Preface

This book is about the general theory of relativity which is concisely labeled as general relativity. The book is based in part on personal notes and tutorials about topics and applications related to modern physics and tensor calculus as well as many insights and analyses that were acquired during my academic and research activities. During the writing of this book, I read and consulted many references (i.e. textbooks and scientific papers) on this subject; the main of these references are listed in the References section in the back of this book. I also read and consulted many general articles, comments, notes, discussions, controversies, etc. which I mostly found on the Internet. The book, therefore, is the result of a rather extensive view to the literature of this theory over most of its lifetime reflecting various stages of its development.

The book contains many solved problems (129) as well as many exercises (606) whose detailed solutions are published in another book that accompanies the present book. The book also includes a detailed index and many cross references (which are hyperlinked in the digital versions) to facilitate the connection between related subjects and parts and hence improve understanding and recollection. The book can be used as an introduction to general relativity at undergraduate and graduate levels.

Unlike most other books on general relativity which are mostly dedicated to the presentation, justification, application and validation of the formalism of the theory (and hence rather minor attention is usually paid to the interpretation and epistemology of the theory), this book is primarily interested in the interpretative and epistemological aspects of the theory. This is obviously inline with the assessment objective of this book, as declared by "Assessed" in the title.

I should announce that unlike almost all other books on general relativity this book does not include any investigation to relativistic cosmology (and indeed to any other application of general relativity like relativistic optics) since it is entirely about the theory of general relativity itself and not about any of its applications and branches. This is partly due to the limitation on the size and scope of the book and partly due to the non-scientific nature of relativistic cosmology in general since it is a topic that is overwhelmingly based on mathematical deliberations and philosophical reflections with very little physical substance or evidence.

I should draw the attention of the readers (and potential readers) of this book that "Simplified" in the title dose not mean "simple". The reader of this book must have a strong background in physics and general mathematics and should be familiar with the basic concepts, notations and techniques of tensor calculus, differential geometry and special relativity. So, it is "Simplified" for the proper reader and not for every reader. Therefore, I strongly advise against acquiring or reading this book by readers who do not have such a suitable background to avoid frustration and disappointment.

Finally, I should announce that it is a great advantage for the reader of the present book to be familiar with my previous book "The Mechanics of Lorentz Transformations" because many topics, discussions, arguments, concepts, notations and techniques in the present book are based on what I established in that book. The familiarity with my other books on tensor calculus and differential geometry (all of which are available on Amazon) should also help the reader to understand and appreciate the contents of this book.
Taha Sochi
London, May 2020

Contents

Preface	1
Table of Contents	2
Nomenclature	6

1 Preliminaries — 9
- 1.1 Synopsis about Gravity and General Relativity — 9
- 1.2 Historical Issues and Credits — 11
- 1.3 General Terminology — 16
- 1.4 General Conventions, Notations and Remarks — 19
- 1.5 Classical Gravity — 21
  - 1.5.1 Planetary Motion — 23
- 1.6 General Relativity versus Classical Gravity — 29
- 1.7 General Relativity versus Special Relativity — 30
- 1.8 General Principles — 31
  - 1.8.1 The Principle of Invariance — 31
  - 1.8.2 The Principle of Equivalence — 34
  - 1.8.3 The Principle of Correspondence — 40
  - 1.8.4 Other Principles — 40
- 1.9 Criteria of Scientific Theory — 41
- 1.10 Reading Too Much in the Equations — 41

2 Mathematical Background — 44
- 2.1 Space, Coordinate System and Transformation — 44
- 2.2 Tensors — 47
- 2.3 Curvature of Space — 51
- 2.4 Variational Principle and Euler-Lagrange Equation — 54
- 2.5 Metric Tensor — 54
  - 2.5.1 Metric Tensor of 4D Spacetime of General Relativity — 57
- 2.6 Christoffel Symbols — 58
- 2.7 Tensor Differentiation — 62
- 2.8 Parallel Transport — 66
- 2.9 Geodesic Path — 68
  - 2.9.1 Geodesic Path in 2D Spaces — 70
  - 2.9.2 Geodesic Equation from Straightness — 72
  - 2.9.3 Geodesic Equation from Variational Principle — 72
  - 2.9.4 Geodesic Equation from Parallel Transport — 75
  - 2.9.5 Geodesic Equation — 76
- 2.10 Riemann-Christoffel Curvature Tensor — 78
- 2.11 Ricci Curvature Tensor — 81
- 2.12 Ricci Curvature Scalar — 82
- 2.13 Einstein Tensor — 83
- 2.14 Energy-Momentum Tensor — 84

3 Formalism of General Relativity — 92
3.1 Rationale of the Field Equation — 92
3.2 The Field Equation of General Relativity — 94
3.3 The Field Equation with Cosmological Constant — 96
3.4 The Linearized Field Equation — 97
3.5 General Relativity as Gravity Theory and as General Theory — 100

4 Solutions of the Field Equation — 102
4.1 Schwarzschild Solution — 103
4.1.1 Derivation of Schwarzschild Metric — 104
4.1.2 Geodesic Equation in Schwarzschild Metric — 111
4.2 Kerr Solution — 119
4.3 Reissner-Nordstrom Solution — 120
4.4 Kerr-Newman Solution — 121

5 Classical Limit of General Relativity — 122
5.1 Convergence to Newtonian Gravity — 122
5.2 Planetary Motion — 127

6 Frames, Coordinates and Spacetime — 130
6.1 Frames in General Relativity — 130
6.2 Coordinates in General Relativity — 132
6.3 Time in Schwarzschild Spacetime — 134
6.3.1 Relation between Coordinate Time and Proper Time — 134
6.3.2 Interpretation of Coordinate Time — 135
6.3.3 Gravitational Time Dilation — 135
6.3.4 Gravitational Frequency Shift — 136
6.3.5 Comparison with Classical Mechanics and Special Relativity — 137
6.4 Length in Schwarzschild Spacetime — 137
6.4.1 Relation between Spatial Coordinates and Proper Length — 138
6.4.2 Interpretation of Spatial Coordinates — 139
6.4.3 Gravitational Length Contraction — 139
6.4.4 Comparison with Classical Mechanics and Special Relativity — 140
6.5 General Relativity and Absolute Frame — 141
6.5.1 Absolute Frame and Mach Principle — 145

7 Physics of General Relativity — 147
7.1 Coordinates of Spacetime — 147
7.2 Time Interval and Length — 148
7.3 Frequency — 149
7.4 Mass — 149
7.5 Velocity, Speed and Acceleration — 150
7.6 Force — 150
7.7 Momentum — 150
7.8 Energy and Work — 151
7.9 Conservation Principles — 151
7.10 Orbital Motion in Terms of Constants of Motion — 151
7.11 Effective Potential in Orbital Motion — 153
7.12 Radial Trajectories in Spacetime — 154
7.13 Geodesic Deviation — 155

8 Consequences and Predictions — 157
- 8.1 Perihelion Precession of Mercury — 157
- 8.2 Light Bending by Gravity — 163
- 8.3 Gravitational Time Dilation — 168
- 8.4 Gravitational Frequency Shift — 171
- 8.5 Gravitational Length Contraction — 174
- 8.6 Gravitational Waves — 174
- 8.7 Black Holes — 175
  - 8.7.1 Schwarzschild Black Holes — 179
  - 8.7.2 Kerr Black Holes — 180
  - 8.7.3 Hawking Radiation — 180
- 8.8 Geodetic Effect — 182
- 8.9 Frame Dragging — 183
- 8.10 Wormholes and Other Fantasies — 184
- 8.11 Cosmological Predictions — 184

9 Tests of General Relativity — 185
- 9.1 Perihelion Precession of Mercury — 187
- 9.2 Light Bending by Gravity — 189
- 9.3 Gravitational Time Dilation — 191
  - 9.3.1 Hafele-Keating Experiment — 192
  - 9.3.2 Shapiro Time Delay Test — 193
  - 9.3.3 Gravity Probe A — 195
  - 9.3.4 Global Positioning System — 195
- 9.4 Gravitational Frequency Shift — 197
  - 9.4.1 Gravitational Red Shift from Astronomical Observations — 198
  - 9.4.2 Pound-Rebka Experiment — 198
  - 9.4.3 Gravity Probe A — 199
- 9.5 Gravitational Length Contraction — 200
- 9.6 Gravitational Waves — 201
  - 9.6.1 Indirect Observation of Gravitational Waves — 201
  - 9.6.2 Direct Observation of Gravitational Waves — 202
- 9.7 Black Holes — 203
- 9.8 Geodetic Effect — 204
  - 9.8.1 Precession of "Moon-Earth Gyroscope" in Motion around Sun — 204
  - 9.8.2 Gravity Probe B — 205
- 9.9 Frame Dragging — 205
  - 9.9.1 LAGEOS Satellites — 205
  - 9.9.2 Gravity Probe B — 205
- 9.10 Wormholes and Other Fantasies — 206
- 9.11 Cosmological Predictions — 206
- 9.12 Tests of the Equivalence Principle — 206
- 9.13 Tests of Special Relativity — 207
- 9.14 Circumstantial Evidence — 208
- 9.15 Evidence for Newtonian Gravity — 208
- 9.16 Final Assessment — 209
- 9.17 Evidence against General Relativity — 211

10 Challenges and Assessment — 212
10.1 Challenges and Criticisms — 212
- 10.1.1 Limitations and Failures of the Equivalence Principle — 212
- 10.1.2 Necessity of Metaphysical Elements — 214
- 10.1.3 Creation Theory — 214
- 10.1.4 Dependence on Special Relativity — 215
- 10.1.5 Triviality of General Invariance — 215
- 10.1.6 Interpretation of Coordinates — 216
- 10.1.7 Strong Dependency of Physical Results on Coordinate System — 216
- 10.1.8 Local Relativity — 217
- 10.1.9 Nonsensical Consequences and Predictions — 218
- 10.1.10 Incompatibility with Quantum Mechanics — 218
- 10.1.11 Gaps, Ambiguities and Question Marks — 219
- 10.1.12 Over-Mathematization of Physics — 221
- 10.1.13 Violation of Sacred Rules — 221
- 10.1.14 Circularity — 222
- 10.1.15 Using Einstein Tensor to Represent Curvature — 222
- 10.1.16 Ambiguity of Vacuum Equation — 223
- 10.1.17 Need for Classical Gravity — 224
- 10.1.18 Failure of Geodesic to Replace Force — 224
- 10.1.19 Limitation of Evidence — 226
- 10.1.20 Absurdities and Paradoxes — 226
- 10.1.21 The Paradox of Absolute Frame and Reality of Spacetime — 227
- 10.1.22 Constancy, Invariance and Ultimacy of the Speed of Light — 228
- 10.1.23 Lack of Practicality and Realism — 228
- 10.1.24 Limited Usefulness — 228

10.2 Overall Assessment — 228
- 10.2.1 Geometric Nature of General Relativity — 229
- 10.2.2 High Complexity — 229
- 10.2.3 Highly Theoretical Nature — 229
- 10.2.4 Publicizing and Politicizing Science — 229
- 10.2.5 Controversies, Conflicts and Uncertainties — 230
- 10.2.6 Practical Value — 231
- 10.2.7 Theoretical Value — 232

Epilogue — 234

References — 236

Index — 237

Author Notes — 247

Nomenclature

In the following list, we define the common symbols, notations and abbreviations which are used in the book as a quick reference for the reader. The list generally excludes what is used locally and casually.

∇	nabla differential operator
∇f	gradient of scalar f
$\nabla \cdot \mathbf{A}$	divergence of vector $\mathbf{A}$
$\nabla \times \mathbf{A}$	curl of vector $\mathbf{A}$
∇^2	Laplacian operator
$\Box$	nabla differential 4-operator
$\Box^2$	d'Alembertian 4-operator
, (subscript)	partial derivative with respect to the following index(es)
; (subscript)	covariant derivative with respect to the following index(es)
nD	n-dimensional
$\delta/\delta\lambda,\ \delta^2/\delta\lambda^2$	absolute derivative operators with respect to λ
$\partial_i,\ \partial^i$	partial differential operators with respect to i^{th} coordinate
$[ij,k],\ [\mu\nu,\omega]$	Christoffel symbol of the 1st kind in 3D (or nD) space, in 4D spacetime
a	mean distance (in planetary motion)
a	semi-major axis of ellipse or transverse semi-axis of hyperbola
$\mathbf{a}$	acceleration vector in 3D space
A	area
A	constant in the orbital equations
a, Δ, ρ^2	parameters in Kerr metric
b	semi-minor axis of ellipse or conjugate semi-axis of hyperbola
B	constant representing magnitude of angular momentum per mass or twice areal speed
B2, B3, B4	author books (see References)
B2X, B3X, B4X	author exercise books (see References)
c	the characteristic speed of light in vacuum ($= 299792458$ m/s)
ct, r, θ, ϕ	Schwarzschild coordinates, Boyer-Lindquist coordinates
C_s	constant equal to Schwarzschild radius
d	distance
$d\mathbf{r}$	differential of position vector
$ds, d\sigma$	infinitesimal line element in 3D (or nD) space, in 4D spacetime
$\mathbf{D}, D^\mu$	displacement (or deviation) 4-vector
diag $[\cdots]$	diagonal matrix with the embraced diagonal elements $\cdots$
e	eccentricity of ellipse
$\mathbf{e}_r, \mathbf{e}_\theta, \mathbf{e}_\phi$	basis vectors of normalized spherical coordinate system
E, E_0	energy, rest energy
$\mathbf{E}_i, \mathbf{E}^i$	i^{th} covariant and contravariant basis vectors
Eq., Eqs.	Equation, Equations
ESA	European Space Agency
f	magnitude of force in 3D space
F	dimensionless factor in Schwarzschild metric ($= 1 - [R_S/r]$)
$\mathbf{f}$	force vector in 3D space
f_t	tidal force
g	magnitude of gravitational acceleration or gravitational force field
g	determinant of covariant metric tensor
$\mathbf{g}$	gravitational acceleration or gravitational force field
G	the gravitational constant ($\simeq 6.674 \times 10^{-11}$ N m^2/kg^2)
G	trace of Einstein tensor

$\mathbf{G}, G_{\mu\nu}, G^{\mu\nu}, G^{\mu}_{\nu}$	Einstein tensor
$g_{ij}, g^{ij}, g^{i}_{j}$	metric tensor of 3D (or nD) space
$g_{\mu\nu}, g^{\mu\nu}, g^{\mu}_{\nu}$	metric tensor of 4D spacetime
GPS	Global Positioning System
h	Planck constant ($\simeq 6.62607 \times 10^{-34}$ J s)
$\hbar$	reduced Planck constant ($= h/2\pi$)
H, K	functions of r in Schwarzschild metric
iff	if and only if
J	magnitude of angular momentum
k	Boltzmann constant ($\simeq 1.38065 \times 10^{-23}$ J/K)
L	length
LAGEOS	LAser GEOdynamics Satellites
LIGO	Laser Interferometer Gravitational-wave Observatory
LISA	Laser Interferometer Space Antenna
m, M	mass
m_g, m_i	gravitational mass, inertial mass
n	number of dimensions of space
n	number of particles inside a given volume
N	number of components or elements
N, N_0	number density, proper number density
N_c	number of planetary revolutions per century
$\mathbf{n}, n_i$	normal unit vector to surface
NASA	National Aeronautics and Space Administration
$\mathbf{p}, \mathbf{P}$	momentum vector in 3D space, in 4D spacetime
Q	electric charge
r	radius, radial distance, radial coordinate
$\mathbf{r}$	position vector [e.g. (x,y,z) or (x^1, x^2, x^3) or $(x^1, \cdots, x^n)$]
r_0, r_1	perihelion distance, aphelion distance
r, ϕ	plane polar coordinates
r, θ, ϕ	spherical coordinates of 3D space
R	radius, radial distance
R	Ricci curvature scalar
R_{ij}, R^{i}_{j}	Ricci curvature tensor of the 1st and 2nd kind in 3D (or nD) space
$R_{\mu\nu}, R^{\mu}_{\nu}$	Ricci curvature tensor of the 1st and 2nd kind in 4D spacetime
R_{ijkl}, R^{i}_{jkl}	Riemann-Christoffel curvature tensor of the 1st and 2nd kind in 3D (or nD) space
$R_{\mu\nu\rho\omega}, R^{\mu}_{\nu\rho\omega}$	Riemann-Christoffel curvature tensor of the 1st and 2nd kind in 4D spacetime
R_S	Schwarzschild radius
s	space interval (or length of arc)
$S_{\mu\nu}, S^{\mu\nu}, S^{\mu}_{\nu}$	electromagnetic field strength tensor
SAO	Smithsonian Astrophysical Observatory
t	time, temporal coordinate
T	periodic time
T	temperature
T	trace of energy-momentum tensor
T (superscript)	transposition
T_c	time of a century
$\mathbf{T}, T^i, T_i$	unit tangent vector
$\mathbf{T}, T_{\mu\nu}, T^{\mu\nu}, T^{\mu}_{\nu}$	energy-momentum tensor
u	$1/r$
u	speed
$\mathbf{u}$	geodesic unit vector
$\mathbf{u}$	velocity vector in 3D space

$\mathbf{U}, U^\mu$	velocity vector in 4D spacetime
v	speed
$\mathbf{v}$	velocity vector in 3D space
V	volume
v_{es}	escape speed
V_{GR}, V_N	general relativistic effective potential, Newtonian effective potential
VLBI	Very Long Baseline Interferometry
$\mathbf{x}$	position vector in 4D spacetime, i.e. (x^0, x^1, x^2, x^3)
x_i	i^{th} Cartesian coordinate (or covariant form of general coordinates)
x^i, x^μ	general coordinates of 3D (or nD) space, of 4D spacetime
x, y, z	coordinates of 3D space (usually orthonormal Cartesian)
α	angle of precession per revolution (in geodetic effect)
γ	Lorentz factor
$\Gamma^k_{ij}, \Gamma^\omega_{\mu\nu}$	Christoffel symbol of the 2nd kind in 3D (or nD) space, in 4D spacetime
$\delta_{ij}, \delta^{ij}, \delta^i_j$	Kronecker delta tensor in 3D (or nD) space
$\delta_{\mu\nu}, \delta^{\mu\nu}, \delta^\mu_\nu$	Kronecker delta tensor in 4D spacetime
$\delta\phi$	angle of deflection of light
$\delta\phi$	extra precession of perihelion per revolution
$\delta\phi_c$	extra precession of perihelion per century
Δ	finite change
ε	trace of perturbation tensor
ε_0	the permittivity of free space
$\varepsilon_{\mu\nu}, \varepsilon^{\mu\nu}, \varepsilon^\mu_\nu$	perturbation tensor
$\eta_{\mu\nu}, \eta^{\mu\nu}$	Minkowski metric tensor
κ	coefficient of the Field Equation ($= -8\pi G c^{-4}$)
κ_g, κ_n	geodesic and normal curvatures
$\mathbf{K}$	curvature vector
$\mathbf{K}_g, \mathbf{K}_n$	geodesic and normal components of curvature vector
λ	natural (or affine) parameter of curve
λ	wavelength
Λ	cosmological constant
μ_0	the permeability of free space
ν	frequency
ρ, ρ_0	density, proper density
ρ, ϕ, z	cylindrical coordinates of 3D space
σ	spacetime interval
σ	Stefan-Boltzmann constant ($\simeq 5.67037 \times 10^{-8}\,\mathrm{W\,m^{-2}\,K^{-4}}$)
τ	proper time
Φ	gravitational potential
$\chi^{\mu\nu}$	perturbation tensor
ω	angular speed

Chapter 1
Preliminaries

In this introductory chapter we present some preliminaries and background materials related mainly to gravity and general relativity. We also discuss general notations, conventions, symbols and terminology that we use in this book. A brief investigation of some general scientific aspects that will be used later in the book is also included.

1.1 Synopsis about Gravity and General Relativity

In simple terms, gravity is an attractive force that acts between two massive bodies. Although it is the weakest of all known forces, it is the most dominant and influential force in the Universe as it is the main factor that governs the physics of the Universe on large scale. The exceptional importance and dominance of gravity arise from its universality and long range since it is a property of matter regardless of its form and shape (unlike electric force for instance which is restricted to charged matter) and regardless of range (unlike strong force for instance which is restricted to short range). For example, gravity is the primary factor in the formation of groups, clusters, galaxies and stars. It is also responsible for the orbital motion of the celestial bodies such as the revolution of the planets in the solar system around the Sun. Therefore, general relativity as a gravity theory finds its natural applications in astronomy, astrophysics and cosmology.

We may note the following characteristics about gravity:[1]

• As indicated above, gravity is an attractive force and hence there is no repulsive gravitational force unlike electric force for instance which is attractive between unlike charges and repulsive between like charges. However, we should note that the subject of this remark is ordinary matter although there are indications that anti-matter also follows the same gravitational rules as ordinary matter and hence the gravitational force that involves anti-matter (i.e. if one or both massive objects are made of anti-matter) is also attractive.

• Gravity is an attribute of mass (compared for instance to electric force which is an attribute of charge). However, since mass is a general attribute of matter, unlike some other restricted attributes of matter like charge, gravity can be regarded as an attribute of matter.

• Gravity is a long range force and hence it dominates the physics of the Universe on large scale where the effects of other forces diminish. This dominance should also be attributed in part to its dependency on matter regardless of any restricted attribute of matter such as charge (see exercise 1).

Regarding general relativity, it is essentially a theory of gravity[2] that can be characterized by the following:

• It is a geometric theory of gravity and hence from a general relativistic perspective gravity is not a physical force but it is an effect of the distortion of the spacetime by the presence of matter and energy.

• The fundamental physical pillar of general relativity is the equivalence principle (see § 1.8.2) which essentially abolishes the difference between acceleration and gravitation, while its fundamental mathematical pillar is differential geometry and tensor calculus where the theory heavily relies in its formalism on the concepts, language, notation and techniques of these branches of mathematics. More specifically, the role of differential geometry is mainly in conceptualizing and formalizing the geometric description

---

[1] The following discussion about gravity and general relativity is for the purpose of setting the scene and hence it can be loose in some parts. However, most of the discussed issues will be revisited in the future in a more formal and rigorous way.

[2] As we will see, general relativity has also non-gravitational generalizations and consequences which we refer to as "General Theory".

of multi-dimensional spaces, while the role of tensor calculus is to provide efficient and compact mathematical notations and techniques for the description of multi-dimensional spaces and their attributes and accessories in an invariant way.

- General relativity has emerged to address certain limitations of the classical Newtonian theory of gravity (see § 1.5). Apart from its direct and common applications that it shares with classical gravity, the theory currently plays an important role in cosmological and astrophysical studies and provides a theoretical framework and motivation for these fields of investigation and research.
- The essence of general relativity is that the presence and movement of matter and energy cause distortion in the surrounding spacetime and hence the spacetime in their vicinity becomes curved. Physical objects (like massive particles and light) under the effect of gravity will then follow geodesic paths[3] in this curved spacetime. Accordingly, objects deviate from these geodesic paths only if forces (i.e. other than gravity which is not a force according to general relativity) do exist. So, if gravity is the only "force" that influences the objects then the objects will be in a state of free fall moving along geodesic trajectories (or world lines).
- Roughly speaking, the effects predicted by general relativity that distinguish it from classical gravity are only significant and detectable in the strong gravitational fields[4] due to the presence of very large or/and very compact aggregates of matter such as black holes and neutron stars where the departure of the predictions of general relativity from the predictions of the classical theory of gravity is significant and observable.
- The curvature of spacetime caused by the presence of matter[5] depends on the amount and density of matter in the spacetime and hence the curvature increases as the amount and density of matter increase. Accordingly, in the vicinity of small and sparse aggregates of matter the spacetime is minimally curved and hence it is virtually flat while in the vicinity of large and dense aggregates of matter the spacetime is significantly curved.
- The curvature of spacetime also depends on the distance from the matter that causes this curvature and hence the curvature of spacetime diminishes as the distance increases. So, the curvature caused even by very large and dense aggregates of matter becomes negligible very far away from these aggregates and hence the spacetime is almost flat in such regions of space.[6]
- General relativity is a highly mathematical theory since it heavily relies on differential geometry and tensor calculus, as indicated earlier. On the observational and experimental side, general relativity largely depends in its verification and support on astronomical observations and related applications although it has also been endorsed by Earth-based and space-based experimental evidence according to its supporters (refer to § 9 for details).

Problems

1. What are the main types of force in physics?
   Answer: According to modern physics there are four types of fundamental forces (or interactions): gravitational, electromagnetic, weak and strong. We note that gravity may not be considered technically a force according to some modern theories (including general relativity) although the term "force" may still be attached to gravity following the common notion and convention. This should also apply to other terms of classical gravity which are based on the paradigm of force (such as gravitational field and potential) where the use of these classical terms is generally tolerated in the literature of general relativity although they are technically incompatible with the theoretical framework and paradigms of general relativity.

---

[3] Geodesic path or curve in a given space is the most straight path connecting two points in that space (see § 2.9).

[4] Some exceptions like perihelion precession of Mercury (see § 8.1) may be considered.

[5] Due to the mass-energy equivalence according to the relation $E = mc^2$, matter may include energy as well as mass. It is noteworthy that there is a convention (mainly in cosmology) that massless objects are called radiation while massive objects are called matter. Also, there are similar other conventions (e.g. in particle physics and relativistic quantum field theories) about these terms. However, we do not strictly follow these conventions. More clarifications about these issues will follow.

[6] In fact, this seems logical if we combine classical gravity (where gravity vanishes far away from the source of gravity due to the inverse square relation between gravitational field and distance) with the general relativistic idea that gravity is demonstrated by the curvature of spacetime and hence weak/strong gravity means weak/strong curvature and vice versa.

2. Why is gravitational force always attractive while electric force can be attractive and repulsive?
   Answer: This may be justified by having only one type of mass but two types of charge. However, the attraction (rather than repulsion) between masses still requires explanation.
3. Is there any way to nullify gravity?
   Answer: There is no way to nullify gravity (or shield it off) in a real sense, i.e. turn it off so it ceases to exist. However, the effect of this force can be neutralized to a certain extent by free fall where the attractive force of gravity is not felt although it is still present as it is the cause of the fall.[7] In fact, the effect of gravity in free fall can be observed by an outside observer, i.e. an observer who is not in the frame of free fall. For example, the planets in their orbits around the Sun and the artificial satellites in their orbits around the Earth are in a state of free fall but the effect of gravity on them can be observed by an outside observer who observes their revolution due to gravity.
4. What is the colloquial idiom that is commonly used to express the essence of general relativity?
   Answer: It is "matter tells spacetime how to curve, spacetime tells matter how to move".
5. Mention a factor (other than those given in the text) that distinguishes gravity from other forces and demonstrates its exceptional nature.
   Answer: It is the ability of gravity to provide some of the most efficient and powerful energy release mechanisms through the accretion of matter onto compact objects (see § 8.7).

Exercises
1. Explain why gravity, unlike electric force, is the dominant force on large scale although they have identical dependency on distance since both are subject to the same inverse square relation (classically at least).
2. Compare mass and electric charge in their relation to gravity and electromagnetism.
3. Compare, briefly, between gravity in classical physics and gravity in general relativity.
4. Outline the need for general relativity and the factors that contributed to the emergence of this theory.
5. According to general relativity, what is the characteristic feature of the world line followed by an object under the influence of gravity alone (and hence it is in a state of free fall)? Consider in your answer massless objects (such as light) as well as massive objects.
6. Establish a connection between special relativity and general relativity with regard to the underlying spaces of these theories and the motion of physical objects in these spaces.
7. Outline the main criticisms that faced Newtonian gravity and contributed to the emergence of general relativity.
8. Briefly discuss the implications of the fact that general relativity is a geometric theory.
9. What are the main pillars of general relativity?
10. Explain the need for tensor calculus and differential geometry in the formulation of general relativity.
11. In general relativity, what is the link between the doctrine of representing gravity by the curvature of spacetime and the redundancy of the concept of force (as applied to gravity)?

1.2 Historical Issues and Credits

The tendency of massive objects to fall to the Earth was seen in the ancient philosophies (which are typically represented by the Aristotelian philosophy) as a natural inclination to come down to the center of the Universe which is believed to be the Earth. Hence, this tendency was seen as an obvious and self-explanatory phenomenon with no need for an elaborate theory to justify. Also, there was no reason to suspect that this tendency has any connection with the tendency of heavenly bodies, such as Mars, to follow certain routes or orbits in their perpetual voyage. With the emergence of the Copernican astronomical model there were some contemplations about possible existence of a "physical force" that pulls down massive objects towards the center of the Earth and may even be responsible for the observed trajectories of celestial bodies. For example, Galileo investigated the acceleration of massive objects under the influence of the "gravity" of the Earth[8] where he observed the independence of the gravitational

---

[7] The purpose of "certain extent" is to exclude tidal forces which are not neutralized by free fall.
[8] Galileo was one of the first scholars (but not the only scholar) in the era of Renaissance to investigate gravity in a systematic way using modern (but rather crude) experimental and observational methods instead of relying on philosophical reasoning

acceleration from the mass of objects and this observation can be seen as the root of the modern principle of equivalence which is one of the pillars of general relativity (see § 1.8.2).

However, the modern scientific notion of gravity and its quantitative formulation did not emerge until the formulation of a mathematically rigorous astronomical model represented by Kepler's laws which are based on the elaborate and systematic observations of Tycho Brahe. These laws paved the way to Isaac Newton to formulate the first modern and scientifically sound theory of gravity (see § 1.5) as represented by his universal law of gravitation which provides qualitative and quantitative explanation not only to the fall of massive objects towards the Earth and the motion of celestial bodies along their trajectories but to many other phenomena that characterize the physics of the Universe on large scale such as the formation of stars and galaxies and the interaction between them.[9]

Although the triumph of the Newtonian gravity theory (as well as its applications such as celestial mechanics) was stunning and hence it was seen by the majority of physicists over more than two centuries as the ultimate theory, there was continuous emergence of problems in the application of this theory to astronomical observations. In fact, these problems can be classified into two main categories: problems based on the limitation of the observation and hence they do not affect the theory itself such as the perturbation of some celestial trajectories from their theoretical predictions due to the presence of unobserved astronomical objects, and problems that seem to threaten the theoretical model itself due to the failure of the theory to provide convincing explanations such as the perihelion precession of Mercury. In fact, the first category endorsed, rather than damaged, the Newtonian theory by providing evidence for its predictive power such as the discovery of Neptune as a result of discrepancies observed in the orbit of other planets with the application of the classical theory of gravity. However, the second category cast a shadow on the integrity of the theoretical model itself although the majority of physicists seemed to believe that these failures of the Newtonian theory do not arise from the theory itself which is essentially sound and reliable but from some other factors and details, such as limitations in the observational and computational techniques, or from minor theoretical limitations and hence there were attempts to find solutions to these problems within the classical theory rather than dismissing this theory altogether and looking for a different theory. A prominent example of these attempts is the retarded gravitational potential theory[10] whose main objective was to fix the problem of Mercury precession that was a major failure to the Newtonian gravity (see § 8.1). However, apart from the apparently undisputed fact that this theory produced the same precession formula as the one based on general relativity we have no sufficient historical records about this theory to reach a conclusion or judgment about its essence, validity and merit.

There are also tentative historical records about contemplations and theories of gravity in the 19th century that were based on the ether assumption or on hypothesized velocity dependent potentials, as well as contemplations and proposals about the finity of the speed of propagation of gravitational effects which should be based on assuming a temporal dependency that is missing in the Newtonian gravitation theory. There were also attempts to model gravity on the style of electromagnetism (represented by Maxwell's equations) as a full field theory with temporal and spatial dependency. Although we cannot reach a decisive conclusion about these contemplations and theories and their contents and significance, we can deduce that many of the basic elements and ideas on which general relativity rests were existent even before the formal and complete development of this theory. In this regard, we should also refer to the history of the development of the equivalence principle, which is a fundamental pillar of general relativity, as one of the precursors of this theory (see § 1.8.2).

Following the development of Lorentz mechanics and Einstein involvement in elucidating the theory of special relativity, which is restricted to inertial frames, he started an effort to generalize this theory

---

which was the standard method of investigation in the middle ages.

[9] In fact, we should also point out to the essential contributions of several predecessors and contemporaries of Newton (e.g. Erigena, Adelard of Bath, Galileo, Bulliadus, Huygens and Hooke) to the development and formulation of the classical gravity law especially the inverse square dependency on distance and its relation to Kepler's laws.

[10] This theory is based on a modified classical approach and is commonly associated with the name of Gerber (who was a German school master) although Gerber may not be the originator of this theory or the main contributor to its development.

to include non-inertial frames.[11] This in essence is a generalization of the invariance with respect to velocity to the invariance with respect to acceleration where inertial frames (of constant velocity) are considered as a special type of accelerating frames (of variable velocity). On contemplating the similarity (which became equivalence later on) between acceleration and gravitation the attempted generalization became an effort to find a theory of gravity where gravitational frames will be considered equivalent to accelerating frames to which the fundamental principles of Lorentz mechanics should apply in some sense. In other words, this generalization will ensure the invariance of physical laws across all frames of reference (i.e. not only across inertial frames as in Lorentz mechanics) where the rules of Lorentz mechanics should apply in a potentially restricted sense (i.e. locally but not globally as in Lorentz mechanics).

Now, Lorentz mechanics as a theory of space and time and their transformation across frames is essentially a geometric theory[12] and hence in this context it seemed natural for any invariant theory of physics to consider the geometry of space where this space was already generalized in Lorentz mechanics to include time, i.e. Minkowski spacetime. So, following several attempts in different directions it was concluded (rightly or wrongly) that the answer to the quest of an invariant general theory of physics lies in the investigation of the geometry of spacetime and its invariant properties. Accordingly, the mathematical tools of differential geometry (whose subject is the analytical investigation of the geometrical properties of multi-dimensional spaces in general) and tensor calculus (whose subject is the formulation of invariant mathematical concepts and relations through the use of tensors) were seen as the ideal choice for developing such a general invariant theory of physics that should include inertial frames and accelerating frames as well as their equivalent gravitational frames. The persistent effort of Einstein and other scholars who worked closely with him was finally culminated by the formulation of the Field Equation of general relativity in 1915-1916 which marks the birth of this theory.

Following the controversial endorsement of general relativity by the 1919 solar eclipse expedition of Eddington and his team, there was strong interest in the first few decades in finding applications to the theory in fields like cosmology and astrophysics. For example, there were many suggestions of general-relativity-based cosmological models for the evolution and destiny of the Universe. This may have been enhanced by the "discovery" of the expansion of the Universe, according to the Hubble's law, and the emergence of the Big Bang theory. There were also proposals of general-relativity-based astrophysical theories related for example to compact objects like neutron stars and black holes. However, most of these investigations and proposals were theoretical contemplations and mathematical models and hence general relativity was largely seen as a mathematically aesthetic but not very practical theory. In fact, in the 1940s and 1950s general relativity seemed to be in its way to die out and become extinct.

This trend was reversed in the late 1950s and in 1960s where general relativity restored its momentum and presence in the mainstream science. This, in large part, is due to the promotional effort by pro-relativity physicists like John Wheeler who revived the interest in this theory and popularized it in the scientific community and even among the wider general public. An auxiliary reason for this stimulated revival is the technological revolution in the computational methods and technology and the advancements in astronomical equipment and techniques. The availability of high computational power associated with very advanced and efficient numerical methods facilitated the solution of the Field Equation of general relativity in many cases where analytical solutions do not exist. Also, the advancement in the technology and methods of astronomical observation, including things like space missions and artificial satellites and the use of radio astronomy, enabled physicists and astronomers to search for applications and verifications to the theory in several fields of interest.

Nowadays general relativity is the dominant theory in many fields of research especially in cosmology and astrophysics. However, general relativity did not get this privileged status by natural selection and fair competition but mostly by promotion and indoctrination where general relativity was hugely favored

---

[11] In fact, generalizing special relativity (or Lorentz mechanics) was a natural objective at that stage in the development of science and hence Einstein was not alone in this effort. So, specifying Einstein is because we are investigating general relativity which Einstein is commonly given the main credit for its development. Otherwise, other physicists were also working on this project (as well as developing new gravitation theories) at that time although their attempts and theories are generally ignored in the modern literature.

[12] This is particularly true when we consider the Minkowskian merge of space and time into spacetime.

while alternative theories were not allowed to survive and thrive. In general, the mainstream scientists (including academic, research and funding bodies) treat alternative theories in principle as inferior to general relativity and even as pseudo science. However, we should distinguish between the alternative theories which are totally independent of general relativity and those which are based on or derived from general relativity and hence in essence they are no more than alterations and variant forms of general relativity and so they generally rest on the same physical and epistemological principles and follow very similar or identical theoretical methods. The latter generally survive and thrive under the umbrella of general relativity and hence they usually enjoy the favoritism that is enjoyed by general relativity. In fact, there are many theories of the latter type although most of them do not deserve particular attention due to their lack of originality and independence. Anyway, this is not the subject of investigation of this book and hence we refer the interested reader to the relevant literature.

Finally, let briefly discuss the credit for the development of general relativity which is generally believed to be the creation of Einstein alone (or at least the contribution of others to this theory is regarded insignificant compared to the contribution of Einstein). According to the available historical records, this view is biased in favor of Einstein because although Einstein was the main contributor to the pursuit of this theory it was impossible for him to develop this theory single handed due to the limitation on his mathematical knowledge and skills.[13] In fact, there are other major contributors to the creation and development of this theory such as Minkowski, Grossmann, Levi-Civita and Hilbert. The contribution of these scholars has particular importance and significance with regard to the development of the formalism of this theory (as opposite to the qualitative and epistemological sides) which is the really scientific part of the theory.

Apart from acquiring fundamental ideas and techniques (e.g. the 4D spacetime of Minkowski and his energy tensor), Einstein got major guidance and advice on essential mathematical and theoretical aspects from the aforementioned scholars (and possibly from others). Moreover, he got substantial help in correcting his numerous blunders (e.g. by Levi-Civita and Hilbert) and reorienting his search into the right direction. In fact, without the essential contributions of these scholars general relativity could not have reached its final conclusion. These contributions include even the formulation of the final version of the Field Equation as will be discussed later. In brief, it was impossible for Einstein to develop general relativity single handed especially in its formal and mathematical part which is the really scientific part of the theory. The reality is that Einstein was mainly skillful in digging, finding and developing qualitative ideas of philosophical and epistemological nature but he did not have the essential mathematical knowledge and skills to develop a highly complex mathematical theory like general relativity. In fact, this should be evident from the many basic formal and logical mistakes and contradictions that can be found in his work (e.g. in his special relativity paper of 1905) especially in that early period of his research career.

It is also important in this context to note that general relativity (in its version which is commonly attributed to Einstein) is not the first metric theory of gravity and hence Einstein (and possibly some of his collaborators) is not the first to come with such an idea and formulate such a theory that links the metric to the energy-momentum tensor. In this regard, major credit should be attributed to Nordstrom for his gravitation metric theory of 1913 (regardless of its experimental validity or invalidity and the alleged contradiction with observation). This theory (like its successor general relativity) also correlates the metric of spacetime $g_{\mu\nu}$ to the energy-momentum tensor $T_{\mu\nu}$. In fact, Nordstrom was not the only one to search for and propose such a metric theory prior to the development of general relativity, as a quick inspection of the history of research in that era should reveal. Moreover, precursors of general relativity (in the form of generic gravitation metric theories) can be traced back even to earlier times where historical indications suggest such speculations and initial formulations in the work of earlier physicists and mathematicians such as Minkowski and Laue. The unfortunate fact is that the history of science of that era (especially with regard to this part of research) is severely distorted and falsified by the followers of Einstein and the enthusiastic propaganda that followed the alleged endorsement of general relativity in

---

[13] In this regard, we refer to the quote by Minkowski (who was a mathematics teacher of Einstein): "The mathematical education of the young physicist [Albert Einstein] was not very solid, which I am in a good position to evaluate since he obtained it from me in Zurich some time ago". We should also refer to the Einstein admission of his mathematical limitation and his struggle with mathematics during his school days.

1919 by Eddington and his team.

Accordingly, we think substantial credit for developing this theory should go to the aforementioned scholars (as well as others) without whom "general relativity" will be no more than a collection of philosophical contemplations, qualitative physical ideas and basic (and potentially inconsistent) mathematical formulation that do not rank to the state of a scientific theory. It is a huge bias to consider the contributions of these scholars as something like consulting a textbook on a mathematical problem or seeking the advice of a mathematics or science teacher to solve a casual difficulty (as it is usually depicted in the literature). Their efforts and contributions in developing the theory are not less important or significant than the effort and contribution of Einstein, and this should be considered when re-writing the history of general relativity which is not less distorted than the history of special relativity (see our book "The Mechanics of Lorentz Transformations"). Regarding the dispute about the priority of developing the final form of the Field Equation and whether the credit should go to Hilbert or Einstein, we question the priority of Einstein in developing this equation and hence we think the credit (or at least the main credit) belongs to Hilbert (see exercise 6).[14]

We should finally note that general relativity in its current state and extent is the product of the persistent effort of several generations of scientists around the world over more than a century. Hence, the theory in its current state is far from being "the theory of Einstein" (as commonly depicted in the literature) even if we accept that Einstein was the major contributor to the original proposition and formulation of the theory. The reality is that large parts of the formulation, interpretation, expansion, generalization, correction and refinement of the current theory belong to other physicists and mathematicians. However, despite this there is a general tendency in the literature to attribute the credit for the entire theory to Einstein, which is very far from reality. This, in fact, is part of the general trend in the literature to magnify and exaggerate the role of Einstein whenever and wherever this seems possible, which is a practice inconsistent with the ethics and spirit of science.[15]

Problems

1. State Kepler's laws.
   Answer: They are:
   (**a**) Planets follow elliptical orbits in their revolution around the Sun with the Sun being at one focus.
   (**b**) The line joining the Sun to the planet sweeps out equal areas in equal periods of time.
   (**c**) The cubes of the mean distances between the Sun and the planets are proportional to the squares of their orbital periods. This law may also be stated as: for any two planets, the ratio of the cubes of the semi-major axes of their orbits is equal to the ratio of the squares of their orbital periods.
2. State Newton's law of gravity.
   Answer: Two bodies attract each other by a force that is directly proportional to the product of their masses and inversely proportional to the square of the distance between their centers.

Exercises

1. State some limitations of the classical gravitation theory that motivated the development of general relativity.
2. Discuss the twists and turns in the development of general relativity.
3. Discuss the failure of Lorentz mechanics to include gravity under the invariance requirement (to be like mechanics and electromagnetism).
4. Outline the main limitations of Lorentz mechanics (or special relativity) with regard to the requirement of the invariance of physical laws and explain and assess the general relativistic methodology in addressing these limitations.

---

[14] Due to the dispute about the contribution of Hilbert and Einstein in developing the final form of the Field Equation of general relativity, as well as the contributions of other scholars to the development of this equation, we prefer to call this equation "the Field Equation", i.e. with uppercase initials and without the "Einstein" or "Hilbert" attachments which are used in the literature.

[15] In fact, this practice of exaggeration and bias is not restricted to Einstein and general relativity (for example we have a similar situation with Newton and his theories) although the situation is more grave in the case of Einstein and general relativity.

5. Briefly outline the role of Einstein in the development of general relativity and in the development of a general theory of space and time (as well as gravitation metric theory).
6. Why we question the priority of Einstein in developing the final form of the Field Equations of general relativity and attribute the main credit to Hilbert?
7. Is general relativity the first metric theory of gravity?

1.3 General Terminology

In this section we provide brief definitions of essential concepts and terms that are frequently used in this book.

- **Massive** or **material** object is a physical object that possesses finite mass (i.e. $m > 0$) such as a particle of sand, while **massless** object is a physical object with no mass (i.e. $m = 0$) such as a photon. However, we should note that massless object does not have "rest" mass but it should have effective (or equivalent) mass according to the mass-energy relation, i.e. $m = E/c^2$ where m is the effective mass of the object, E is its energy and c is the characteristic speed of light. So, massive object has rest mass while massless object has no rest mass although it has effective mass.

- **Spacetime** is a combination of space and time in which objects do exist and physical events take place. In general relativity, spacetime is not seen as space plus time but as a single manifold in which temporal and spatial dimensions are treated equally as dimensions of a 4D manifold. The concept and formal treatment of spacetime in general relativity is as in special relativity apart from the fact that the spacetime of special relativity is flat (i.e. Minkowskian) while the spacetime of general relativity is curved (i.e. Riemannian).

- **Event** is a physical occurrence in the spacetime manifold and hence it is represented by a single point in this manifold. Accordingly, an event is identified by a set of 4 spacetime coordinates: 1 temporal and 3 spatial, e.g. (x^0, x^1, x^2, x^3) where x^0 represents the temporal coordinate ct (with c being the characteristic speed of light in free space and t is time) while x^1, x^2, x^3 represent the spatial coordinates. We may call (x^0, x^1, x^2, x^3) the 4-position in accord with the upcoming terminology (e.g. 4-velocity).

- **Riemannian space** is a manifold characterized by the existence of a symmetric rank-2 tensor called the metric tensor that describes the geometry of the space (see § 2.5). More accurately, Riemannian space is characterized by having an invariant line element ds of the following form: $(ds)^2 = g_{ij}dx^i dx^j$, i.e. $(ds)^2$ is quadratic in the coordinate differentials as well as being dependent on a symmetric rank-2 metric tensor g_{ij} which is generally a function of coordinates.[16]

- **Pseudo-Riemannian space** is like Riemannian space in having a symmetric rank-2 metric tensor but, unlike Riemannian space, its metric (or rather the quadratic form of its line element) is not positive definite. This distinction is a mathematical convention and hence physicists generally do not follow this distinction because it is irrelevant to their purpose. Accordingly, the spacetime of general relativity is commonly described as Riemannian (in disregard to this convention) although it is actually pseudo-Riemannian. In fact, this is a matter of terminology more than an abolishment of the difference between Riemannian and pseudo-Riemannian which is a real difference (as we will see later).

- In Riemannian space, the differential **line element** ds is defined by $(ds)^2 = g_{ij}dx^i dx^j$ where g_{ij} is the metric tensor, dx^i and dx^j are coordinate differentials and i and j range over the space dimensions (noting that we are using the summation convention, as will be declared later). In this book, the line element of the spacetime of general relativity is symbolized with $d\sigma$ to make it specific to general relativity.[17] The square of the line element [i.e. $(ds)^2$ and $(d\sigma)^2$] is called the **quadratic form**.

- In Riemannian space, the **space interval** is the length of a general path (whether straight or curved) of a finite size connecting two points in the space and hence it is the integral of the line element ds (as defined in the previous point) along the path, that is:

$$s = \int_P ds \qquad (1)$$

---

[16] As we will see, other conditions (e.g. being invertible) are also imposed on the metric tensor of Riemannian space.
[17] In fact, this also applies to the line element of the spacetime of special relativity and hence it is specific to the relativity theories.

where s is the space interval and the integral is evaluated over the path P which connects the two points. This definition applies to the spacetime of general relativity[18] (which is a pseudo-Riemannian space) and hence the **spacetime interval** σ is defined as the integral $\sigma = \int d\sigma$. However, we may stretch the term "spacetime interval" to include even its infinitesimal form $d\sigma$ (which is the line element).

- A space is **flat** if it is possible to find a coordinate system for the space with a diagonal metric tensor whose all diagonal elements are ± 1; otherwise the space is **curved**. More formally, an nD space is flat *iff* it is possible to find a coordinate system in which the quadratic form $(ds)^2$ is given by:

$$(ds)^2 = \sum_{i=1}^{n} \zeta_i (dx^i)^2 \tag{2}$$

where the indexed ζ are ± 1 while the indexed x are the coordinates of the space.

- **Minkowski space**, or Minkowski spacetime, is the 4D spacetime manifold that underlies special relativity and hence it is flat and void of matter and energy. In fact, it is void of matter and energy as a source of gravity (i.e. in considerable quantities that distort the metric properties of the spacetime) but it could (and usually should) contain some matter and energy.
- **4-vector** means a vector (in its technical tensorial sense) in the 4D spacetime. Similarly, **3-vector** means a vector in the 3D ordinary "spatial" space. Terms like 3-tensor and 4-tensor may also be used to refer to tensors in 3D space and 4D spacetime. This also applies to terms like 3-operator, 4-operator, 3-*quantity* and 4-*quantity* where *quantity* is a given physical quantity (e.g. 3-velocity and 4-momentum). It should be noted that the "4-" label generally means invariance under the presumed transformations in the spacetime (i.e. Lorentz transformations in the Minkowskian spacetime of special relativity and general transformations in the Riemannian spacetime of general relativity although in the latter case the label usually refers to the local Lorentzian origin of the quantity).
- **Geodesic** curve is the most "straight" or most "direct" route connecting two points in a given space. Geodesic curves are normally, but not necessarily, the curves of shortest length.
- **Null geodesic** is the geodesic path that is followed by a light signal (or any massless object to be more general) in the spacetime (assuming the signal is not under non-gravitational influence). The line element $d\sigma$ (and hence the spacetime interval σ) vanishes identically along null geodesic and that is why it is "null".
- **Free particle** (or object) is a particle that moves in the spacetime under no influence at all (as in the spacetime of special relativity) or under the influence of gravity alone (as in the spacetime of general relativity). In fact, if we consider the inertia as well then we may say: a particle is free (according to general relativity) if it is under the influence of only inertia or/and gravity (i.e. it is not under the influence of any force). This should be appreciated in the context of the principle of equivalence (see § 1.8.2) where inertia and gravity are unified.
- **Inertial mass** is a quantitative measure of the tendency of massive objects to resist any attempt by a foreign agent to change their state of rest and motion (i.e. being accelerated). Accordingly, we need to apply a physical force to change the state of a massive object from rest to motion or from motion to rest or change its speed or/and direction of motion. The inertial mass is formally represented (and defined in a sense) by Newton's second law of motion and hence it is associated with acceleration.
- **Gravitational mass** is a quantitative measure of the pulling force (or attraction) that a massive object experiences in the presence of another massive object. Accordingly, in the presence of a given massive object A, object B which is made of certain amount of matter (i.e. it has a given gravitational mass) will experience a stronger attractive force than object C that is made of a smaller amount of matter (i.e. it has a smaller gravitational mass) assuming that the distance between A and B and the distance between A and C are equal. The gravitational mass is formally represented (and defined in a sense) by Newton's law of gravitation and hence it is associated with gravitational field.
- **Tidal force** in the context of gravity means a force that originates from a gradient in the gravitational field due to its non-uniformity and hence different parts of an object in such a field experience forces of

---

[18] In fact, this also applies to the spacetime of special relativity (see exercise 5). Also, due to physical considerations the sensibility of this may be restricted to timelike intervals (or at least lightlike intervals are excluded because they are null).

different strength or/and direction and this results in physical effects that tend to distort the object such as stretching the object along a given orientation. In fact, we may define tidal force more generally and succinctly as differential force across a distance that spans one (or more) physical object.
- **Gravitating object** or body means a source of gravitational field while **gravitated object** or body means a body affected by (or being under the influence of) the gravitational field of the gravitating object. We usually use the "gravitating object" term to refer to the larger (i.e. the more massive) of the two objects while we normally use the "gravitated object" term to refer to the smaller object. So, in the gravitational relation between the Sun and the Earth the Sun is the gravitating object while the Earth is the gravitated object. We may also refer to such two objects as gravitating objects or gravitating system.
- We commonly use the term "**other forces**" to exclude gravity although gravity is not a force in a technical sense according to general relativity.
- We may use **free fall gravitation** to mean motion in the absence of any force other than gravity. However, there are some details about this issue which will be discussed in the Exercises.
- **Test particle** (or test object) is a massive object whose mass is so tiny that it does not have tangible effect on the gravitational field that influences the test particle.
- **Weak gravitational field** may be defined as a field in which the gravitational potential energy of a test particle is negligible in comparison to its mass energy while **strong gravitational field** is a field in which the gravitational potential energy is comparable to the mass energy (also see exercise 19 of § 4.1.1). As we see, there is no sharp border between the two and hence what is weak and what is strong may depend on the case and context. It is generally accepted that Newtonian gravity is a valid approximate theory in weak fields but not in strong fields although this should depend on the case and context as well as other factors (as will be investigated later).
- **Local** means a confined part of space (or spacetime) over which the effects of curvature, non-linearity, non-uniformity, non-homogeneity, non-isotropy, etc. are negligible. In this context, we should note that terms like "locally flat" (which frequently occur in the literature of relativity including this book) may be used in two meanings: as approximation (and hence it applies at any location even in curved spaces assuming they are Riemannian) and as strict condition and hence it applies only to the strictly flat parts of space (e.g. plane that is smoothly connected to a semi-sphere). The technical distinction between the two (within the context of the relativity theories) is that in the first case the metric of the spacetime region is given by $g_{\mu\nu} \simeq \eta_{\mu\nu} + \varepsilon_{\mu\nu}$ (where $\eta_{\mu\nu}$ is the Minkowski metric tensor and $\varepsilon_{\mu\nu}$ is a perturbation tensor such that $|\varepsilon_{\mu\nu}| \ll 1$) while in the second case the metric of the spacetime region is given by $g_{\mu\nu} = \eta_{\mu\nu}$.

Problems
1. What is the relation between Minkowskian space and Riemannian space?
    Answer: Minkowskian space is an instance (or special case) of Riemannian space (or rather pseudo-Riemannian space). In fact, Minkowskian space is a flat 4D pseudo-Riemannian space (or rather spacetime).
2. Define inertial and gravitational mass in a few words.
    Answer: Inertial mass is a measure of the resistance to kinematical change (i.e. the state of rest and motion) while gravitational mass is a measure of the strength of gravitational attraction.
3. Classify space intervals in pseudo-Riemannian space.
    Answer: In pseudo-Riemannian space, space intervals are classified (according to the sign of their quadratic form) as timelike when $(ds)^2 > 0$, lightlike when $(ds)^2 = 0$, and spacelike when $(ds)^2 < 0$. We note that the classification with regard to the sign of the quadratic form may be reversed [i.e. timelike when $(ds)^2 < 0$ and spacelike when $(ds)^2 > 0$] depending on the adopted convention about the sign pattern of the metric like $+---$ or $-+++$ (similar to the situation in special relativity where the Minkowski metric may be given as $\text{diag}[+1,-1,-1,-1]$ or as $\text{diag}[-1,+1,+1,+1]$). In this book we follow the $+---$ sign pattern and hence $(ds)^2 > 0$ for timelike and $(ds)^2 < 0$ for spacelike.
    Note: an issue related to the above classification is the classification of vectors where a vector (e.g. **V**) is classified (according to the sign of its square which is given by the inner product $\mathbf{V} \cdot \mathbf{V}$) as timelike when $\mathbf{V} \cdot \mathbf{V} > 0$, lightlike when $\mathbf{V} \cdot \mathbf{V} = 0$, and spacelike when $\mathbf{V} \cdot \mathbf{V} < 0$. For example, the velocity 4-vector **U** is timelike because we always have $\mathbf{U} \cdot \mathbf{U} = c^2 > 0$ (see our book "The

Mechanics of Lorentz Transformations"). Also, a trajectory (or curve or world line) is described as timelike/lightlike/spacelike when its tangent vector is timelike/lightlike/spacelike.

Exercises

1. Make a brief comparison between the space of special relativity and the space of general relativity.
2. Clarify the issue of line element and quadratic form in general nD space and in the spacetime of general relativity.
3. What is the essence of the definition (which is given in the text) of flat space?
4. State some of the distinctive properties of pseudo-Riemannian spaces.
5. Give a detailed and formal definition of "spacetime interval" in special relativity.
6. What "free particle" means in special relativity and in general relativity? What is common to both? Also, link this to the concepts of inertia and gravity and hence find a more fundamental common factor.
7. Give examples of massive free objects (i.e. under the influence of inertia or/and gravity alone) in the spacetime of general relativity.
8. Is there a difference between "free" and "freely falling"?
9. Compare inertial mass and gravitational mass.
10. What is the relation between inertial mass and acceleration?
11. Give brief definitions of the following terms: pseudo-Riemannian space, null geodesic, tidal force, free fall gravitation, and test particle.
12. Give a mathematical condition that distinguishes between Riemannian space and pseudo-Riemannian space.
13. Give some examples of the significance of the locality condition in various physical contexts.
14. Give common examples of how curved spaces of various dimensionality look flat locally. Also justify your findings.
15. Give mathematical definitions for "locally Euclidean space" and "locally flat spacetime". Also explain "local Cartesian coordinate system" and "local inertial frame".
16. Use the answer of exercise 15 to give mathematical conditions for local flatness in Riemannian and pseudo-Riemannian spaces.

1.4 General Conventions, Notations and Remarks

In the following points we provide some notes about the conventions and notations that we use in this book. We also include general remarks that we need for future investigation.

• Due to the restrictions on the size of the book, some additional materials (like some non-essential proofs) are not included in this book. However, these additional materials are mostly given in our other books. So, for completeness and to satisfy the need of the readers who are interested in these details, we make frequent references to our books using the abbreviations B2, B3, B4, B2X, B3X, B4X where we linked these abbreviations to the books in the References of this book.

• We use the summation convention in the tensor formulations and hence a twice-repeated index in a tensor term like i in $A_i A^i$ means summing over the range of i.

• The Latin indices generally range over $1, 2, 3$ while the Greek indices generally range over $0, 1, 2, 3$. The Latin indices usually represent the three spatial coordinates while the Greek indices represent the four spacetime coordinates (one temporal indexed with 0 and three spatial indexed with $1, 2, 3$). However, the Latin indices may also be used for general nD spaces of non-specific dimensionality (see for example § 2).

• The mathematical quantities in this book are generally real, i.e. not imaginary or complex. Accordingly, all the arguments of real-valued functions that are not defined for negative quantities, like square roots and logarithmic functions, are assumed non-negative by taking the absolute value, if necessary, without using the absolute value symbol. This is to simplify the notation and avoid potential confusion with other notations. So, $\sqrt{g}$ means $\sqrt{|g|}$ and $\ln(g)$ means $\ln(|g|)$.

• We use "classical mechanics" in a more general sense than that we used in our book "The Mechanics of Lorentz Transformations" (i.e. B4) by including classical gravity as represented by Newton's gravitation

1.4 General Conventions, Notations and Remarks

theory. We also use "classical gravity" (or similar tags) to refer to the Newtonian gravity. Also, "classical" may be used occasionally as opposite to quantum mechanical (and hence even the relativity theories are classical in this sense).

- Although we frequently express personal opinions, the book in general is based on the common views held by the mainstream scientists and presented in the mainstream literature of general relativity (or at least they are obtained from an objective analysis of the theory and its formalism). Therefore, the views and style of presentation in the book should not be interpreted as personal views. In fact, some of these contradict our views as expressed in the present book and in our previous books.
- We provided a Nomenclature list (in the front of the book) which the reader is advised to consult when needed for interpreting the mathematical symbols and abbreviations. Therefore, we do not provide such explanations within the text systematically although we usually explain the symbols at the place of their first occurrence or when there is a potential confusion with other symbols.
- Because there are many symbols in this book, due to its highly mathematical nature, and because we prefer to use the conventional symbols that are commonly used in the literature of this subject, some symbols are used in more than one meaning, e.g. G for the gravitational constant and for the trace of Einstein tensor. Therefore, to avoid confusion we generally explain the appropriate meaning of such symbols within the text although we might rely on the context for clarification.
- We generally use "space" in its technical mathematical sense and hence the 4D spacetime is a space in this sense. However, "space" may also be used occasionally to mean the ordinary 3D "spatial" space. The meaning should be generally obvious from the context.
- "Matter" is usually a label for massive objects. However, for brevity "matter" may include energy in some contexts where the meaning is obvious. This is justified by the equivalence between mass and energy and the fact that the source of gravity in general relativity includes energy and is not restricted to mass as it is the case in classical gravity.
- We use "gravitational frame" or "gravitating frame" to refer to a frame of reference in which gravitational fields do exist, i.e. the spacetime in which the frame exists is not void of matter and energy and the frame is at a finite distance from the source of gravity. These tags may also be restricted to frames in gravitational fields but they are not in a state of gravitational free fall.
- Unlike our book "The Mechanics of Lorentz Transformations" where we used "Lorentz mechanics" systematically instead of "special relativity", we commonly use special relativity in the present book to refer to this subject due to the familiarity of the readers with this term as well as the correspondence with general relativity. Moreover, we made our point in that book and there is no strong reason to do the same in this book.
- Unlike our previous books where we used different notations for the coordinates of Cartesian and general coordinate systems, in this book we use contravariantly-indexed x (e.g. x^i) for both and the distinction (if needed) will rely on the context if no explicit statement is made. We may also distinguish the Cartesian coordinates by indexing them with lower indices (e.g. x_i) since in Cartesian systems the covariant and contravariant types are equivalent. However, in a few cases the lower index is used to distinguish the covariant form of general coordinates.
- We use "the Field Equation" as singular or plural depending on the context and convenience where the singular refers to its symbolic or tensorial form while the plural refers to its components. This also applies to other tensorial equations like the geodesic equation.
- Physical quantities (such as energy and momentum) in this book are generally relativistic (or what we call Lorentzian) unless it is stated or indicated otherwise. However, the context should be considered to determine the ultimate meaning.
- "Radiation" in general relativity and cosmology is commonly used to refer to massless particles and hence it is not restricted to electromagnetic radiation. We may use this without clarification relying on this convention as well as context.
- For simplicity and convenience, in the case of rank-2 tensors we commonly use tensor and matrix (which represents the tensor) interchangeably. However, the matrix symbol is usually distinguished by square brackets, e.g. the tensor g_{ij} is represented by the matrix $[g_{ij}]$.
- We generally follow the common convention of using the indicial notation of tensors to symbolize and

represent the components of tensors as well as the tensors themselves. For example, A^i can represent the contravariant type of a vector $\mathbf{A}$ as well as its ith component. The meaning should be obvious from the context if no explicit explanation is given (e.g. the vector A^i or the component A^i). The rules of index pattern may also be used as a good intuitive indicator (although not always definite) to the nature of expressions, equations and relations, e.g. $A^i{}_{jk} = B^i{}_{jk}$ should suggest a tensor equation while $A_{ij} = B_{ji}$ or $A^j_i = B^i_j$ must be a component relation between different tensors.

- Regarding the mathematics of tensors, we generally follow the old school approach where tensors and their main characteristics are generally distinguished by the type and pattern of indices and hence we talk for example about covariant and contravariant basis vectors and tensors avoiding the mathematics and terminology of differential forms and dual vectors. This is to avoid unnecessary complications and needless overheads. Moreover, in our view the old approach is more intuitive and physical in spirit than the mathematically-oriented modern approach.
- The common convention in the literature of relativity theories is to write the quadratic forms as ds^2 and $d\sigma^2$. However, to avoid confusion we do not follow this convention. Instead, we write them as $(ds)^2$ and $(d\sigma)^2$.
- The term "acceleration" (and its alike like accelerating) may be used in its general sense to mean a change of the state of a physical object from rest to motion or from motion to rest or change of its speed or/and direction of motion. It may also be used to mean temporal rate of increase of velocity in translational motion specifically and hence it excludes for example deceleration and rotation. To avoid complications in the phrasing and presentation, we commonly rely on the context to determine the intended meaning.
- We use "relativity theories" in this book to mean the theories of special and general relativity specifically.
- The formulae, physical constants and calculations are generally given and performed in standard SI units (Système Internationale d'Unités).

1.5 Classical Gravity

The classical concept and formalism of gravity are embedded in the Newton's law of gravitation which states that any two massive objects attract each other by a force that is directly proportional to the product of their masses and inversely proportional to the square distance between them. Formally, this law is expressed as:

$$f = \frac{Gm_1m_2}{d^2} \tag{3}$$

where f is the magnitude of the attractive force, G ($\simeq 6.674 \times 10^{-11}$ N m^2/kg^2) is the gravitational constant,[19] m_1 and m_2 are the masses of the two objects and d is the distance between them. We note that "attract each other" indicates that the direction of the force is along the line joining the two objects.

The above gravitation law represents classical gravity in its simplest scalar form assuming, for instance, that the masses are point-like (or particles) with no spatial extension. For more general and technical form of classical gravity, where gravity is formulated as a field theory with the use of the concepts and techniques of calculus, we re-cast the above equation in a vector form where the gravity is assumed to be between a spherically-shaped gravitating body of mass M of spherical symmetric distribution and a gravitated test particle of mass m with the use of a normalized spherical coordinate system whose origin is at the center of M, that is:

$$\mathbf{f} = -\frac{GmM}{r^2}\mathbf{e}_r \tag{4}$$

where $\mathbf{f}$ is the gravitating force (i.e. the force that M exerts on m),[20] r is the radial distance of m from the origin of coordinates (which is at the center of M) and $\mathbf{e}_r$ is a unit radial vector. The minus sign is because the force of gravity is attractive and hence it points towards the origin.

---

[19] It is noteworthy that G (according to some theories) may vary in space and time and hence it is a function of spacetime coordinates although such variations (if they really exist) should be negligible for most common purposes and hence it is essentially constant.

[20] For simplicity, we use M and m to refer to the gravitating and gravitated objects (i.e. not only to their masses).

1.5 Classical Gravity

Next, we define the "gravitational force field" $\mathbf{g} \equiv \mathbf{g}(\mathbf{r})$ (with $\mathbf{r}$ being the position vector) as the gravitating force per unit mass of the test particle (and hence $\mathbf{g}$ is a vector field), that is:

$$\mathbf{g} \equiv \frac{\mathbf{f}}{m} = -\frac{GM}{r^2}\mathbf{e}_r \tag{5}$$

We then define the "flux" of the gravitational field as its area integral over a spherical surface S that encloses the entire M, that is:

$$\text{Flux} \equiv \int_S \mathbf{g} \cdot \mathbf{n}\, dA \tag{6}$$

where $\mathbf{n}$ is a unit vector that is normal to the surface and pointing outward and dA is an infinitesimal area element. Now, if we choose this surface to be the surface of M, which is a sphere of radius R, then we have:

$$\begin{aligned}
\int_S \mathbf{g} \cdot \mathbf{n}\, dA &= \int_S \left(-\frac{GM}{R^2}\mathbf{e}_r\right) \cdot \mathbf{e}_r\, dA \\
&= \int_S -\frac{GM}{R^2}\, dA \\
&= \int_{\theta=0}^{\theta=\pi}\int_{\phi=0}^{\phi=2\pi} -\frac{GM}{R^2} R^2 \sin\theta\, d\theta\, d\phi \\
&= \int_{\theta=0}^{\theta=\pi}\int_{\phi=0}^{\phi=2\pi} -GM \sin\theta\, d\theta\, d\phi \\
&= \int_{\theta=0}^{\theta=\pi} -2\pi GM \sin\theta\, d\theta \\
&= -4\pi GM
\end{aligned} \tag{7}$$

where in line 1 we use Eq. 5 (with $r = R$), line 2 is because $\mathbf{e}_r$ is a unit vector, in line 3 the generic area integral over the sphere of radius R is converted to its explicit form in terms of spherical coordinates, and in lines 5 and 6 we perform and evaluate the integrals over ϕ and θ respectively. It should be obvious that M inside the integral is independent of θ and ϕ.

Now, by the divergence theorem we have:

$$\int_S \mathbf{g} \cdot \mathbf{n}\, dA = \int_\Omega \nabla \cdot \mathbf{g}\, dV \tag{8}$$

where Ω is the region of space occupied by the gravitating object and dV is an infinitesimal volume element. On comparing the last two equations we get:

$$\int_\Omega \nabla \cdot \mathbf{g}\, dV = -4\pi GM \tag{9}$$

To eliminate the integral (which can be annoying in manipulation and calculation) we express M as an integral and hence we have:

$$\int_\Omega \nabla \cdot \mathbf{g}\, dV = -4\pi G \int_\Omega \rho\, dV \tag{10}$$

where ρ (which is a function of r alone) is the mass density of the gravitating object. Now, since this equation is valid regardless of the form of dV and the system that we use, then we can conclude:

$$\nabla \cdot \mathbf{g} = -4\pi G\rho \tag{11}$$

The last equation, which states that the divergence of the gravitational field is proportional to the mass density of the gravitating body, is valid in general although it is derived under certain restricting conditions and assumptions, e.g. spherical shape of gravitating body and spherical symmetry of its mass distribution. This also applies to the following equations that are derived from this equation.[21]

Now, although Eq. 11 is essentially a scalar equation (since the divergence of vector is scalar) it still involves vector quantities and operations. It is advantageous in applications to have a purely scalar form of this equation and hence we will do that to Eq. 11. We note first that the gravitational field is conservative and hence it can be expressed as a gradient of a potential field Φ (which is the gravitational potential energy per unit mass), that is:

$$\mathbf{g} = -\nabla \Phi \qquad (12)$$

Accordingly, Eq. 11 becomes:

$$\nabla \cdot (-\nabla \Phi) = -4\pi G \rho \qquad (13)$$
$$\nabla^2 \Phi = 4\pi G \rho \qquad (14)$$

where in line 2 we use the definition of the Laplacian operator ∇^2 as the divergence of gradient. The last equation, which is the Poisson equation for gravity, is the scalar form (or rather purely scalar form) of the Newtonian gravity as a field theory.

Problems

1. Find the magnitude of the Newtonian gravitational force between two point masses with $m_1 = 2.5$ kg and $m_2 = 10^5$ kg which are separated by a distance $d = 10^5$ km.
   Answer: We have:
   $$f = \frac{G m_1 m_2}{d^2} \simeq \frac{6.674 \times 10^{-11} \times 2.5 \times 10^5}{10^{16}} \simeq 1.669 \times 10^{-21}\,\mathrm{N}$$

2. What "conservative field" means?
   Answer: It is a field whose line integral is independent of the path, i.e. the value of its line integral depends only on the value of the field at the initial and final points of the path.

Exercises

1. Find the gravitational force field on the surface of the Earth (assuming it is a perfect sphere with spherically symmetric mass distribution).
2. Obtain a mathematical expression for the gravitational potential Φ.[22]
3. Give two properties of conservative vector field.
4. Prove that the gravitational field is conservative.
5. Verify the divergence theorem for the gravitational field of a gravitating body assuming it is a perfect sphere with spherically symmetric mass distribution.

1.5.1 Planetary Motion

Planetary motion in classical physics is part of the celestial mechanics which is based on Newton's laws of motion and Newton's law of gravity. However, from a historical perspective planetary motion is described by Kepler's laws which can be loosely stated as follows:
(a) The orbits of planets have elliptical shape with the Sun being at one focus.
(b) The areal velocity of planets is constant.[23]
(c) The squares of the orbital periods of planets are proportional to the cubes of the distances of the planets from the Sun.
It can be shown (see Problems) that these laws are direct consequences of Newton's laws (supported by certain definitions, conventions and assumptions as well as simple mathematical and physical facts).

---

[21] The generality may be inferred from the linearity (and hence superposition) plus the fact that any gravitating body can be considered as an assembly of particles each of which obeys the above equation (and hence the following equations).
[22] In questions like this we assume a single gravitating object with spherically symmetric (or point like) mass distribution.
[23] The areal (or sectorial) velocity (or speed) is the temporal rate of change of the area swept by a line joining an object to a given point in space. In planetary motion, the object is the planet while the given point is the center of the Sun.

1.5.1 Planetary Motion

In fact, planetary motion in classical mechanics is described by a system of two equations which are derived from Newton's laws (see Problems).[24] These equations are:

$$\frac{d^2 u}{d\phi^2} + u = \frac{GM}{B^2} \qquad (15)$$

$$\frac{d\phi}{dt} = Bu^2 \qquad (16)$$

where $u = 1/r$ with r being the radial distance in plane polar coordinates, ϕ is the polar angle in these coordinates, G is the gravitational constant, M is the mass of the gravitating object (i.e. the Sun in our case assumed to be at the origin of coordinates), B is a constant (which represents the magnitude of angular momentum per unit mass or twice the areal speed), and t is the time.

Using the above system of equations, it can be shown (see Problems) that the shape of the planetary orbits is described by the following equation:

$$r = \frac{e/A}{e \cos\phi + 1} \qquad (17)$$

where A is a positive constant (with physical dimensions of reciprocal length) and $e = \frac{AB^2}{GM}$. This is a standard equation of an ellipse (with eccentricity e) in polar coordinates; a fact that is inline with the statement of Kepler's first law (see Problems for details).[25]

Problems

1. Use Newton's law of gravity (supported by Newton's second law) to develop a system of equations that describe planetary motion (and hence show that it is the system given by Eqs. 15-16).
   Answer: We use a plane polar coordinate system (r, ϕ) where the Sun (i.e. the gravitating body to be more general) is at the origin of the coordinate system, r is the radial distance from the Sun to the planet (i.e. gravitated body to be more general), and ϕ is the polar angle (with $\phi = 0$ corresponding to the perihelion position).[26] Accordingly, the radial position $\mathbf{r}$ (noting that we are dealing with radial force), the velocity $\mathbf{v}$ and the acceleration $\mathbf{a}$ of the planet are given by (refer to mathematical textbooks):

$$\begin{aligned} \mathbf{r} &= r\mathbf{e}_r \\ \mathbf{v} &\equiv \dot{\mathbf{r}} = \dot{r}\mathbf{e}_r + r\dot{\phi}\mathbf{e}_\phi \\ \mathbf{a} &\equiv \dot{\mathbf{v}} = \left(\ddot{r} - r\dot{\phi}^2\right)\mathbf{e}_r + \left(r\ddot{\phi} + 2\dot{r}\dot{\phi}\right)\mathbf{e}_\phi \end{aligned} \qquad (18)$$

where the radial distance r and the unit vector $\mathbf{e}_r$ are functions of ϕ (and hence functions of time due to revolution) and the overdot symbolizes derivative with respect to time t (i.e. d/dt).[27] Now, the gravitational force is given by Eq. 4, that is:

$$\begin{aligned} \mathbf{f} &= -\frac{GmM}{r^2}\mathbf{e}_r \\ m\mathbf{a} &= -\frac{GmM}{r^2}\mathbf{e}_r \\ \mathbf{a} &= -\frac{GM}{r^2}\mathbf{e}_r \end{aligned} \qquad (19)$$

---

[24] Eq. 15 can also be derived from the conservation principles of energy and angular momentum.
[25] We are considering the case of ellipse based on physical considerations related to the limits on the eccentricity of the planetary orbits; otherwise the above formulation is more general.
[26] This coordinate system can be seen as the 2D section of a 3D spherical coordinate system in the plane $\theta = \pi/2$ (refer to § 1.5 and § 5.2). In fact, we are assuming that the planetary orbits are planar; a fact that is justified classically because gravity is a central force field (which is also inline with the conservation of angular momentum in the absence of torques).
[27] In fact, we can write: $r = r(\phi(t))$ and similarly $\mathbf{e}_r = \mathbf{e}_r(\phi(t))$.

1.5.1 Planetary Motion

where m and M stand for the mass of planet and Sun respectively and where in line 2 we use Newton's second law of motion. On comparing Eq. 18 with Eq. 19 we obtain:

$$\ddot{r} - r\dot{\phi}^2 = -\frac{GM}{r^2} \tag{20}$$

$$r\ddot{\phi} + 2\dot{r}\dot{\phi} = 0 \tag{21}$$

Now, if we multiply Eq. 21 by r we get:

$$r^2\ddot{\phi} + 2r\dot{r}\dot{\phi} = 0$$

$$\frac{d}{dt}\left(r^2\dot{\phi}\right) = 0$$

where in line 2 we use the product rule of differentiation. So, if we integrate the last equation with respect to t we get:

$$r^2\dot{\phi} = B \tag{22}$$

where B is a constant (with physical dimensions of angular momentum per unit mass or areal speed).[28] Accordingly, the system of Eqs. 20 and 21 becomes:

$$\ddot{r} - r\dot{\phi}^2 = -\frac{GM}{r^2} \tag{23}$$

$$r^2\dot{\phi} = B \tag{24}$$

Now, if we introduce a new dependent variable $u = 1/r$ (and hence $r = 1/u$) then we have:

$$\frac{dr}{dt} = -\frac{1}{u^2}\frac{du}{dt} = -\frac{1}{u^2}\frac{du}{d\phi}\frac{d\phi}{dt} = -\frac{du}{d\phi}r^2\dot{\phi} = -B\frac{du}{d\phi}$$

where in step 1 we use the power and chain rules (or composite rule), in step 2 we use the chain rule, step 3 is just a notational modification, and in step 4 we use Eq. 24. Similarly:

$$\frac{d^2r}{dt^2} = \frac{d}{dt}\left(\frac{dr}{dt}\right) = \frac{d}{dt}\left(-B\frac{du}{d\phi}\right) = -B\frac{d}{dt}\left(\frac{du}{d\phi}\right)$$

$$\frac{d^2r}{dt^2} = -B\frac{d}{d\phi}\left(\frac{du}{d\phi}\right)\frac{d\phi}{dt}$$

$$\frac{d^2r}{dt^2} = -B\frac{d^2u}{d\phi^2}\frac{d\phi}{dt}$$

$$\frac{d^2r}{dt^2} = -B\frac{d^2u}{d\phi^2}\frac{B}{r^2}$$

$$\ddot{r} = -B^2u^2\frac{d^2u}{d\phi^2} \tag{25}$$

where in line 2 we use the chain rule, in line 4 we use Eq. 24, and in line 5 we use the dot notation and $u = 1/r$. On substituting from Eq. 25 into Eq. 23 (and using $u = 1/r$) we get:

$$-B^2u^2\frac{d^2u}{d\phi^2} - \frac{1}{u}\dot{\phi}^2 = -GMu^2$$

$$-B^2u^2\frac{d^2u}{d\phi^2} - \frac{1}{u}B^2u^4 = -GMu^2$$

$$\frac{d^2u}{d\phi^2} + u = \frac{GM}{B^2}$$

---

[28] With a proper choice of coordinates, $\dot{\phi}$ is positive (as well as r^2) and hence B is positive.

1.5.1 Planetary Motion 26

where in line 2 we use Eq. 24, and in line 3 we divide by $-B^2 u^2$. Accordingly, the system of Eqs. 23 and 24 becomes:

$$\frac{d^2 u}{d\phi^2} + u = \frac{GM}{B^2} \qquad (26)$$

$$\frac{d\phi}{dt} = Bu^2 \qquad (27)$$

which is the same as the system of Eqs. 15-16. The last two equations represent the classical system of equations that describe the orbital motion of planets (and indeed the classical orbital motion in general) and hence the classical solution of the problems of planetary motion requires solving this system of equations (refer to the next problems and see § 5.2).

2. From the answer to problem 1, show that Kepler's first law is a consequence of Newton's laws.
Answer: To show that Kepler's first law is a consequence of Newton's laws it is sufficient to show that Kepler's first law can be derived from the system of equations that we developed in problem 1 (i.e. Eq. 26 to be more specific) since this system is obtained from Newton's laws. The solution of Eq. 26 is:[29]

$$u = A\cos\phi + \frac{GM}{B^2} \qquad (28)$$

(with A being a positive constant) as can be easily verified by substituting from the last equation into Eq. 26. From the last equation we get:

$$\frac{1}{r} = A\cos\phi + \frac{GM}{B^2}$$

$$r = \frac{1}{A\cos\phi + \frac{GM}{B^2}} \qquad (29)$$

$$r = \frac{\frac{B^2}{GM}}{\frac{AB^2}{GM}\cos\phi + 1} \qquad (30)$$

$$r = \frac{e/A}{e\cos\phi + 1} \qquad (31)$$

where in line 1 we use $u = 1/r$, in line 2 we take the reciprocal of both sides, in line 3 we multiply the numerator and denominator by $\frac{B^2}{GM}$, and in line 4 we symbolize $\frac{AB^2}{GM}$ with e. The last equation is a standard form of an ellipse with eccentricity e in polar coordinates where the origin of coordinates is located at one focus (refer to mathematical textbooks). This shows that the planetary orbit has elliptical shape with the Sun being at one focus (as stated by Kepler's first law) and hence Kepler's first law is a consequence of Newton's laws.
Note: the condition "with the Sun being at one focus" is implicit since the above derivation is based on choosing a coordinate system whose origin is located at the Sun. We should also note that with some minor amendments to the above equations and conditions the case of circular orbit may be included as a special case of elliptical orbit corresponding to $e = 0$ where the orbital equation becomes $r = \frac{B^2}{GM}$ (which corresponds to $A = 0$ in Eq. 28).

3. Show that the perihelion distance r_0 and the aphelion distance r_1 are given respectively by:[30]

$$r_0 = \frac{B^2}{GM(1+e)} \qquad \text{and} \qquad r_1 = r_0 \frac{1+e}{1-e}$$

Answer: As seen in problem 2, the equation of the planetary orbit is (see Eq. 29):

$$r = \frac{1}{A\cos\phi + \frac{GM}{B^2}}$$

---

[29] In fact, this is "a solution" and not "the solution" although it is chosen as "the solution" due to physical and practical considerations.

[30] Perihelion (aphelion) is the point in the orbit where the planet is closest to (farthest from) the Sun. Hence, the perihelion (aphelion) distance is the minimum (maximum) distance between the planet and the Sun.

1.5.1 Planetary Motion

Now, since $A > 0$ and $\frac{GM}{B^2} > 0$ then r is minimum (i.e. $r = r_0$) when $\cos\phi = 1$ (since the denominator is maximum when $\cos\phi = 1$) and r is maximum (i.e. $r = r_1$) when $\cos\phi = -1$ (since the denominator is minimum when $\cos\phi = -1$).[31] So, if we use another given form of the equation of planetary orbit (see Eq. 30), i.e.

$$r = \frac{\frac{B^2}{GM}}{\frac{AB^2}{GM}\cos\phi + 1} = \frac{\frac{B^2}{GM}}{e\cos\phi + 1} = \frac{B^2}{GM(1 + e\cos\phi)}$$

then we have:

$$r_0 = \frac{B^2}{GM(1+e)}$$

$$r_1 = \frac{B^2}{GM(1-e)} = \frac{B^2}{GM(1-e)}\frac{(1+e)}{(1+e)} = \frac{B^2}{GM(1+e)}\frac{(1+e)}{(1-e)} = r_0\frac{1+e}{1-e}$$

4. From the answer to problem 1, show that Kepler's second law is a consequence of Newton's laws.
   Answer: As in the answer of problem 2, we use the system of equations that we developed in problem 1 (i.e. Eq. 27 to be more specific) to show that Kepler's second law is a consequence of Newton's laws since that system is obtained from Newton's laws. Now, the infinitesimal area dA swept by the line joining the planet to the Sun in an infinitesimal time interval dt is given by:

$$dA = \frac{1}{2}r^2 d\phi$$

where $d\phi$ is the infinitesimal polar angle swept during dt. So, if we divide the last equation by dt we get:

$$\begin{aligned}\frac{dA}{dt} &= \frac{1}{2}r^2\frac{d\phi}{dt} \\ \frac{dA}{dt} &= \frac{1}{2u^2}\frac{d\phi}{dt} \\ \frac{dA}{dt} &= \frac{1}{2}B\end{aligned} \quad (32)$$

where in line 2 we use $r = 1/u$, and in line 3 we use Eq. 27. Now, since B is constant then the last equation means that the areal velocity of planet is constant (as stated by Kepler's second law) and hence Kepler's second law is a consequence of Newton's laws.
   Note: Eq. 32 shows that the areal speed is constant. The constancy of the areal velocity can then be concluded from the fact that the orbit is planar as indicated earlier. In fact, we are assuming that the original statement of the Kepler's second law is about areal velocity rather than areal speed; otherwise we do not need to show the constancy of direction.

5. Show that Kepler's second law is a demonstration of the principle of conservation of angular momentum.
   Answer: As demonstrated in problem 4, Kepler's second law can be derived from Eq. 27. So, to show that Kepler's second law is a demonstration of the principle of conservation of angular momentum we just need to demonstrate that Eq. 27 implies the conservation of angular momentum. Now, if we multiply Eq. 27 with m (i.e. the mass of planet) we get:

$$\begin{aligned}m\frac{d\phi}{dt} &= mBu^2 \\ mr^2\frac{d\phi}{dt} &= mB\end{aligned}$$

---

[31] In fact, this originates from our choice of the coordinate system where the solution $u = A\cos\phi + \frac{GM}{B^2}$ is based on the condition that $\phi = 0$ corresponds to the perihelion position (as stated in problem 1). We should also remark that the condition $A > 0$ (which we stated previously with no justification) can be justified by the fact that $r > 0$ and hence from Eq. 31 we get $\frac{e/A}{e\cos\phi+1} > 0$. Now, since for ellipse $0 < e < 1$ then $(e\cos\phi + 1) > 0$ and hence A must be greater than zero if r to be greater than zero.

1.5.1 Planetary Motion

where in line 2 we use $u = 1/r$. Now, according to classical mechanics $mr^2 \frac{d\phi}{dt}$ is the magnitude of the planet angular momentum, and since m and B are constants then the last equation expresses the conservation of angular momentum (at least its magnitude). Again, the constancy of direction can be concluded from the planar nature of the orbit.

6. From the answer to problem 1, show that Kepler's third law is a consequence of Newton's laws.

Answer: In the answer to the present problem we use a few elementary mathematical facts as well as the results that we obtained already in the previous problems which are derived from the system of equations that we developed in problem 1, and since that system is ultimately based on Newton's laws then what we will obtain here is a consequence of Newton's laws.

It is well known that for ellipse we have:

$$b = a\sqrt{1 - e^2} \tag{33}$$

where a is the semi-major axis, b is the semi-minor axis and e is the eccentricity. It is also obvious that the sum of the perihelion distance r_0 and the aphelion distance r_1 is twice the semi-major axis, that is:

$$\begin{aligned}
2a &= r_0 + r_1 \\
a &= \frac{r_0 + r_1}{2} \\
a &= \frac{1}{2}\left(\frac{B^2}{GM(1+e)} + \frac{B^2}{GM(1-e)}\right) \\
a &= \frac{B^2}{2GM}\left(\frac{1}{1+e} + \frac{1}{1-e}\right) \\
a &= \frac{B^2}{2GM}\left(\frac{1-e}{1-e^2} + \frac{1+e}{1-e^2}\right) \\
a &= \frac{B^2}{2GM}\left(\frac{2}{1-e^2}\right) \\
a &= \frac{B^2}{GM(1-e^2)}
\end{aligned} \tag{34}$$

$$\tag{35}$$

where line 3 is from the results of problem 3 while the other lines are based on simple algebraic manipulation. Now, the area of ellipse is πab, so if the time period of a complete revolution is T then the areal speed of planet (which is constant) should be given by:

$$\begin{aligned}
\frac{dA}{dt} &= \frac{\pi ab}{T} \\
\frac{B}{2} &= \frac{\pi ab}{T} \\
T &= \frac{2\pi ab}{B} \\
T &= \frac{2\pi a^2 \sqrt{1-e^2}}{B} \\
T &= \frac{2\pi a^2}{\frac{B}{\sqrt{1-e^2}}} \\
T &= \frac{2\pi a^2}{\sqrt{GM}\left(\frac{B}{\sqrt{GM}\sqrt{1-e^2}}\right)} \\
T &= \frac{2\pi a^2}{\sqrt{GM}\sqrt{a}}
\end{aligned}$$

$$T = \frac{2\pi}{\sqrt{GM}} a^{3/2}$$
$$T^2 = \frac{4\pi^2}{GM} a^3 \qquad (36)$$

where in line 2 we use Eq. 32, in line 4 we use Eq. 33, and in line 7 we use Eq. 35. Now, since $\frac{4\pi^2}{GM}$ is constant then the last equation means that the square of the planetary time period is proportional to the cube of the mean distance (which is represented by a since a is the mean of r_0 and r_1 as seen in Eq. 34). This result is no more than Kepler's third law and hence this law is a consequence of Newton's laws.

Note: for circular orbit, the derivation of the Kepler's third law from Newton's laws can be obtained more easily by equating the magnitude of centripetal acceleration $\omega^2 r$ (where $\omega \equiv \frac{2\pi}{T}$ is the angular speed and r is the orbital radius) to the magnitude of gravitational acceleration $\frac{GM}{r^2}$, that is:

$$\omega^2 r = \frac{GM}{r^2}$$
$$\frac{4\pi^2}{T^2} = \frac{GM}{r^3}$$
$$T^2 = \frac{4\pi^2}{GM} r^3$$

which is the same as Eq. 36 with $r \equiv a$.

Exercises
1. What is the relation between Kepler's laws and Newton's law of gravity?
2. Show that for a binary system whose objects revolve around their center of mass Kepler's third law keeps its form with a standing for the semi-major axis of the ellipse that represents the orbit of one object in the rest frame of the other object while M stands for the sum of the masses of the two objects. (Hint: you can simplify the analysis by assuming circular motion.)

1.6 General Relativity versus Classical Gravity

The classical theory of gravitation is based on considering gravity as a force originating from matter (mass) while general relativity is essentially a geometric theory in which gravity is seen as an effect of the distortion (or curvature) of spacetime due to the presence and movement of matter and energy. In brief, classical gravity is a physical force like electromagnetic force while general relativistic gravity is a geometric effect of spacetime distortion. In the following points we briefly compare the two theories:
• Classical theory of gravity is a physical theory while general relativity is a geometric theory of a physical phenomenon.
• Gravity in classical theory is a physical force while gravity in general relativity is an effect of the spacetime curvature.
• Classical gravity has no temporal dependency and hence its space is the familiar Euclidean 3D space while general relativity is a theory of spacetime and hence it belongs to a 4D Riemannian space which includes a temporal dimension as well as spatial dimensions.
• General relativity converges to classical gravity in the classical limit, i.e. weak, time independent gravitational field at low speed with the source of gravity being restricted to matter (see § 5).

We should remark that in the above comparison we essentially restrict our attention to general relativity as a gravity theory, i.e. we do not consider its status as a "General Theory" with its extended formalism. In fact, from this perspective this should make another difference between classical gravity and general relativity since general relativity is more general than classical gravity because it claims to provide general rules of transformation between reference frames and hence it formulates the laws of physics in an invariant form in all types of frame.

Exercises

1. What are the manifolds that underlie the classical theory of gravity and general relativity?
2. Outline the different philosophical views which the classical theory of gravity and general relativity rest upon.

1.7 General Relativity versus Special Relativity

In brief, both special relativity and general relativity are theories about spacetime but the spacetime of special relativity is void of matter and energy[32] (and hence it is flat) while the spacetime of general relativity is not void (and hence it is curved). In the following points we briefly compare the two theories:
• Special relativity is a theory about spacetime coordinate transformations (i.e. Lorentz transformations) while general relativity is essentially a theory about gravity. However, because gravity in general relativity is an effect of the curvature of spacetime due to the distortion caused by the distribution of matter and energy, the two theories share the common factor of being theories about spacetime.[33]
• While the spacetime in special relativity is flat Minkowskian, the spacetime in general relativity is curved Riemannian.
• General relativity supposedly converges to special relativity in weak and vanishing gravitational fields because in these cases the spacetime is approximately flat (or flat).
• Special relativity applies locally in the spacetime of general relativity (i.e. even in strong gravitational fields) since this spacetime is locally flat although it is globally curved.

Regarding the relation between special relativity and general relativity and if special relativity is contained in general relativity or they are two different theories, it is not a clear cut issue and hence both opinions have followers and supporters. The opinion that special relativity is contained in general relativity may be supported by the following:
(a) The domain and subject matter of the two theories is spacetime where this spacetime is curved in general relativity and flat in special relativity. Hence, considering the domain of the two theories, the special-general relation is justified since flat space is a special case of curved space.
(b) Special relativity applies locally in the spacetime of general relativity and hence in this sense it is a special case of general relativity and contained in it. Similarly, general relativity supposedly converges to special relativity in weak and vanishing gravitational fields and hence special relativity is contained (as a special case) in general relativity.
(c) The invariance principle of general relativity is more general than the invariance principle of special relativity (see § 1.8.1).
(d) According to the equivalence principle (see § 1.8.2), accelerating frames (which are supposedly more general than inertial frames of special relativity) are equivalent to gravitational frames which are the primary subject of general relativity.

On the other hand, the opinion that special relativity and general relativity are different theories may be supported by the following:
(A) Special relativity is not contained in general relativity in the sense that the formalism of special relativity can be obtained from the formalism of general relativity[34] like classical mechanics in its relation to Lorentz mechanics (see B4).
(B) The local application of special relativity in the spacetime of general relativity is because this condition is imposed rather arbitrarily in general relativity (based on the local flatness of the spacetime of general

---

[32] In fact, it is void of matter and energy as source of gravity.

[33] As indicated earlier and will be discussed further later on, general relativity is also a "General Theory" and not only a gravity theory. So, from the "General Theory" perspective it may be seen as a generalization to special relativity due to the local application of special relativity in the spacetime of general relativity and hence it is also a theory of spacetime from this perspective.

[34] The formalism of general relativity here refers to the gravitational formalism (as represented by the Field Equation and its direct consequences) although the formalism of general relativity supposedly includes the formalism of special relativity in its local application. In brief, the gravitational component of the formalism of general relativity (i.e. the Field Equation and its consequences) and the Lorentzian (or special relativistic) component of the formalism of general relativity (i.e. Lorentz transformations and their consequences) are two independent formalisms that are pieced together (rather arbitrarily) in general relativity where no one of these formalisms can be obtained from the other (like obtaining classical formalism from Lorentzian formalism as a special case which we demonstrated in B4).

relativity) and hence it is not a natural convergence of the formalism of the theory itself. In fact, the local application of special relativity may qualify special relativity to be the more general theory from this perspective. Similarly, the alleged convergence to special relativity in weak and vanishing gravitational fields (also based on the approximate flatness and flatness of spacetime in these cases) is a convergence of the underlying spacetime and not a convergence of one formalism to the other. Yes, the geometric paradigm of general relativity (i.e. gravity is an attribute of spacetime curvature) implies such a convergence but it is not the same as the actual convergence of one formalism to the other. Moreover, this requires physical evidence to be fully established unlike the convergence of the formalism which is a fact that can be established by purely theoretical means. To demonstrate the nature of this alleged convergence (whether the convergence locally or the convergence in weak and vanishing fields) we need just to compare the convergence of the formalism of general relativity to the formalism of Newtonian gravity in the classical limit (see § 5) with the alleged convergence of general relativity to special relativity to see the obvious difference between the types of these convergences. We may also consider in this comparison the convergence of the formalism of special relativity to the formalism of classical mechanics in the low-speed limit.

(C) Special relativity is about transformation of spacetime coordinates which should apply (when its conditions are met) to any physical subject and theory[35] while general relativity is primarily about gravity and hence special relativity is the more general theory from this perspective (see exercise 6). The requirement of the strong equivalence principle (see § 1.8.2) which extends the theory of general relativity to all physical laws is arbitrary; moreover it requires validation by physical evidence.[36]

Exercises
1. Give examples of common features between special relativity and general relativity.
2. Give examples of features that distinguish special relativity and general relativity from each other.
3. Compare the fundamental principles which special relativity and general relativity rest upon.
4. List some of the features (such as foundations, implications and predictions) of special relativity and the features of general relativity.
5. What "special" and "general" mean with regard to the two relativity theories? Justify these labels.
6. It is commonly claimed that special relativity is less general than general relativity. Can we claim that in a sense special relativity is more general than general relativity?
7. Assess the relation between special relativity and general relativity and if special relativity is contained in general relativity or they are two different theories.

1.8 General Principles

In this section we discuss the principles which general relativity is based on. We note that these principles are mostly epistemological in nature with scientific implications although some have prime and explicit scientific content.

1.8.1 The Principle of Invariance

The essence of this principle is that observers in different frames of reference between which certain coordinate transformations apply should observe and record the same laws of physics, i.e, the laws of physics take the same form regardless of the frame of reference.[37] This principle may also be stated as:

---

[35] As explained earlier (see § 1.2), special relativity (or rather Lorentz mechanics) should have been extended to include even gravity if Lorentz transformations are the right transformations.

[36] We should note that general relativity (even as a gravity theory) also provides a sort of transformations between local frames in a given spacetime (as demonstrated for example by time dilation where the time in one local frame is correlated to the time in another local frame). However, these "local transformations" are not for the purpose of transforming the laws invariantly between frames.

[37] In the literature of general relativity, the principle of invariance is commonly known as the principle of covariance (or general covariance). However, we mostly avoid using this term to avoid confusion with covariance of tensors. This should be further justified by our intention (in this section at least) to investigate the principle of invariance in general whether in general relativity or elsewhere. We note that covariance in this context means form invariance. We should

1.8.1 The Principle of Invariance

all frames of reference (possibly of a certain category such as inertial) are suitable for formulating the laws of physics. For example, in classical physics the laws of mechanics take the same form in all inertial frames under the Galilean transformations. This similarly applies to Lorentz mechanics (or special relativity), since the domain of this mechanics is also inertial frames, but under the Lorentz transformations. In general relativity, this invariance principle is extended to include all frames and observers (although usually locally) and hence the "inertial" restriction is lifted. Accordingly, the principle of invariance in general relativity is more general than that in classical mechanics and in Lorentz mechanics.

The justification of the principle of invariance is that objective laws of physics should not depend on our conventions and how we choose our coordinate systems and frames of reference. In other words, there should be no privileged frame of reference in which the laws of physics take a particular form. Accordingly, real physical laws that truly describe the behavior of Nature in an objective and impartial way should be expressed in a form that is independent of observers and frames of reference.

The principle of invariance of physical laws can be characterized by the following:

1. This principle is an epistemological principle rather than a scientific principle. In other words, it is a sort of philosophical convention about what should be considered a valid physical law and what should not be considered so. Accordingly, this principle is not a scientific principle that is based on experiment and extracted from observation of Nature (such as the principle of energy conservation). Instead, we choose a certain set of transformations and we label those rules of Nature that remain invariant under these transformations (according to our observations) as laws of physics.
2. This principle is about the form of physical laws, i.e. any valid physical law should take the same form in different frames of reference under certain transformations. In our book B4 we distinguished between form invariance and value invariance. Accordingly, the principle of invariance is about form invariance and not about value invariance.
3. This principle requires a domain of validity which determines the type of frames to which this principle applies. This domain may be a particular category of reference frames or all reference frames. For example, the domain of this principle in classical mechanics and in Lorentz mechanics is inertial frames while the domain in general relativity is reference frames in general whether inertial or accelerating or gravitational.
4. This principle requires a set of coordinate transformations under which the laws are required to keep their form. For example, in classical mechanics the principle of invariance requires the form of legitimate physical laws to be invariant under the Galilean transformations, while in Lorentz mechanics it requires this form to be invariant under the Lorentz transformations. In general relativity, it supposedly requires the form to be invariant under "general transformations".
5. The validity of the invariance principle is restricted to the determination of which rule is a legitimate candidate for becoming a physical law and hence the invariance principle has no validity in determining the actual physical laws. The latter validity belongs to the available experimental and observational evidence in support of the rule. In other words, form invariance is a necessary but not sufficient condition for a rule to be a law of physics, i.e. all laws of physics should be form invariant but not all form invariant rules should be laws of physics. In fact, this restriction is particularly important in the framework of general relativity (and especially in assessing the validity of the principles of equivalence and *General Covariance*) where we will see that the existence of a valid formal transformation does not guarantee the validity of the transformation physically in the real world.

We should remark that from a historical perspective, the principle of invariance started with the laws of mechanics which are considered invariant by Galileo and the pioneers of classical mechanics under what is subsequently called the Galilean transformations. It was then extended to include electromagnetism (and later to all laws of physics) by Poincare and the pioneers of Lorentz mechanics where the Galilean transformations were replaced by the Lorentz transformations (with some amendments to the laws of mechanics).[38] The principle was then applied with further elaboration and generalization during the

---

also draw the attention of the readers to the existence of a technically-specific *Principle of General Covariance* which will be discussed in § 1.8.2.

[38] Some historical records seem to indicate that the generalization of the principle of invariance belongs to Maxwell who was quoted to talk about "the doctrine of relativity of all phenomena" and "a formula of the same type" (although seemingly

1.8.1 The Principle of Invariance

development of general relativity and incorporated in its more general form in this theory.[39] So, if we include this historical factor in the characterization of the principle of invariance then we can say: the principle of invariance is characterized by its domain (e.g. inertial frames), type of transformations (e.g. Galilean transformations) and subject (e.g. mechanics).

Another remark is that since general relativity is essentially a theory of gravity, there is no actual coordinate transformations between frames in this theory. Yes, it is claimed in this theory that the results and rules of Lorentz mechanics, which are based on the validity of Lorentz transformations, apply locally but this is not the same as applying transformations between frames as it is the case in classical mechanics with regard to the Galilean transformations or in Lorentz mechanics with regard to the Lorentz transformations.[40] This difference arises from the fact that, unlike Lorentz mechanics for example which is a theory about coordinate transformations, general relativity is fundamentally a theory of gravity and not a theory of coordinate transformations of spacetime although it is based on formulating gravity in terms of spacetime and its geometric properties. Accordingly, we may question the general relativistic claims about the generality of the invariance principle with regard to frames of reference (i.e. domain) and if this is implemented and realized in a meaningful way by the theory. Yes, it is claimed that the same laws of physics apply in all frames of reference, and hence the laws should be cast in an invariant tensorial form and they are subject to the most general type of transformations, even though we may not actually apply coordinate transformations as we do in Lorentz mechanics. In fact, we can question even the generality of the invariance principle in general relativity with regard to the laws of physics (i.e. subject) because even if we accept the claims that general relativity passed all the gravitational tests (see § 9), there is no concrete evidence for the invariance of all physical laws in gravitational and accelerating frames. The reality is that the strong equivalence principle (see § 1.8.2) is not experimentally verified and hence the verified gravitational results may establish the weak, but not the strong, equivalence principle. In other words, the claimed evidence in support of general relativity is related to the theory as a gravity theory and not as a "General Theory". In brief, as a gravity theory general relativity is not a theory of transformation and hence it is not concerned with the invariance of physical laws, while as a "General Theory" there is no actual implementation or demonstration or verification of the alleged invariance and hence this invariance is a trivial claim with no substantial evidence in its support (and possibly there is actual evidence against it as it might be claimed based on the evidence against the general validity of the strong equivalence principle). In fact, even the logical essence of the strong equivalence principle (as well as the *Principle of General Covariance*) is questionable. As indicated earlier, the existence of a valid formal transformation does not guarantee the validity of the transformation physically because the invariance is a necessary but not sufficient condition for determining what should be a law of physics and what should not be.

Problems
1. Justify the demand for tensor calculus in formulating the laws of physics.
   Answer: This demand may be justified by the requirement of the principle of invariance because tensor formulation is invariant under coordinate transformations. So, the equations that represent physical laws are guaranteed to be form invariant under coordinate transformations if they are cast in tensor form.
2. What is the implication of the premise that the principle of invariance is an epistemological principle rather than a scientific principle?

---

without suggesting a specific type of transformations). The invariance of the speed of light (which is related to the principle of invariance and the principle of relativity) may also be inferred from his writings and work. In fact, even the roots of what we now call special relativity (or rather Lorentz mechanics) can be detected in some of the writings of Maxwell.

[39] Our view (which is based on the available historical evidence) is that the principle of invariance in its general form (i.e. the requirement that *all* physical laws should be form invariant under certain coordinate transformations) was coined by the pioneers of Lorentz mechanics (like Poincare) although there seems to be no actual work on implementing or demonstrating or verifying this principle outside mechanics and electromagnetism. As we will see, the alleged implementation of this principle in general relativity is trivial and hence we do not see any priority to Einstein in generalizing this principle. This should also apply to the generalization of this principle with regard to the type of frames.

[40] In fact, Lorentz transformations are not valid globally in general relativity as in special relativity.

Answer: The obvious implication is that its value is rather restricted unlike the value of fundamental physical principles (e.g. the conservation of energy) which acquire their legitimacy and validity from the authority of Nature itself and hence they cannot be violated or modified by our needs or conventions. In other words, the principle of invariance determines what type of science we should create and construct and this can be susceptible to changes or modifications depending on our needs and conventions which are largely materialized in our philosophical and epistemological framework. For example, if we find it useful to use non-invariant set of "laws" in a certain frame or category of frames then there is nothing that can prevent us from doing this because there is nothing that is fundamentally wrong with this. In fact, large parts of physics are cast and used in non-invariant form (inconsistent with the demand of general relativity).[41]

Exercises

1. Discuss the "covariance" label that is attached in general relativity to the principle of "invariance".
2. Our assertion that the principle of invariance is an epistemological rather than a scientific principle may be challenged by the fact that certain laws at least (e.g. Newton's second law) are actually invariant regardless of any convention. How do we reply?
3. How is the principle of invariance implemented in general relativity?
4. Question, briefly, the applicability and generality of the principle of invariance in general relativity and if it is really implemented in general under any sort of coordinate transformations.
5. What is the difference between the invariance principle of special relativity (i.e. the first postulate) and the invariance principle of general relativity?
6. Briefly discuss the historical development of the principle of invariance with regard to the subject of this principle.
7. Lorentz tensor in the frames of 4D spacetime of Lorentz mechanics is the equivalent of the Jacobian matrix in spatial coordinate systems (see B4). So, what is the tensor that corresponds to the Lorentz tensor (i.e. Jacobian matrix) in general relativity?
8. Argue against the invariance principle of physical laws and demonstrate its limited value due to its epistemological and conventional nature.
9. Discuss the relation between the principle of relativity and the principle of invariance.
10. According to the principle of invariance there should be no privileged frame of reference in which the laws of physics take a particular form. Discuss this briefly.

1.8.2 The Principle of Equivalence

The principle of equivalence is essentially a scientific principle although it has epistemological contents and implications as well. In simple terms, the principle of equivalence states that a uniformly accelerated reference frame is locally indistinguishable from a frame in a gravitational field, i.e. acceleration in a given direction is equivalent in effect to gravitation of suitable magnitude in the opposite direction. This is inline with the classical view that acceleration introduces fictitious forces from the perspective of the accelerating frame and hence it may be seen as a justification of the general relativistic view that gravity is a "fictitious" force due to its equivalence to the effect of acceleration. As we will see, there are different forms of this principle in the literature of general relativity. Some of these forms are essentially equivalent or they differ in phrasing and presentation but others are fundamentally different, e.g. weak and strong forms. The roots of the equivalence principle can be traced back to the days of Galileo with his observation of the independence of the gravitational acceleration from the mass of the accelerated object which cannot be explained without assuming the equivalence between inertial mass which is related to acceleration and gravitational mass which is related to gravity (see questions). So, the idea of a connection or similarity between acceleration and gravity is not a novelty or an invention of general relativity.

The equivalence principle may by stated generically in a more formal way as: "there is no experiment conducted in a sufficiently small region of spacetime that can distinguish between a gravitational field and

---

[41] As we will see, we distinguish between the invariance of the essence of the laws and the invariance of the form of the laws. Accordingly, the above discussion is mainly about the latter.

1.8.2 The Principle of Equivalence

an equivalent uniform acceleration".[42] Accordingly, no experiment can distinguish between the effects of gravity and the effects of acceleration at local level (where "local" in this context means local in spacetime and not only local in space). As we see, there are several limitations in the equivalence principle, e.g. "sufficiently small region of spacetime" and "uniform acceleration". It may also be challenged by claiming that the equivalence principle can apply to any force and hence gravity is not special in this regard. For example, how can we distinguish between the electric force that is subject to the above restrictions and conditions and the effect of acceleration? In other words, what distinguishes gravity from other forces in its equivalence to acceleration which eventually leads to stripping the "force" qualification from gravity and reducing it to a mere effect of spacetime curvature while keeping this qualification to other types of force? The only potential distinction seems to be the association of gravity with mass while other forces are associated with other attributes of matter like charge. However, this distinction is virtually circular because we are essentially distinguishing gravity by gravity when we take mass into account since gravity is an attribute of mass. We note that this challenge to the equivalence principle is mainly directed to the weak form of this principle in its general relativistic sense. We should also note that in the literature of general relativity there are justifications for the difference between gravity and other forces from this perspective, e.g. the acceleration of gravitated object depends on the ratio of gravitational mass to inertial mass which is universal while the acceleration of charged object depends on the ratio of electric charge to inertial mass which is object-dependent. However, these justifications may be questioned because they do not seem to address the fundamental issue raised in this challenge.

The above criticism to the equivalence principle in its general relativistic sense may be consolidated by the existence of field gradients and tidal forces in the gravitational frames but not in the accelerating frames. In other words, acceleration may partially account for gravity and its effects but it cannot substitute gravity and account for all its effects. The condition of locality may avoid some unwanted consequences (by virtually annihilating the effect of field gradients and tidal forces) but it does not guarantee the equivalence. Moreover, even the locality is conditioned by the limitations on resolution in any real physical experiment (i.e. the actual resolution may not eliminate the effects of field gradients and tidal forces locally) and this should be specially obvious in strongly non-uniform gravitational fields.[43] Therefore, we may conclude that gravity and acceleration are fundamentally different although the similarity between their effects, as embedded and expressed in the equivalence principle of general relativity under certain conditions like locality, may be used to formulate an approximate or phenomenological theory of gravity. However, if this conclusion is valid in principle then it is possible to run the risk of ill-formulating and misinterpreting gravity and drawing illegitimate consequences and conclusions from such formulation.

It is noteworthy that some relativists believe that the equivalence principle is completely wrong or at least redundant and hence this principle provides motivation and starting point (rather than valid theoretical foundation) for the theory of general relativity (see for example Synge in the References). As we will see, the rejection of the equivalence principle with the acceptance of general relativity may be logically consistent with regard to certain forms of the equivalence principle (e.g. the weak form which essentially underlies the gravity theory) but not with regard to other forms (e.g. the strong form which essentially underlies the "General Theory"). In fact, the topic of the equivalence principle, its classification, definition and interpretation is very messy in the literature (although this is a characteristic feature of the relativity theories in general). This mess also extends to the invariance (or covariance) principle due in part to the close relation between the two principles. The mess also includes the relation between these principles (which may be combined into or linked by the *Principle of General Covariance*) as well as their relation to the "principle of relativity".

Regarding the different forms of the principle of equivalence in general relativity, we can distinguish two

---

[42] We use terms like "acceleration equivalent or corresponding to gravity" or "gravity equivalent or corresponding to acceleration" to indicate the aforementioned meaning (i.e. suitable magnitude in the opposite direction) with the locality condition being observed.

[43] In fact, the limitations on the locality are not restricted to the observational limits on resolution but they extend to the intrinsic limitations on the phenomenon itself because as the local region shrinks other effects (e.g. microscopic and quantum) will get involved and this should invalidate the equivalence principle in its conventional interpretation. In such cases, a different theory (or a modified theory such as "quantum general relativity") should be sought where the conventional equivalence principle can (at the best) play only a part in the physical formulation.

1.8.2 The Principle of Equivalence

main forms: weak and strong.[44] These forms may be stated as follows:
- **Weak equivalence principle**: in a sufficiently confined region of spacetime, the mechanical (or kinematical) effects caused by gravity cannot be experimentally distinguished from the mechanical effects of corresponding uniform acceleration.
- **Strong equivalence principle**: in a sufficiently confined region of spacetime, the physical effects caused by gravity cannot be experimentally distinguished from the physical effects of corresponding uniform acceleration. In other words, the physical laws in confined gravitational frames are the same as the physical laws in the corresponding accelerating frames.[45]

So, what distinguishes the two forms of the equivalence principle is that the weak form is about the mechanical effects of acceleration and gravity (where these effects may be demonstrated for example by motion) while the strong form is about any physical effect and hence it is more general than the weak. This should explain why they are "weak" and "strong".

In the following points we discuss some of the implications and consequences of these two forms:
1. The weak equivalence principle is essentially about gravitational free fall and hence it may be called the principle of universality of free fall.
2. Although the weak equivalence principle is similar to the classical equivalence principle, they are not identical (as might be suggested in the literature). Our view is that the essence of the classical equivalence principle is the equivalence (or equality or identicality) between inertial mass and gravitational mass and this is not the same as the weak equivalence principle of general relativity although the classical form is embedded (in a sense) in the weak form of general relativity and could lead to the experimental consequences of the weak form (see questions). In fact, the weak form contains a theoretical or epistemological component (which is experimentally unverifiable) that is not contained in the classical form. So, while the classical form is a verifiable experimental fact the weak form (as such and in its entirety) is not.[46]
3. The equivalence principle (especially the weak form) is based on the assumption that the gravitated object does not modify the gravitational field created by the gravitating object tangibly, i.e. the gravitated object is a test particle. Otherwise, the gravitational field will be more complicated for the equivalence principle to apply since there is no such a simple acceleration that can simulate the effect of the combined gravitational fields. In fact, this should put more question marks on the validity of the equivalence principle since there are situations in which no acceleration can be equivalent to gravity. This should be added to non-uniform acceleration which has no physically-viable gravitational equivalent. Also, decelerating frames and rotating frames (which are accelerating frames in the general sense of acceleration) have no physically-viable gravitational equivalent.[47] In brief, even if we assume that the equivalence between acceleration and gravity is valid in some cases, it is not valid in general and this should degrade the equivalence principle and diminish its value and significance (and even its validity as a general physical principle) especially with regard to its objective of extending the laws of physics to all types of frame (noting that the strong equivalence principle underlies the "General Theory").
4. If we couple the strong equivalence principle with the premise that the spacetime of general relativity is

---

[44] We should note that we have a third main form of the equivalence principle which is the classical form (i.e. the equivalence between inertial mass and gravitational mass). This third from is not included here because it is not specifically general relativistic since it essentially belongs to classical physics although it represents the root of the general relativistic forms of the equivalence principle.

[45] As we will see, the strong form of the equivalence principle also contains generalizations with regard to the local application of the rules of special relativity and hence the physical laws as represented by these rules equally apply in all types of reference frame locally. In fact, this should be implied by the locality condition in Riemannian space (which is locally flat) plus the alleged validity of special relativity in flat spacetime in general (whether global or local). This validity should be based on the fundamental geometric doctrine of general relativity which correlates the physical laws to the geometry of spacetime.

[46] In our view, the distinction between the classical and weak forms should also consider the difference in the nature of the relation between the inertial mass and gravitational mass (i.e. the difference between "identical" and "equivalent") as will be discussed in the questions.

[47] In fact, if we have to accept the distinction between various types of mass (as it is the case in general relativity), then even the equivalence between inertial accelerating mass and inertial rotating mass can be questioned.

1.8.2 The Principle of Equivalence

locally flat (and hence the laws of special relativity apply locally in the spacetime of general relativity), then we may claim that the strong form is more than an "equivalence" principle between gravity and acceleration but it contains an extra claim (i.e. the equivalence of all frames in the local application of special relativity) that has no counterpart in the weak form as stated above.[48] If so, then the essence of the strong equivalence principle is that *all* the laws of physics apply in *all* frames of reference (whether inertial, accelerating or gravitational) where these frames coordinate locally flat patches of spacetime and hence they are subject to the rules of special relativity at local level (noting that any non-special-relativistic effect introduced by gravity or/and acceleration should be accounted for automatically by the coordinate transformations between the locally inertial frames and the corresponding non-inertial frames). Accordingly, the equivalence principle in its strong form contains an element of invariance.[49] As we will see, the strong equivalence principle is what elevates general relativity from the state of being a mere gravity theory to the state of being a "General Theory".

Problems

1. Briefly discuss the history of the experimental investigations of the equivalence principle.
   Answer: As stated in the text, the roots of the equivalence principle can be traced back to the days of Galileo with his alleged Leaning Tower of Pisa experiments or his rolling balls experiments. In fact, some historical records trace the roots of this principle (as represented by the independence of gravitational acceleration from mass and composition) to the Byzantine scholar John Philoponus in the 6th century AD. There are also some historical records of experiments similar to the alleged Leaning Tower experiments of Galileo conducted (prior to Galileo) by the Dutch scholar Simon Stevin around 1580s. Similar experiments have also been attributed to Newton where he used pendulums of equal length with bobs made of different masses and materials and observed that they swing with identical speeds and periods. There are also reports that the German physicist Friedrich Bessel conducted similar experiments to those of Newton around 1830s using pendulums where he confirmed the previous results. The Hungarian physicist Lorand Eotvos conducted a series of highly precise experiments in the late 19th century and early 20th century using torsion balance techniques and verified the equivalence between inertial mass and gravitational mass to unprecedented level of accuracy. More accurate experiments were conducted recently (e.g. the Eot-Wash group series of gravitational experiments in the late 20th century and early 21st century) all of which verified the equivalence between inertial mass and gravitational mass to very high precision and hence confirmed the equivalence principle. We should remark that some of the aforementioned experiments included testing potential variations and correlations not only with mass but even with other factors such as the type of materials used and the location on the Earth or space. Also, some of those experiments are based on celestial (rather than terrestrial) tests and techniques.
   It is very important to note that these investigations are related (at least primarily) to the classical equivalence principle and hence they are actually verifications to the classical form rather than the weak or strong forms (despite the common claim among general relativists that these experiments are verification and evidence in support of general relativity and its equivalence principle; see § 9.12 and § 10.1.1).

2. Show formally, using classical concepts and formalism and employing the classical equivalence principle, that the gravitational acceleration is independent of the mass of the gravitated object. How can this explain the fact that the roots of the equivalence principle are embedded in classical physics?
   Answer: If we symbolize the inertial mass with m_i and the gravitational mass with m_g then from Newton's second law we have $m_i = \frac{f}{a}$ where f and a symbolize the magnitude of force and acceleration, while from Newton's law of gravity we have $m_g = \frac{fd^2}{GM}$ where d is the distance from the Earth center, G is the gravitational constant and M is the mass of the Earth. So, if these two types of mass are

---

[48] We deliberately phrased the strong equivalence principle above in a simple and rather non-comprehensive form to highlight the correspondence between the weak and strong forms.

[49] This should be inline with the *Principle of General Covariance* (and similar interpretations) which will be investigated later.

1.8.2 The Principle of Equivalence

equivalent (i.e. $m_i = m_g$)[50] then we should have:

$$\frac{f}{a} = \frac{fd^2}{GM} \qquad \rightarrow \qquad a = \frac{GM}{d^2} \equiv g \qquad (37)$$

which justifies the fact that the "gravitational acceleration" g of the Earth is independent of the mass of the gravitated object. In fact, the gravitational acceleration of the Earth is just an example and hence the result is general.

As we see, the independence of gravitational acceleration from the mass of the gravitated object is based on the equivalence between inertial mass and gravitational mass and this explains the fact that the roots of the equivalence principle are embedded in classical physics.

3. Discuss briefly the issue of the relation between the principle of invariance (or general covariance) and the principle of equivalence.

   Answer: This very important issue is not clear cut and hence we can find various opinions and interpretations in the literature not only about the relation between these two principles but even about the meaning and significance of each of these principles individually (where the proposed meaning should affect the supposed relation). We expressed our opinion about the meaning of these principles and the relation between them in several places in this book (earlier and later) and hence we do not need to repeat. Regarding the opinions of other authors, we cannot go through these in detail but it is important to highlight a particularly significant opinion which essentially unifies (or combines or links) the two principles and expresses this in a technically-specific principle which is labeled as the *Principle of General Covariance*. The following quote from Weinberg (see References) eloquently represents this view:

   This method is based on an alternative version of the Principle of Equivalence, known as the *Principle of General Covariance*. It states that a physical equation holds in a general gravitational field, if two conditions are met:

   1. The equation holds in the absence of gravitation; that is, it agrees with the laws of special relativity when the metric tensor $g_{\alpha\beta}$ equals the Minkowski tensor $\eta_{\alpha\beta}$ and when the affine connection $\Gamma^\alpha_{\beta\gamma}$ vanishes.

   2. The equation is generally covariant; that is, it preserves its form under a general coordinate transformation $x \rightarrow x'$.

   To see that the Principle of General Covariance follows from the Principle of Equivalence, let us suppose that we are in an arbitrary gravitational field, and consider any equation that satisfies the two above conditions. From (2), we learn that the equation will be true in all coordinate systems if it is true in any one coordinate system. But at any given point there is a class of coordinate systems, the locally inertial systems, in which the effects of gravitation are absent. Condition (1) then tells us that our equation holds in these systems, and hence in all other coordinate systems. (End of quote)

   The implications of this particular *Principle of General Covariance* will be discussed later in the text and questions (see for instance § 10.1.5) where we will use the capitally-initialized italic form "*Principle of General Covariance*" to refer to this technically-specific meaning of this principle (which combines the invariance and equivalence principles).

4. Make a brief comparison between the weak and strong forms of the equivalence principle of general relativity.

   Answer: We note the following:
   - The weak form is about the mechanical effects (of acceleration and gravity) specifically while the strong form is about the physical effects in general.
   - The weak form underlies the gravity theory while the strong form underlies the "General Theory".

Exercises

1. Discuss the following quote which is attributed to Einstein: "The equality of these two masses [i.e. inertial mass and gravitational mass], so differently defined, is a fact which is confirmed by experiments

---

[50] In fact, what is needed is proportionality between these two types of mass, the equivalence then follows by choosing a proper unit system in which the two become equal.

1.8.2 The Principle of Equivalence

of very high accuracy (experiments of Eotvos), and classical mechanics offers no explanation for this equality. It is, however, clear that science is fully justified in assigning such a numerical equality only after this numerical equality is reduced to an equality of the real nature of the two concepts".

2. Distinguish between the different forms of the principle of equivalence and the scientific evidence in their support.
3. What is the difference between the weak and strong forms of the equivalence principle of general relativity?
4. Discuss briefly the importance of the strong equivalence principle for the theoretical framework of general relativity.
5. Discuss the significance of the weak and strong forms of the equivalence principle and their relation to the nature of general relativity.
6. Summarize your main observations about the strong equivalence principle.
7. The equivalence principle of general relativity may be seen as a bridge between geometry and physics. How?
8. The equivalence principle of general relativity may be seen as having two sides (or meanings): static and dynamic. How?
9. Interpret the equivalence principle of general relativity in the light of the locality of frames and hence determine the role of general relativity and the procedure that should be followed in its application.
10. Examine the significance of the locality restriction in the statement of the equivalence principle.
11. Analyze the condition "sufficiently small region of spacetime" in the statement of the equivalence principle.
12. Analyze a freely falling frame in a gravitational field.
13. Show formally, using the classical equivalence principle (and assuming locality), that a freely falling frame in a gravitational field is effectively inertial (i.e. a frame in which Newton's laws of motion are valid).
14. Analyze a frame that is uniformly accelerating in the absence of any gravitational field.
15. Discuss and assess the equivalence principle in the context of free fall situation.
16. Derive a formula for the gravitational field gradient (which estimates tidal forces) along the radial direction in the neighborhood of a gravitating object using the classical formalism of gravity (see § 1.5).
17. Calculate the tidal force f_t (i.e. force per unit mass)[51] along the radial direction over a distance $d = 1$ nm in the vicinity of the Earth (**a**) at its surface and (**b**) at distance $2R$ from its surface (where R is its radius). Comment on the results.
18. Estimate the tidal force f_t (i.e. force per unit mass) along the radial direction over a distance $d = 1$ nm in the vicinity of a black hole of 1 Earth mass (**a**) at the event horizon[52] and (**b**) at distance $2R_S$ (where R_S is its Schwarzschild radius) from the event horizon. Comment on the results.
19. Assess the validity of the equivalence principle (and hence the validity of general relativity which is based on this principle) in very strong gravitational fields.
20. Are tidal forces restricted to the radial direction?
21. Make a brief assessment to the weak and strong forms of the equivalence principle.
22. Use the existence of tidal forces in gravitational systems as a basis for a potentially fundamental challenge to the strong equivalence principle.
23. Discuss the issue of decelerating frames (as a subset of accelerating frames in its general sense) and to which gravity they should correspond according to the equivalence principle. Also, discuss the implication of this on the existence of absolute frame.
24. To what type of gravitation rotating frames correspond?
25. Try to challenge our criticism to the equivalence principle for its failure to include decelerating and rotating frames (as well as non-uniformly accelerating of all types) and assess this challenge.
26. What is the relation between the invariance principle and the equivalence principle in general relativity?

---

[51] What is actually required is its magnitude with the sign indicating its nature.
[52] The event horizon of a black hole is an imaginary sphere whose center is the singularity (or center) of the black hole and whose radius is the Schwarzschild radius ($R_S = \frac{2GM}{c^2}$) of the black hole (see § 8.7).

27. Outline the technical meaning of the *Principle of General Covariance*. What is the procedure for implementing this *Principle* and what is the logic behind this procedure?
28. Assess the *Principle of General Covariance*.
29. Justify the following statement using the invariance and equivalence principles: "physical objects that are not under the influence of non-gravitational forces follow geodesic trajectories".
30. Identify the potentially different types of invariance as embedded in the general principles of general relativity.
31. Discuss how the principle of relativity in special relativity is extended by the equivalence principle to the "principle of relativity" in general relativity.[53]

1.8.3 The Principle of Correspondence

The essence of the principle of correspondence[54] is that any new theory of physics should agree with the predictions of the corresponding classical theory in its domain of validity. For example, the predictions of quantum mechanics should agree with the predictions of classical mechanics in the limit of large quantum number where classical predictions are verified. Similarly, the predictions of Lorentz mechanics (or special relativity) should agree with the predictions of classical mechanics in the limit of low speed where the classical predictions are verified (see B4). This also applies to general relativity where its predictions (as a gravity theory) should agree with the verified predictions of Newtonian gravity in the classical limit (see § 5).

The validity of the correspondence principle should be obvious because no correct theory should contradict the experimentally verified predictions and results of classical physics. It should also be obvious that the principle of correspondence is an epistemological principle. In fact, it is no more than an expression of the principles of reality and truth, i.e. uniqueness of reality and truth (refer to B4). As we will see in § 3 and § 5, the formalism of general relativity is partly based on the principle of correspondence where the required convergence to the classical limit is assumed to find certain parameters in the formulation of general relativity. Moreover, it is shown that the formalism of general relativity converges to the formalism of classical gravity in certain classical cases.

Exercises

1. Define and discuss the correspondence principle.
2. Is the correspondence principle restricted to physics?
3. Discuss the applicability of the correspondence principle in the alleged convergence of general relativity to special relativity at local level.

1.8.4 Other Principles

It is important to note that the principles that we investigated in the last three subsections represent the formally nominated and commonly recognized principles; otherwise the theory of general relativity relies on other principles as well. For example, we can nominate "the principle of metric gravity" (which we may also call "the principle of geometric gravity" or "the principle of curvature") whose essence is the representation of gravity by the geometry of spacetime and its curvature which is determined by the mass-energy distribution. We may also nominate "the principle of geodesic motion" whose essence is the geodesic nature of the spacetime trajectories of free objects (although this principle may be derived from the formalism of the theory and may even be implied by the former principle).[55] In fact, general relativity (like any other scientific theory) relies on many other general principles such as the principle of

---

[53] In fact, this question is about the relation between the principle of relativity and the general principles of general relativity. The restriction to special relativity is because there is no such explicit "principle of relativity" in general relativity (although its essence is embedded in the framework of the theory).

[54] This may also be called the consistency principle to be distinguished from the correspondence principle of quantum mechanics, i.e. the consistency principle is a generalization of the correspondence principle of quantum mechanics.

[55] We note that the principle of metric gravity is essentially epistemological while the principle of geodesic motion is essentially physical.

causality (i.e. any effect has a cause whose existence precedes the existence of the effect). However, we do not go through these principles in detail due to their simplicity or/and generality (and hence they do not require elaborate explanation or special attention as general relativistic principles).

1.9 Criteria of Scientific Theory

For any theory to be qualified as an acceptable scientific theory it should comply with the following criteria:
(a) It should be consistent with the rules of logic because logic is the most fundamental component of any rational intellectual product.
(b) It should be consistent with the principles of reality and truth (i.e. the existence and uniqueness of reality and the uniqueness of truth) because otherwise the theory will defeat itself.
(c) It should be entirely about the physical world because science is about this world.
(d) It should be consistent with other facts and verified theories.
(e) It should be practical to use and apply because science (unlike philosophy and mathematics for example) is about physical reality and hence it should have a practical (as well as theoretical) value and significance. An important instance of the practicality criterion is that it should be possible to verify and falsify the theory, i.e. it can be proved to be right or wrong experimentally.

Problems

1. What criterion (c) in the above list actually means?
   Answer: This criterion does not only require that the subject of the theory should be a natural phenomenon that belongs to the physical world but it also requires that the theory should not involve any non-physical element, entity or concept in its framework and therefore the theory should not include any metaphysical or supernatural element or component.
2. Link criterion (d) in the above list to another criterion in the list. Also, link criterion (d) to the principles that we investigated in § 1.8.
   Answer: Criterion (d) is obviously based on criterion (b) as a requirement for the uniqueness of truth. Criterion (d) is also linked to the principle of correspondence which was investigated in § 1.8.3.

Exercises

1. Can a creation theory be a scientific theory?
2. Assess general relativity in the light of the above criteria.
3. Give broad criteria for ideal scientific theory and assess the theories of modern physics from this perspective.
4. Clarify criterion (e) in the list that we stated in the text giving some examples.
5. Are the criteria for acceptable scientific theory (as stated in points a-e in the text) sufficient or/and necessary conditions (or not) for the truthfulness of a presumed scientific theory?
6. Why metaphysical entities cannot enter in the formulation of a physical theory?
7. Give more justification to the requirement of consistency with logic (i.e. criterion a).

1.10 Reading Too Much in the Equations

In this section we discuss the validity and value of mathematical models and why observations and experiments should lead the mathematical development of science rather than the opposite. Many equations and mathematical models that are developed to describe physical phenomena have limited applicability although from a formal perspective they do not look restricted. Despite the fact that mathematical equations are legitimate and necessary tools for describing physical phenomena, the two are not equal. This means that the mathematical models are not equivalent to the physical phenomena and hence although they agree in general they do not necessarily agree in every aspect and detail. For example, the mathematical models may be based on certain assumptions, some of which can be implicit or hidden, and hence their extension and generalization cannot be justified. Moreover, mathematics is a highly abstract, symbolic and rather simplistic tool while the physical phenomena are usually very complex and messy.

1.10 Reading Too Much in the Equations

Hence, it is very unlikely that a mathematical model can be so comprehensive that it captures and includes all the physical factors involved in a given phenomenon and this is particularly true in highly complex physical phenomena like gravity and highly abstract scientific theories like general relativity. In brief, to construct realistic scientific theories that reflect the reality of the physical world the observations, rather than mathematical models and modeling, should lead the scientific research or at least the observations should go hand in hand with the mathematical modeling and theoretical reasoning.

The conclusion from this preamble is that the reality and truthfulness of many of the claimed consequences and implications of general relativity that are based on analyzing the formalism of this theory are questionable. This is particularly true with respect to modern cosmology which is largely guided by the theoretical framework and formalism of general relativity. This also applies to many astrophysical theories and speculations that are based entirely on the formalism of general relativity and hence they represent mathematical and theoretical curiosities rather than physical reality. Accordingly, many of the consequences of general relativity that have been inferred from purely theoretical models are questionable. Typical examples of this sort of questionable issues are the contemplative details about black holes and their physics (see § 8.7) even if we agree on the possibility of the existence of black holes and even if this existence is supported (or will be supported) by genuine scientific evidence from astronomical observations. In brief, equations (which represent mathematical models and modeling) in science should be read and analyzed wisely and sensibly in association with observation, rationality and proper epistemology.[56]

Problems

1. Define singularity and distinguish between mathematical singularity and physical singularity. Also, discuss the relation between the two.

   Answer: We may define singularity roughly as a situation or condition that leads to the divergence of a given quantity and hence it becomes infinite at certain point or region of space. Singularity is also used to mean the point in space where the singularity occurs.

   We distinguish between mathematical singularity and physical singularity by considering the former as a mathematical requirement obtained from a valid mathematical formulation that leads to the singularity while considering the latter as a physical realization of this mathematical requirement which means that a given physical property goes to infinity in the real world. For example, we may infer from a valid mathematical formulation that the density (or curvature) at the center of black holes is infinite and hence we have a mathematical singularity. We may then conclude that the density of the actual black holes is infinite (i.e. they represent a real mass concentration at a geometric point in space with zero volume) and hence we have a physical singularity.

   We believe that mathematical singularity should not mean or imply physical singularity. The reason is that real world has no infinite quantities and hence no singularity should exist in reality. The truth is that such quantities do exist only in theoretical and hypothetical worlds, e.g. the abstract spaces of mathematics. In fact, this is one of the widespread examples of "Reading Too Much in the Equations" where it is common in modern physics to conclude physical singularities from mathematical singularities. The undisputed existence of accidental (or coordinate) singularity is a clear evidence for the misleading ability of mathematics when it applies to physical phenomena since the accidental singularity is a mathematical singularity with no corresponding physical singularity because it can be removed by replacing the coordinate system. In fact, the conclusion of physical singularities from mathematical singularities may be challenged more formally and technically by denying the validity of the singularity even mathematically (especially when mathematics is used to describe and quantify the real world) because any valid mathematical formulation should exclude such singular points and conditions and hence we do not have legitimate and valid mathematical singularities to conclude physical singularities from them.

---

[56] Unfortunately, the majority of modern physicists (unlike classical physicists) are poor in logic, poor in epistemology and poor in philosophy of science and they are indoctrinated to follow the equations literally and make the mathematical models lead the observations and that is why modern physics is full of absurdities and mythical objects such as wormholes, time machines, strings, membranes, dark matter, dark energy, etc. In fact, we propose including the above subjects (i.e. logic, epistemology and philosophy of science) in the curriculum of science education (at least starting from college or undergraduate level if not earlier).

Note: the reader should note that we use "physical singularity" in a different meaning to its common use in the literature. In fact, we use accidental singularity for coordinate singularity, essential singularity for what is called by some authors physical singularity, and we use physical singularity to mean a singularity in the real physical world.

Exercises

1. Give some reasons for our rejection of over-mathematization of science.

Chapter 2
Mathematical Background

In this chapter we summarize the main mathematical concepts and techniques that we need for the presentation and formulation of general relativity in a rigorous mathematical form. As we will see, these concepts and techniques are mostly imported from tensor calculus and differential geometry of abstract multi-dimensional manifolds. The reader, however, should note that the present chapter includes the minimum of the most important parts of the required mathematical background and hence many aspects of the mathematical background that is needed in the formulation of general relativity are not discussed in this chapter (at least in sufficient details) due to the restriction on the size of the book and because most of these aspects are treated in my previous books. The reader should therefore consult those books (or other books such as those in the References) for the missing materials and details about differential geometry, tensor calculus and Lorentz mechanics (see the References in the back of this book). However, the reader should note that some of the missing details and supplementary materials are included in the exercise book that accompanies this book.

It is noteworthy that because the materials in this chapter are mostly about abstract spaces in general regardless of dimensionality we do not strictly follow the convention about the Latin indices (see § 1.4). In fact, we generally use Latin indices in our mathematical formulations even for non-3D spaces unless the subject or context dictates otherwise and hence the use of Latin indices in this chapter does not necessarily mean that the given formulation is restricted to 3D spaces. This may also extend to some other symbols (like ds which is primarily used for 3D spaces) where in certain contexts they are assumed to belong to an nD space and hence they are not specific to a space of certain dimensionality. We should also note that apart from general mathematical and physical terminology which should be known to the readers at this level, the basic terminology in this chapter is mostly explained in the previous chapter while some other terminology is explained within its proper context.

2.1 Space, Coordinate System and Transformation

In mathematics, **space** or manifold is an abstract concept whose origin comes from the physical space that we live in. So, we can imagine mathematical space as a container of abstract mathematical objects. The main property of space is the number of its dimensions (or dimensionality). So, a space with 1 dimension is a 1D space, a space with 2 dimensions is a 2D space, and so on. The space can be regarded as an assembly of points and hence we need an abstract device to identify and distinguish these points; this device is called "coordinate system" which will be investigated later in this section.

A space is described as Euclidean when it can be coordinated by a Cartesian coordinate system[57] and described as flat when it can be coordinated by a Cartesian or quasi-Cartesian system (see Problems). If the space cannot be coordinated by a Cartesian or quasi-Cartesian system it is described as curved. A Riemannian space is a space characterized by having a symmetric rank-2 tensor called the metric tensor that represents the geometry of the space. Riemannian space may also be characterized (and hence restricted) by the condition that it is possible to construct a locally Cartesian coordinate system at any of its points, i.e. it is locally Euclidean or flat.[58] Riemannian space may similarly be characterized by having a line element ds whose square is defined by the quadratic form $(ds)^2 = g_{ij}dx^i dx^j$ where g_{ij} is the metric tensor,

---

[57] This is equivalent to the condition that there is a permissible coordinate transformation that takes the general quadratic form $(ds)^2 = g_{ij}dx^i dx^j$ of the space to the Cartesian quadratic form $(ds)^2 = \sum_{i=1}^{n}(dx^i)^2$. In other words, there is a permissible coordinate transformation that takes the metric tensor g_{ij} to the Kronecker delta tensor δ_{ij}.

[58] We note that there are some abstract metric spaces which are not Euclidean or flat even locally and accordingly they are not Riemannian.

dx^i and dx^j are coordinate differentials and i and j range over the space dimensions.[59] Conventionally, the quadratic form (which is based on the metric tensor) of Riemannian space is positive definite. Hence, a space whose characteristics are similar to the characteristics of Riemannian space but its quadratic form is not positive definite (i.e. it is Riemannian apart from this mathematical nicety) is commonly known as pseudo- or quasi-Riemannian. As indicated earlier, physicists generally do not discriminate between Riemannian and pseudo-Riemannian and hence both might be described as Riemannian.

Coordinate system of a given space is a mathematical device used to identify the location of points and objects in the space. Coordinate systems are also needed to define non-scalar quantities (mainly tensors) in a specific form and identify their components in reference to the basis vector set of the system. This applies in particular to the metric tensor (which represents the geometry of the space) where a coordinate system is needed to obtain a specific form of this exceptionally important tensor. An nD space requires a coordinate system with n mutually independent coordinates to be fully determined so that any point in the space can be uniquely identified by n independent coordinates. Coordinate systems are characterized by having coordinate curves and coordinate surfaces. Coordinate curve is a curve along which exactly one coordinate varies while the other coordinates are held constant, whereas coordinate surface is a surface over which exactly one coordinate is held constant while the other coordinates vary. So, at any point in the space we have n independent coordinate curves and n independent coordinate surfaces.

The n coordinate curves uniquely identify a set of n mutually independent covariant basis vectors $\mathbf{E}_i$ ($i = 1, \cdots, n$) which are defined as the tangents to the coordinate curves. Similarly, the n coordinate surfaces uniquely identify a set of n mutually independent contravariant basis vectors $\mathbf{E}^i$ ($i = 1, \cdots, n$) which are defined as the gradients of the coordinate surfaces (or rather gradients of coordinates and hence they are perpendicular to the coordinate surfaces). Mathematically, the covariant and contravariant basis vectors are given by:

$$\mathbf{E}_i = \frac{\partial \mathbf{r}}{\partial x^i} \qquad \mathbf{E}^i = \nabla x^i \qquad (38)$$

where $\mathbf{r}$ is the position vector (of the point on the ith coordinate curve) in Cartesian coordinates $(x_1, \ldots, x_n)$, ∇ is the nabla operator in Cartesian coordinates (of the point on the ith coordinate surface), and x^i represents general coordinates.[60] Accordingly, the covariant and contravariant basis vectors are generally variable functions of coordinates. A coordinate system whose all basis vectors (whether covariant or contravariant) are constant throughout the space is described as rectilinear or affine; otherwise it is described as curvilinear. A coordinate system (whether rectilinear or curvilinear) is described as orthogonal if the basis vectors (whether covariant or contravariant) are mutually perpendicular at every point in the space.

In general terms, a **transformation** from an nD space to another nD space[61] is a correlation that maps the points of the first space to the points of the second space where each point in the first and second spaces is identified by n independent coordinates. To distinguish between the two sets of coordinates in the two spaces, the coordinates of the points in the second space may be notated with primed symbols like $(x'^1, \ldots, x'^n)$, while the coordinates of the points in the first space are notated with unprimed similar symbols like $(x^1, \ldots, x^n)$. Barred symbols like $(\bar{x}^1, \ldots, \bar{x}^n)$ may also be used instead of primed symbols. For the transformation to be useful and versatile, it should be bijective and hence an inverse transformation from the second space to the first space is also defined, i.e. we have direct transformation from first space to second space and inverse transformation from second space to first space.

In many cases (especially in the theories of relativity), the transformation is not between different spaces but it is between different coordinate systems of the same space. However, this does not introduce

---

[59] To be more clear, $(ds)^2$ is the square of the line element ds while $g_{ij}dx^i dx^j$ is the quadratic from (since it is quadratic in dx) although it is common to call $(ds)^2$ the quadratic form (referring to its mathematical expression $g_{ij}dx^i dx^j$).

[60] Cartesian in this definition can be local and hence it applies to curved spaces as well as to flat spaces noting that the spaces are presumably Riemannian and hence they can be coordinated locally (at least) by a Cartesian system. It should be noted in this context that the basis vectors actually belong to the tangent (or cotangent) space (where they represent a link between the Riemannian space at a given point and the tangent space at that point). The reader should be aware of the fact that the tangent space at any point of a Riemannian (or pseudo-Riemannian) space is a Euclidean (or flat) space of the same dimensionality that represents the local geometry of the space. The existence of the tangent space is guaranteed by the Riemannian nature of the space since Riemannian space is locally flat.

[61] Transformations can also be between spaces of different dimensionality, but this is of no interest to us.

a fundamental difference on the aforementioned facts and procedures and hence it could be seen as a mere conceptual difference. Accordingly, in this book we generally talk about transformations between coordinate systems (or reference frames which are the coordinate systems of spacetime) rather than transformations between spaces.

Mathematically, each one of the direct and inverse transformations can be regarded as a mathematical correlation expressed by a set of equations in which each coordinate in one system is considered a function of the coordinates in the other system. Hence, the transformations between the two sets of coordinates in the two systems can be expressed mathematically in a generic form by the following two sets of independent relations:

$$x'^i = x'^i(x^1, \ldots, x^n) \tag{39}$$
$$x^i = x^i(x'^1, \ldots, x'^n) \tag{40}$$

where $i = 1, \ldots, n$ and n is the space dimension. The first equation represents the direct transformation while the second equation represents the inverse transformation.

Problems

1. What we mean by Cartesian and quasi-Cartesian coordinate systems?
   Answer: We generally mean by Cartesian orthonormal rectilinear coordinate system of nD space with a positive definite quadratic form. So, all the coordinate curves of the system are straight lines and all the coordinate surfaces of the system are planes. Moreover, the coordinate curves (as well as the coordinate surfaces and the basis vectors) of the system are mutually perpendicular and the axes have identical length scale. An obvious example of Cartesian system is the familiar (x, y, z) system of 3D Euclidean space (which is the space of classical physics and Newtonian gravity).
   Quasi-Cartesian (or pseudo-Cartesian) is the same as Cartesian but it, unlike Cartesian, has a quadratic form that is not positive definite. So, the quadratic form of Cartesian is $(ds)^2 = \sum_{i=1}^{n}(dx^i)^2$ while the quadratic form of quasi-Cartesian is $(ds)^2 = \sum_{i=1}^{n} \zeta_i (dx^i)^2$ with some ζ_i being $+1$ and some ζ_i being -1. An example of quasi-Cartesian is the coordinate system (or frame) of the Minkowski spacetime.
   We note that because the difference between Cartesian and quasi-Cartesian is not of primary interest to physicists in most contexts (although it is of primary interest to mathematicians due to its mathematical nature) we may use Cartesian to mean Cartesian or quasi-Cartesian (and may similarly use Euclidean to mean flat).

2. Give examples of common spaces of different dimensionality and characteristics.
   Answer:
   Plane: 2D flat Riemannian.
   Cylinder: 2D flat Riemannian.[62]
   Sphere: 2D curved Riemannian.
   Space of Euclidean geometry: 3D flat Riemannian.
   Space of special relativity: 4D flat pseudo-Riemannian.
   Space of general relativity: 4D curved pseudo-Riemannian.

3. Discuss the need for using local coordinate systems. Also, give examples of commonly used local coordinate systems.
   Answer: Local coordinate systems are used to have a closer look (or zoom in) into a particular region or point of space or to obtain simplified formulation by exploiting the local properties of the space which are simpler than its global properties (e.g. flatness of space at local level).
   A prominent example of local coordinate system that is commonly used in the proofs and arguments of differential geometry (and hence in general relativity) is the "geodesic coordinates" which can be defined as a local coordinate system constructed at and in the neighborhood of a given point P (which is called the pole of the geodesic coordinate system) of the space such that all the Christoffel symbols vanish at P. Another prominent example is the local inertial frame which is widely used in general

---

[62] We note that cylinder is intrinsically flat. Therefore, it can be coordinated intrinsically by a Cartesian system since it is isometric to plane and hence the Cartesian system of the plane is mapped onto the intrinsic Cartesian system of the cylinder.

relativity and its applications (as we saw and will see throughout the book).

Note: regarding the relation between geodesic coordinate system and local Cartesian (or pseudo-Cartesian) system, we note that according to the given definitions and conventions in the literature local Cartesian system is a geodesic coordinate system but the opposite is not true in general. Hence, geodesic coordinate system is more general than local Cartesian system.

Exercises

1. What is the relation between space, coordinate system and transformation?
2. Characterize the spaces of special relativity and general relativity.
3. What is the difference between Riemannian space and pseudo-Riemannian space? Link this to the spaces of the relativity theories.
4. Should Riemannian space be curved?
5. List the main characteristics of Riemannian space.[63]
6. What is the coordinate system of the spaces of special relativity and general relativity?
7. In the special and general theories of relativity, what transformation from one coordinate system to another coordinate system means?
8. The covariant basis vectors of a given space are transformed from an unprimed coordinate system O to a primed coordinate system O' by the relation $\mathbf{E}'_i = \frac{\partial x^j}{\partial x'^i} \mathbf{E}_j$. Justify this relation and do the same for contravariant basis vectors.
9. Which space can be coordinated by rectilinear coordinate systems and which space cannot?
10. Show that $\mathbf{E}_j \cdot \mathbf{E}^i = \mathbf{E}^i \cdot \mathbf{E}_j = \delta^i_j$ (where $\mathbf{E}^i$ and $\mathbf{E}_j$ are the basis vectors and δ^i_j is the Kronecker delta).
11. Why should we invent (and hence use) different types of coordinate systems which leads for example to the complications of transformation and to potential confusion?
12. Discuss the relation between the invariance principle and coordinate transformations and link this to the nature of the coordinate systems and physical laws.

2.2 Tensors

Tensors are mathematical entities that have certain transformation rules. Tensors are required to be invariant under certain transformations across coordinate systems, although this property of invariance may not be general. For example, a tensor may be invariant (and hence it is a tensor in a technical sense) with respect to certain types of transformations (e.g. proper transformations or Lorentz transformations) but not with respect to other types of transformations (e.g. improper transformations or Galilean transformations). It can also be invariant only across certain types of coordinate system (e.g. affine) but not across other types of coordinate system (e.g. general curvilinear).

Tensors are characterized by a number of attributes; the main ones are:

- **Indices** which refer to the coordinates of the underlying coordinate system and its basis vectors.[64] For example, in the tensor A^i_j the index i refers to the ith coordinate (or i^{th} covariant basis vector) while the index j refers to the jth coordinate (or j^{th} contravariant basis vector).
- **Order** which is represented by the total number of indices whether repetitive (known as dummy) or non-repetitive (known as free). For example, the order of the tensor A is zero, the order of the tensor B^j_i is two and the order of the tensor C^a_{ijka} is five.
- **Rank** which is represented by the number of non-repetitive (or free) indices. For example, the rank of the tensor A is zero, the rank of the tensor B^j_i is two and the rank of the tensor C^a_{ijka} is three. The tensors of rank-0 are commonly known as scalars while the tensors of rank-1 are commonly known as vectors.
- **Dimension** (or dimensionality) which is represented by the number of dimensions of the space which the tensor belongs to. For example, in a 2D space (or surface) the dimension of the tensor A_{ij} is 2 while in a 4D space (e.g. the spacetime of general relativity) the dimension of A_{ij} is 4.

---

[63] This question is about Riemannian as opposite to non-Riemannian and not as opposite to pseudo-Riemannian and hence it includes pseudo-Riemannian.

[64] As we will see, rank-0 tensors (or scalars) have no indices. However, they are still characterized by indices because they have 0 indices.

2.2 Tensors

- **Variance type**, i.e. covariant or contravariant or mixed. The variance type is determined by the position of the indices, i.e. being lower (like A_i) or upper (like B^{ij}) or both (like C^i_j). So, a tensor whose all indices are lower is covariant (e.g. A_{ijk}) and a tensor whose all indices are upper is contravariant (e.g. A^{ijk}) while a tensor that has both lower and upper indices is mixed (e.g. A^k_{ij}). It is obvious that rank-0 tensors (or scalars) have no variance type because they have no indices, while the rank of the mixed tensors should be 2 or higher so that it is possible to have some upper and some lower indices. In fact, the use of upper and lower indices is just a notational convention that refers to a more fundamental issue that is upper indices are associated with covariant (or tangent) basis vectors while lower indices are associated with contravariant (or gradient) basis vectors. Accordingly, the tensor A_i means $A_i \mathbf{E}^i$, the tensor B^{ij} means $B^{ij} \mathbf{E}_i \mathbf{E}_j$ and the tensor C^i_{jk} means $C^i_{jk} \mathbf{E}_i \mathbf{E}^j \mathbf{E}^k$. It is common to indicate the variance type of a tensor by (n_u, n_l) where n_u is the number of upper (or contravariant) indices and u_l is the number of lower (or covariant) indices. For example, the type of A_{ij} is $(0, 2)$, the type of A^{ijk} is $(3, 0)$ and the type of A^{ij}_k is $(2, 1)$.
- **Symmetry properties**, i.e. being symmetric or anti-symmetric or neither. A symmetric tensor does not change if two of its indices are exchanged while an anti-symmetric tensor reverses its sign following such an exchange. For example, if A_{ij} is a symmetric tensor then its components satisfy the relation $A_{ij} = A_{ji}$ while if it is an anti-symmetric tensor then its components satisfy the relation $A_{ij} = -A_{ji}$. A tensor that does not meet any one of these criteria is neither symmetric nor anti-symmetric. The symmetry properties are restricted to tensors of rank-2 and higher and hence scalars and vectors have no symmetry properties because they do not have sufficient number of indices to meet the criteria of symmetry properties. A tensor whose rank is greater than 2 can be totally symmetric or anti-symmetric and it can be partially symmetric or anti-symmetric and hence the symmetry properties apply with respect to some (but not all) of its indices. Such a tensor may also be symmetric with respect to some indices and anti-symmetric with respect to other indices. For example, if the components of the tensor A_{ijk} satisfy the relation $A_{ijk} = A_{jik}$ but not the relation $A_{ijk} = A_{ikj}$ then it is partially symmetric (i.e. with respect to its ij indices). Similarly, if the components of the tensor A_{ijk} satisfy the relation $A_{ijk} = -A_{jik}$ but not the relation $A_{ijk} = -A_{ikj}$ then it is partially anti-symmetric (i.e. with respect to its ij indices). Some tensors may have block symmetry properties when the symmetry properties apply not to individual indices but to sets (or blocks) of indices. For example, a tensor whose components satisfy the relation $A_{ijkl} = A_{klij}$ has block symmetry with respect to its first two indices and last two indices because the exchange of the position of these blocks does not change the tensor. It should be remarked that the symmetry properties of tensors are invariant under coordinate transformations and hence a tensor that is symmetric/anti-symmetric/neither in a given coordinate system remains so following its transformation to other coordinate systems.
- **Number of components** (or elements or entries) which is given by the simple formula n^r where n is the dimension and r is the rank. Accordingly, a scalar has only one component in any space because its rank is 0 and hence $n^0 = 1$, while a vector has n components (e.g. 2 in 2D space, 3 in 3D space, etc.) because its rank is 1 and hence $n^1 = n$. Similarly, a rank-2 tensor has n^2 components (e.g. $2^2 = 4$ in 2D space, $3^2 = 9$ in 3D space, etc.).
- **Number of independent (or distinct) components** which is the number of non-correlated components. Since the components of symmetric and anti-symmetric tensors are correlated (i.e. by relations like $A_{ij} = A_{ji}$ and $A_{ij} = -A_{ji}$), the number of independent components is less than the number of components. It can be easily shown (see exercise 2) that the number of independent components of rank-2 symmetric tensors is $\frac{n(n+1)}{2}$ while the number of independent non-zero components of rank-2 anti-symmetric tensors is $\frac{n(n-1)}{2}$ where n is the dimension.
- **Transformation rules** which are the mathematical formulae used to transform tensors from one coordinate system to another coordinate system. An arbitrary tensor of type (m, n) which is symbolized with $A^{i \cdots m}_{j \cdots n}$ in an unprimed coordinate system O and symbolized with $A'^{I \cdots M}_{J \cdots N}$ in a primed coordinate

2.2 Tensors

system O' is transformed from O to O' according to the following rule:[65]

$$A'^{I\cdots M}_{J\cdots N} = \frac{\partial x'^I}{\partial x^i} \cdots \frac{\partial x'^M}{\partial x^m} \frac{\partial x^j}{\partial x'^J} \cdots \frac{\partial x^n}{\partial x'^N} A^{i\cdots m}_{j\cdots n} \qquad (41)$$

This rule can be obtained as a generalization of the transformation rules of low-rank tensors with the help of the transformation rules of the basis vectors (see exercise 3). It should be noted that the type (m,n) includes the case $m=0$ (i.e. covariant tensor like A_{kl}) and the case $n=0$ (i.e. contravariant tensor like A^{kl}). In fact, it includes even the case $m=n=0$ (i.e. scalar) where the transformation rule becomes $A' = A$.

- **Algebraic rules** which determine the tensorial nature of the entities obtained by simple algebraic operations on tensors. These rules include:[66]
    1. Linear combinations of tensors (such as $a\mathbf{C} \pm b\mathbf{D}$ where a and b are scalars and $\mathbf{C}$ and $\mathbf{D}$ are tensors) are tensors.
    2. Outer and inner products of tensors (e.g. $\mathbf{CD}$ and $\mathbf{C}\cdot\mathbf{D}$ where $\mathbf{C}$ and $\mathbf{D}$ are tensors) are tensors.
    3. Contractions of tensors (e.g. C^i_{ji} where C^i_{jk} is a tensor) are tensors.
    4. Permutations of tensors (e.g. C_{ji} which is a permutation of a tensor C_{ij}) are tensors.
- **Differential rules** which determine the tensorial and operational nature of the entities obtained by differential operations (mainly partial and total derivatives and covariant and absolute derivatives) on tensors. Some of these rules are discussed in § 2.7.
- **Metrical rules** which determine the tensorial nature of the entities obtained by metrical operations on tensors. The main of these rules is that entities obtained by raising and lowering the indices of a tensor are tensors, i.e. all variance types of a tensor are tensors (see § 2.5). For example, if C_{ij} is a tensor then C^i_j and C^{ij} are also tensors.

Problems

1. Distinguish between general tensors and special (or qualified) tensors.
    Answer: General tensors transform invariantly (according to the tensor transformation rules) under all permissible (i.e. differentiable invertible) coordinate transformations, while special (or qualified) tensors transform invariantly under special (or qualified) type of transformations or across certain types of coordinate systems. For example, the tensors in general relativity are supposed to be general tensors, while the tensors in special relativity are tensors under Lorentz transformations and hence they are not required to be tensors under general transformations. For this reason the tensors of special relativity may be labeled as Lorentzian (or Lorentz) tensors. We can also find in the literature tensors that are labeled as Cartesian tensors or affine tensors because they transform as tensors only across Cartesian or affine coordinate systems (e.g. the tensors of classical physics and the Christoffel symbols which will be investigated in § 2.6).
    Note: the tensors of special relativity are more commonly known as "4-tensors" (although "4-tensors" is also used to label the tensors of general relativity which are supposed to be general tensors). So, "4-tensors" are special tensors in special relativity and are general tensors in general relativity.[67] We should also note that the tensors that transform invariantly across a certain type of coordinate systems should also be characterized by a certain type of transformations (which characterize that type of coordinate systems) although they are labeled by the type of coordinate systems.
2. What is the relation between the indices of a tensor and its dimension?
    Answer: Each index of a tensor ranges over its dimension. For example, A^j_i in nD space means $i = 1, \cdots, n$ and $j = 1, \cdots, n$. This should explain why the number of components of tensor are given by n^r where n is the dimension and r is the rank which represents the number of free indices.

---

[65] The use of uppercase letters to index primed symbols in this equation and its alike is for the purpose of more clarity although this is not necessary because the indices in the primed and unprimed symbols are totally independent of each other, e.g. i in x'^i is independent of i in x^i and hence they are totally different like i and k.

[66] In curved spaces, the legitimacy (and sensibility) of the operations that involve more than one tensor may depend on the co-locality of the tensors involved in these operations (i.e. the involved tensors correspond to the same point in space). For example, adding a tensor at one location to another tensor at another location may not be legitimate and hence it is meaningless to determine the tensorial nature of such illegitimate operation.

[67] As indicated earlier in § 1.3, the "4-" label in general relativity may also refer to the local Lorentzian origin of the quantity.

3. Show that the algebraic sum of symmetric tensors is a symmetric tensor.
   Answer: If A^{ij} and B^{ij} are symmetric tensors then we have $A^{ij} = A^{ji}$ and $B^{ij} = B^{ji}$ and their algebraic sum C^{ij} will be $C^{ij} = A^{ij} \pm B^{ij}$. On shifting the indices of the last equation we get:

   $$C^{ji} = A^{ji} \pm B^{ji} = A^{ij} \pm B^{ij} = C^{ij}$$

   and hence C^{ij} is also symmetric. We note that equality 2 is justified by the symmetry of A^{ij} and B^{ij} while equality 3 is justified by the definition of C^{ij}. The above applies similarly if the tensors are covariant.
   Note: although we used rank-2 contravariant tensors in the above demonstration, the result can be easily generalized to tensors of any rank and covariant type. We should also note that the above demonstration can also be easily applied to anti-symmetry, that is if A^{ij} and B^{ij} are anti-symmetric tensors and C^{ij} is their algebraic sum then we have:

   $$C^{ji} = A^{ji} \pm B^{ji} = \left(-A^{ij}\right) \pm \left(-B^{ij}\right) = -\left(A^{ij} \pm B^{ij}\right) = -C^{ij}$$

   In fact, we can even generalize "algebraic sum" to "linear combination".
4. What is the quotient rule of tensors?
   Answer: The quotient rule of tensors states: if the inner product of a suspected tensor by a known tensor is a tensor then the suspect is a tensor. For example, if **A** is a suspected tensor and **B** and **C** are known tensors and we have $\mathbf{A} \cdot \mathbf{B} = \mathbf{C}$ (where the dot represents inner product) then **A** is a tensor.
5. What is the important property of the zero tensor? What is the implication of this?
   Answer: It is the value-invariance (or constancy) of its components across all coordinate systems (i.e. its components are zero in any coordinate system). In other words, if a tensor vanishes (i.e. it is a zero tensor) in one coordinate system then it vanishes in any other coordinate system.[68]
   The implication of this is that a tensorial relation (or equation) that is valid in one coordinate system should hold in all other coordinate systems due to its tensorial (invariant) nature since it can be expressed as a tensorial relation of the zero tensor. For example, if we have the tensorial relation $\mathbf{A} = \mathbf{B}$ in a given coordinate system then we can put it in the form $\mathbf{A} - \mathbf{B} = \mathbf{0}$ and hence it should be valid across all coordinate systems due to the value-invariance of the zero tensor. In fact, the invariance of tensorial relations can also be easily deduced from the transformation rules of tensors (as will be seen next).
6. Show that if two tensors of the same type (i.e. same rank, variance type, etc.) are equal (i.e. have identical corresponding components) in one coordinate system then they should also be equal in any other coordinate system.
   Answer: This may be "shown" easily by noting that the supposedly "two tensors" are actually identical (because they are equal) and hence they represent the same tensor.[69] Therefore, the statement is self-evident (or even trivial) without need of a technical proof. However, in the following we show this formally (for the sake of clarity and rigor).
   Let $A^{i \cdots m}_{j \cdots n}$ and $B^{i \cdots m}_{j \cdots n}$ be two such tensors in coordinate system O (unprimed) and hence we have $A^{i \cdots m}_{j \cdots n} = B^{i \cdots m}_{j \cdots n}$. On transforming this relation to another coordinate system O' (primed) by multiplying the two sides by the Jacobian factors $\frac{\partial x'^I}{\partial x^i} \cdots \frac{\partial x'^M}{\partial x^m} \frac{\partial x^j}{\partial x'^J} \cdots \frac{\partial x^n}{\partial x'^N}$ (see Eq. 41) we get:

   $$\frac{\partial x'^I}{\partial x^i} \cdots \frac{\partial x'^M}{\partial x^m} \frac{\partial x^j}{\partial x'^J} \cdots \frac{\partial x^n}{\partial x'^N} A^{i \cdots m}_{j \cdots n} = \frac{\partial x'^I}{\partial x^i} \cdots \frac{\partial x'^M}{\partial x^m} \frac{\partial x^j}{\partial x'^J} \cdots \frac{\partial x^n}{\partial x'^N} B^{i \cdots m}_{j \cdots n}$$

---

[68] This may be expressed by saying: a non-zero tensor cannot be transformed to a zero tensor and vice versa, (i.e. a non-zero tensor cannot be transformed away and a zero tensor cannot be transformed to a non-zero tensor). We should also note that the value-invariance property of the zero tensor also applies to its various variance types, i.e. if one variance type of a tensor is zero then all its variance types are also zero (e.g. if $A_{ij} = 0$ then $A^i_{\ j} = g^{ik} A_{kj} = 0$).

[69] In fact, this may be questioned because the equality may not imply identicality (especially when considering the difference between mathematical and physical perspectives). For example, two objects may be mathematically equal but they are physically (or even mathematically) different, e.g. they represent realistically-different physical entities or conceptually-different mathematical entities. Anyway, this is a trivial matter and should not concern us.

As we see, each side of the last equation represents the transformation of a tensor from system O to system O' and hence we should have:
$$A'^{I...M}_{J...N} = B'^{I...M}_{J...N}$$
i.e. the two tensors are also equal in system O'. Now, since O' is an arbitrary system then the two tensors should be equal in all systems.

Note: the presentation of the above proof may be simplified by the use of the zero tensor as follows: Since $A^{i...m}_{j...n}$ and $B^{i...m}_{j...n}$ are equal (i.e. $A^{i...m}_{j...n} = B^{i...m}_{j...n}$), then $A^{i...m}_{j...n} - B^{i...m}_{j...n} = 0$ and hence this zero tensor can be transformed invariantly to any other system and accordingly $A'^{I...M}_{J...N} - B'^{I...M}_{J...N} = 0$ which leads to $A'^{I...M}_{J...N} = B'^{I...M}_{J...N}$ as required. However, this requires the applicability of the transformation to the individual terms and hence it is the same as the above demonstration.

7. How are index shifting and index replacement operations conducted?
   Answer: Index shifting operations are conducted by contracting an index of the metric tensor with an index of another tensor (i.e. by inner product). So, in lowering a contravariant index the covariant metric tensor is used (e.g. $A_i = g_{ij} A^j$) while in raising a covariant index the contravariant metric tensor is used (e.g. $A^i = g^{ij} A_j$).
   Index replacement operations are conducted by contracting an index of the Kronecker delta tensor with an index of another tensor, e.g. $A^i = \delta^i_j A^j$ and $A_i = \delta^j_i A_j$ where the index j in tensor **A** is replaced by the index i.
   Note: index replacement operation is conducted by using the Kronecker delta tensor in its mixed form (i.e. δ^j_i) which is the same as the mixed metric tensor g^j_i (see § 2.5). Accordingly, we can say that all the above operations (i.e. index shifting and index replacement) are conducted by using the metric tensor, i.e. the covariant metric tensor for lowering index, the contravariant metric tensor for raising index and the mixed metric tensor for replacing index.

Exercises
1. Justify the fact that the number of components of a rank-r tensor in an nD space is given by n^r.
2. Justify the fact that the number of independent components of rank-2 symmetric tensors is $\frac{n(n+1)}{2}$ while the number of independent non-zero[70] components of rank-2 anti-symmetric tensors is $\frac{n(n-1)}{2}$.
3. Find the transformation rules of a vector **A** between unprimed and primed coordinate systems and hence generalize these rules to tensors of any rank.
4. Distinguish between covariant and contravariant tensors using their transformation rules.
5. Show that the zero tensor is value-invariant (i.e. constant) across all coordinate systems.
6. What you notice about the tensor relation $A^{i...m}_{j...n} = 0$?
7. Show that linear combinations of tensors are tensors.
8. Show that outer products of tensors are tensors.
9. Show that contractions of tensors (i.e. mixed tensors of rank > 1) are tensors.
10. Show that inner products of tensors are tensors.
11. Show that permutations of tensors are tensors.
12. Show that the symmetry properties (i.e. being symmetric/anti-symmetric/neither) of tensors are invariant under coordinate transformations.
13. Why in the contraction operation of tensors the contracted indices should be of opposite variance type (i.e. one upper and one lower)?
14. Show that all the variance types of a tensor are tensors.
15. Discuss the nabla and Laplacian operators in 4D spacetime.

2.3 Curvature of Space

The curvature of space can be an intrinsic property and hence it can be detected and measured by an inhabitant of the space. An example of this is the Gaussian curvature of a surface (such as a sphere)

---

[70] "Non-zero" here means they are not necessarily zero due to the anti-symmetry requirement although they could be zero accidentally or for another reason.

which is an intrinsic property that can be detected and measured by a 2D inhabitant of the surface. The curvature of space can also be an extrinsic property and hence it is detected and measured by an inhabitant of the embedding space. For example, cylinder has no intrinsic curvature and hence a 2D inhabitant of the cylinder will see it flat like a plane with no curvature. However, from the 3D embedding space that contains this cylinder it is a curved surface, unlike plane, and hence a 3D inhabitant can detect and measure its curvature.

We note the following about the curvature of space:
• A space can be flat intrinsically and extrinsically (e.g. plane), or flat intrinsically and curved extrinsically (e.g. cylinder), or curved intrinsically and extrinsically (e.g. sphere). In fact, these are the common and familiar cases of space curvature. So, we may also find in the literature examples of spaces that are curved intrinsically but not extrinsically (although some of these examples are questionable). Also, there are examples of mathematical spaces that cannot be embedded in an embedding space and hence there is no meaning for them to have (or not have) extrinsic curvature although it is still meaningful to characterize them as curved spaces (i.e. intrinsically).
• The curvature of interest in special relativity and in general relativity is the intrinsic curvature and not the extrinsic curvature.[71] In fact, we lack the ability to envisage the 4D spacetime of special relativity and general relativity let alone embedding this spacetime in a space of higher dimensionality.
• The intrinsic curvature is mainly quantified by the Riemann-Christoffel curvature tensor (see § 2.10) although it may also be quantified by other parameters that are related to this tensor (e.g. the Gaussian curvature in the case of 2D spaces). Therefore, in general relativity the Riemann-Christoffel curvature tensor is the main building block for the formulation of spacetime curvature (see § 2.13). In fact, we will see that the Einstein tensor which defines the curvature of spacetime in the Field Equation of general relativity can be defined by the Christoffel symbols directly without resort to the Riemann-Christoffel curvature tensor. However, the Ricci curvature tensor and scalar (which enter in the definition of the Einstein tensor) are originally defined by the Riemann-Christoffel curvature tensor and hence the connection between the curvature of spacetime in general relativity and the Riemann-Christoffel curvature tensor is established.
• A curved space may have constant curvature all over the space (e.g. sphere), or have variable curvature and hence the curvature is position dependent (e.g. torus).
• There are several indicators and conditions for testing the intrinsic curvature and flatness of space. Some of these indicators and conditions will be briefly discussed in the questions and some will be investigated in detail later.

Problems

1. Briefly discuss the commonly used indicators and conditions for testing the intrinsic curvature and flatness of space.
   Answer: These include:
   • Special geometric tests such as the sum of the angles θ_s of triangles in the space and the relation between the circumference C and the diameter D of circles in the space. So, in curved space $\theta_s \neq \pi$ and $C \neq \pi D$ while in flat space $\theta_s = \pi$ and $C = \pi D$. In fact, these conditions reflect the local curvature and flatness (i.e. where the triangle and circle in our examples are located). The space is then qualified as curved if it is curved somewhere and as flat if it is flat everywhere. We should also remark that although the conditions in flat space (i.e. $\theta_s = \pi$ and $C = \pi D$) are obviously valid everywhere, the validity of the conditions in curved space (i.e. $\theta_s \neq \pi$ and $C \neq \pi D$) at every locally curved location are not obvious and hence they need to be established rigorously (e.g. in the case where the location is negatively curved in part and positively curved in other part).
   • Coordination by Cartesian system (see § 2.1). So, curved space cannot be coordinated by Cartesian system while flat space can be coordinated by Cartesian system. It is noteworthy that Cartesian in this context includes quasi-Cartesian (and may even be generalized to rectilinear). We should also note that coordination here means globally since curved space can be coordinated by Cartesian system locally but not globally.
   • Riemann-Christoffel curvature tensor (see § 2.10). So, this tensor does not vanish identically in

---

[71] So, flatness in special relativity means intrinsic flatness, i.e. vanishing of intrinsic curvature.

2.3 Curvature of Space

curved space, while it does vanish identically in flat space. In some cases this test also applies to other parameters of curvature, such as the Gaussian curvature of 2D spaces, that are based on the Riemann-Christoffel curvature tensor. It should be remarked that the Gaussian curvature in differential geometry is defined for 2D spaces although the concept may be extended to manifolds of higher dimensionality in the form of Riemannian curvature (see B2). We should also note that the Riemann-Christoffel curvature tensor in 2D space has only one independent component and hence the vanishing of Gaussian curvature is equivalent to the vanishing of this tensor.

- Parallel transport (see § 2.8). So, in flat space a parallel-transported vector around a closed curve returns to its initial position in the same direction while in curved space it does not necessarily return so.
- Metric tensor (see § 2.5). So, if all the components of the metric tensor are constants then the space is flat but the opposite is not true, i.e. the metric tensor of flat space may not be constant (e.g. Euclidean space coordinated by spherical coordinate system). In brief, the constancy of the metric tensor is a sufficient but not necessary condition for the space to be flat (and hence the value of this test is limited).[72]
- Christoffel symbols (see § 2.6). So, if all the Christoffel symbols of a given coordinate system of a space vanish identically then the space is flat; however the opposite is not true, i.e. the Christoffel symbols may not vanish in a flat space if the coordinate system is curvilinear (e.g. Euclidean space coordinated by spherical coordinate system). In brief, the vanishing of Christoffel symbols is a sufficient but not necessary condition for the space to be flat (and hence the value of this test is limited). In fact, this test is based on the previous test since the Christoffel symbols are synthesized from the components of the metric tensor and their derivatives.

We should remark that a space is flat only if it is globally flat (i.e. it is flat everywhere) while it is curved even if it is partially curved (i.e. it is curved somewhere). Hence, all these indicators and conditions should comply with this provision (or rather convention) when they are used to verify the flatness or curvature of space. For example, a space whose Riemann-Christoffel curvature tensor vanishes at certain location (but not globally) is not flat. In fact, adopting this convention should depend on the purpose and hence we may adopt a convention that is based on the local state of the space (i.e. curved and flat have only local significance) if it suits us. We may also use "curved" and "flat" only in global sense and hence we may have "partially flat and partially curved spaces".

Exercises

1. Justify the indicators and conditions (for testing the intrinsic curvature and flatness of space) that we stated in the Problems.
2. Give an example of an indicator for testing the intrinsic curvature and flatness of space other than those given in the Problems.
3. Give an example of a special geometric test (other than those given in the Problems) for testing the intrinsic curvature and flatness of space.
4. How do you define triangle and circle in curved spaces?
5. In special and general relativity theories, what we mean by the curvature of space and why?
6. Give simple examples of how a 2D inhabitant of a surface can detect the intrinsic curvature of his 2D space by conducting simple measurements.
7. Briefly discuss the coordination by a Cartesian system as a sign for the curvedness and flatness of space.
8. How can the Riemann-Christoffel curvature tensor be used for testing the curvedness and flatness of space?
9. Flat spaces may be metricized by non-constant metric tensor (i.e. tensor whose components are functions of coordinates). For example, 3D Euclidean space can be coordinated by a spherical coordinate system whose metric g_{ij} is diag $\left[1, r^2, r^2 \sin^2 \theta\right]$. Comment on this.

---

[72] The constancy of the metric tensor means the constancy of its individual elements although these elements are generally different.

2.4 Variational Principle and Euler-Lagrange Equation

In many mathematical and physical situations we need to find the extremum (or optimum) points of a given function[73] where the function reaches its maximum or minimum values (possibly locally), and this is achieved by using the variational principle. The essence of the variational principle is that the variation of the function at its extremum points vanishes. This principle is intuitive because at an extremum point the function should change its trend (i.e. from increasing to decreasing or from decreasing to increasing) and hence at the extremum point the function should cease to vary abruptly. In other words, a positive/negative variation trend followed by a negative/positive variation trend should be separated by a zero variation "trend".

The Euler-Lagrange equation is a relation that implements the variational principle in a specific mathematical form when the function that should be optimized is a functional. More specifically, the Euler-Lagrange equation is a mathematical relation whose objective is to minimize or maximize a certain functional $F(f)$ which depends on a function f. It is represented mathematically by a differential equation whose solutions optimize the particular functional F. The Euler-Lagrange equation in its generic, simple and most common form is given by:

$$\frac{\partial f}{\partial y} - \frac{d}{dx}\left(\frac{\partial f}{\partial y_x}\right) = 0 \qquad (42)$$

where $f(x, y, y_x)$ is a function of the given variables that optimizes the functional F, y is a function of x, and y_x is the derivative of y with respect to x.

Exercises
1. What is the relation between the variational principle and the Euler-Lagrange equation?
2. What is the role of the Euler-Lagrange equation in implementing the variational principle?

2.5 Metric Tensor

The metric tensor is a rank-2 symmetric differentiable invertible tensor that reflects and represents the geometric properties of the space. The main objective of the metric is to generalize the concept of length to general coordinate systems and hence maintain the invariance of length in different coordinate systems (see next paragraph). The metric tensor is also used to raise and lower indices and thus facilitate the transition between the covariant and contravariant types. As a tensor, the metric has geometric significance regardless of any coordinate system although it requires a coordinate system to be represented in a specific form. So in brief, the coordinate system and the space metric are independent but correlated entities. More specifically, the metric uniquely determines the space and identifies its geometric properties, but the opposite is untrue (i.e. the geometric properties of the space do not uniquely identify its metric) because the metric also depends on the choice of coordinate system, and the coordinate system for a given space is not unique.

In an orthonormal Cartesian coordinate system of an nD space the length of infinitesimal element of arc, ds, connecting two neighboring points in the space, one with coordinates x_i and the other with coordinates $x_i + dx_i$ ($i = 1, \cdots, n$), is given by:

$$(ds)^2 = dx_i dx_i = \delta_{ij} dx_i dx_j \qquad (43)$$

where δ_{ij} is the Kronecker delta. This definition of length is the key to the introduction of a rank-2 tensor, g_{ij}, called the metric tensor which for a general coordinate system is defined by:

$$(ds)^2 = g_{ij} dx^i dx^j \qquad (44)$$

where the indexed x represent general coordinates. Accordingly, we may consider the metric tensor as a generalization of the Kronecker delta tensor that is associated with orthonormal Cartesian coordinate

---
[73] We note that "function" in this context is more general than "simple function" or "function of function" where the latter is usually known as functional. So, $f(x)$ is a simple function while $F(f(x))$ is a functional.

2.5 Metric Tensor

systems (see Problems). The metric tensor in the last equation is of covariant form, however it has also a contravariant form (which is notated with g^{ij}) and a mixed form (which is notated with g_i^j).

The components of the metric tensor in its covariant, contravariant and mixed forms are closely related to the basis vectors of the coordinate system, that is:

$$g_{ij} = \mathbf{E}_i \cdot \mathbf{E}_j = \frac{\partial x_k}{\partial x^i}\frac{\partial x_k}{\partial x^j} \tag{45}$$

$$g^{ij} = \mathbf{E}^i \cdot \mathbf{E}^j = \frac{\partial x^i}{\partial x_k}\frac{\partial x^j}{\partial x_k} \tag{46}$$

$$g_j^i = \mathbf{E}^i \cdot \mathbf{E}_j = \frac{\partial x^i}{\partial x_k}\frac{\partial x_k}{\partial x^j} = \frac{\partial x^i}{\partial x^j} = \delta_j^i \tag{47}$$

where the indexed $\mathbf{E}$ are the covariant and contravariant basis vectors and where in the second equalities we used the definition of these basis vectors as given by Eq. 38 (noting that x^i and x^j represent general coordinates while x_k represents Cartesian coordinates whose variance type is irrelevant). As we see, the mixed metric tensor g_j^i is the same as the identity tensor δ_j^i (i.e. Kronecker delta).

The metric tensor is necessarily symmetric with non-zero diagonal elements but not necessarily diagonal.[74] Moreover, for the subject of this book (i.e. general relativity as well as special relativity to some extent) the diagonal elements of a diagonal metric tensor are not necessarily positive.[75] The last two statements obviously apply to the covariant and contravariant metric tensor but not to the mixed metric tensor which is the identity tensor and hence it is symmetric, diagonal with positive diagonal elements.

The covariant and contravariant forms of the metric tensor are inverses of each other and hence we have the following relations:

$$[g_{ij}] = [g^{ij}]^{-1} \qquad [g^{ij}] = [g_{ij}]^{-1} \tag{48}$$

where $[\cdots]$ symbolizes matrix while $[\cdots]^{-1}$ symbolizes its inverse. Therefore:

$$[g^{ik}][g_{kj}] = [g_{jk}][g^{ki}] = [\delta_j^i] \tag{49}$$

Since the metric tensor is invertible, its determinant should not vanish at any point in the space. Also, being invertible means that the metric tensor has always covariant and contravariant forms.

Apart from its conceptual significance (as representative of the space geometry) and apart from its practical and theoretical use in computing the line element (i.e. ds), the metric tensor is commonly used as an operator for lowering and raising indices (i.e. covariant metric tensor lowers indices and contravariant metric tensor raises indices) and hence facilitating the transition between the covariant and contravariant types. Consequently, in Riemannian space tensors can be cast into covariant or contravariant or mixed form. This means that the different forms of a tensor are equivalent and hence they represent the same tensor but in reference to different basis vector sets (i.e. covariant refers to contravariant basis and contravariant refers to covariant basis as seen in § 2.2).

As indicated earlier, the metric tensor is a variable function of coordinates in general. However, in rectilinear coordinate systems the metric tensor is constant because the basis vectors in these systems are constants. As we will see, a sufficient and necessary condition for the metric tensor to be constant in a given coordinate system is that the Christoffel symbols of the first or second kind vanish identically (refer to exercise 7 of 2.6). The metric tensor in general (i.e. whether variable or constant) behaves like a constant with respect to tensor differentiation and hence the covariant and intrinsic derivatives of the metric tensor vanish identically (see § 2.7).

Problems

1. Compare the concept of "transformation" with the concept of "shifting variance type".
   Answer: We note the following:

---

[74] The necessary and sufficient condition for the metric tensor to be diagonal is that the coordinate system is orthogonal.

[75] As indicated earlier (see § 1.3 and § 2.1), being necessarily positive (i.e. having positive definite quadratic form) or not depends on the space being Riemannian or pseudo-Riemannian.

- Transformation is an operation for shifting from one coordinate system to another coordinate system, while shifting variance type (by using the metric tensor as an index shifting operator) is an operation for shifting from one type of basis vectors (i.e. covariant or contravariant) of a given coordinate system to another type of basis vectors of that coordinate system. So in brief, transformation means changing coordinate system while shifting variance type means changing basis vectors of the same coordinate system. Hence, in shifting variance type both the covariant and contravariant types belong to the same coordinate system. However, we should notice the significance of the index shifting operation of covariant and contravariant metric tensor (see the next two problems).
- From a procedural perspective, in transformation we use a set of equations (e.g. Cartesian-to-spherical transformations or Lorentz transformations),[76] while in shifting variance type we use the metric tensor.
- The subject of transformation is coordinate system, while the subject of shifting variance type is tensor (or tensor-like entity).

2. Show that the metric tensor can be regarded as a transformation of the Kronecker delta tensor in its different variance types from an orthonormal Cartesian coordinate system to a general coordinate system.
   Answer: From the definition of the metric tensor in its different variance types, we have:

$$g_{ij} = \frac{\partial x^k}{\partial x^i}\frac{\partial x^k}{\partial x^j} \qquad g^{ij} = \frac{\partial x^i}{\partial x^k}\frac{\partial x^j}{\partial x^k} \qquad g^i_j = \frac{\partial x^i}{\partial x^k}\frac{\partial x^k}{\partial x^j}$$

where x^i and x^j represent general coordinates while x^k represents Cartesian coordinates (whose variance type is irrelevant). These equations can be written as (noting that the indices k and l refer to Cartesian coordinates):

$$g_{ij} = \frac{\partial x^k}{\partial x^i}\frac{\partial x^l}{\partial x^j}\delta_{kl} \qquad g^{ij} = \frac{\partial x^i}{\partial x^k}\frac{\partial x^j}{\partial x^l}\delta^{kl} \qquad g^i_j = \frac{\partial x^i}{\partial x^k}\frac{\partial x^l}{\partial x^j}\delta^k_l \qquad (50)$$

where they are justified by the fact that the Kronecker delta tensor is an index replacement operator plus the fact that the corresponding components of all the variance types of the Kronecker delta are the same, i.e. $\delta_{ij} = \delta^{ij} = \delta^i_j$. As seen, the relations in Eq. 50 are the transformation equations of the Kronecker delta tensor in its different variance types from an orthonormal Cartesian coordinate system to a general coordinate system, as required. In other words, the relations in Eq. 50 are the transformation rules of the metric tensor in orthonormal Cartesian coordinate system (which is the Kronecker delta tensor $\delta_{kl}, \delta^{kl}, \delta^k_l$) to the metric tensor in general coordinate system (i.e. g_{ij}, g^{ij}, g^i_j). We note that in curved space this transformation applies only locally.

3. What is the significance of raising an index of the covariant metric tensor g_{jk} and lowering an index of the contravariant metric tensor g^{jk}?
   Answer: Using g^{ij} as an index raising operator, we have:

$$g^{ij}g_{jk} = g^i_k = \delta^i_k$$

Similarly, using g_{ij} as an index lowering operator, we have:

$$g_{ij}g^{jk} = g^k_i = \delta^k_i$$

Hence, these operations mean taking the metric tensor back to its original Cartesian form (i.e. Kronecker delta) which the tensor originates from (according to the previous problem). In other words, we are reversing the transformation of the previous problem. This should be seen as a link between changing variance type (although restricted here to the metric tensor) and transforming between coordinate systems (although restricted here to Cartesian and locally), and hence it is related to the first problem.

---

[76] In fact, in transformation we use the Jacobian matrix which is derived from the transformation equations (see B3 and B3X).

In fact, this should be more appreciated if we note that the covariant and contravariant forms of the metric tensor are inverses of each other and hence if one of these forms corresponds to a transformation (of the Kronecker delta) from an orthonormal Cartesian coordinate system to a general coordinate system then the other form (when applied to the first form) should correspond to an opposite transformation from a general coordinate system to an orthonormal Cartesian coordinate system (i.e. taking the metric back to its origin which is the Kronecker delta). Also, see question 4.52 of B3X.

Exercises

1. Explain the relation between space, coordinate system and metric tensor.
2. What are the physical dimensions of the metric tensor?
3. Show that $g_{ij} = \mathbf{E}_i \cdot \mathbf{E}_j$.
4. Show that the metric is a rank-2 tensor.
5. Show that the Kronecker delta is a rank-2 tensor.
6. Why is the metric tensor symmetric?
7. Summarize the main properties of the line element ds in Riemannian space highlighting its relation to the metric tensor.
8. Show the invariance of the quadratic form $(ds)^2$ across coordinate systems.
9. Give the metric tensor of the 3D Euclidean space for the following coordinate systems: orthonormal Cartesian, cylindrical and spherical.
10. Give the metric tensor of the 4D flat spacetime of special relativity using the following 3D spatial coordinate systems: orthonormal Cartesian, cylindrical and spherical.
11. Justify the use of the Kronecker delta as an index replacement operator.
12. Show that the covariant and contravariant components of a vector $\mathbf{v}$ can be obtained from the relations: $\mathbf{v} \cdot \mathbf{E}_i = v_i$ and $\mathbf{v} \cdot \mathbf{E}^i = v^i$.
13. Justify the use of the metric tensor as an index shifting operator.
14. Show that in orthonormal Cartesian systems of Euclidean spaces there is no difference between the covariant and contravariant types of tensors.
15. Show that the metric tensor is diagonal *iff* the coordinate system is orthogonal.
16. Show that the constancy of the metric tensor is a sufficient but not necessary condition for the space to be flat.[77]

2.5.1 Metric Tensor of 4D Spacetime of General Relativity

Unlike the metric tensor of the flat spacetime of special relativity which has a general form (assuming a particular underlying 3D spatial coordinate system), the metric tensor of the curved spacetime of general relativity has no such general form because it depends on the distribution and flow of matter and energy as described by the energy-momentum tensor (see § 2.14 and § 3.2).[78] Accordingly, we should search for the metric tensor for each individual case and this depends on the physical setting and the adopted assumptions (as well as the employed coordinate system). We will see a few examples of the metric tensor of general relativity for some special and simple cases, e.g. Schwarzschild metric (see § 4.1).

Exercises

1. What we mean by "general form" when we say "the metric tensor of the flat spacetime of special relativity has a general form"?
2. Compare the metrics of special relativity and general relativity.
3. What is the implication of not having a unique and general form of the metric in general relativity?

---

[77] As indicated earlier, the constancy of the metric tensor means the constancy of its individual elements.

[78] In fact, this is a physical argument based on the validity of general relativity and its Field Equation. We can also have a purely geometric argument based on the fact that curved geometry, unlike flat geometry, is not unique.

2.6 Christoffel Symbols

The Christoffel symbols are rank-3 affine tensors but they are not general tensors (see exercise 12). As a consequence of not being general tensors, if all the Christoffel symbols of either kind vanish in a particular coordinate system of a given space they will not necessarily vanish in other coordinate systems of that space. For instance, all the Christoffel symbols vanish in Cartesian coordinate systems of 3D Euclidean space but not in cylindrical or spherical coordinate systems of this space (see B3). As we will see, the Christoffel symbols solely depend on the metric tensor and hence they are intrinsic properties of their space. The Christoffel symbols are classified as those of the first kind and those of the second kind. These two kinds are linked through the index raising and lowering operators. Both kinds of the Christoffel symbols are variable functions of coordinates since they depend on the metric tensor which is coordinate dependent in general.

The Christoffel symbols of the first kind are given by:

$$[ij,l] = \frac{1}{2}\left(\partial_j g_{il} + \partial_i g_{jl} - \partial_l g_{ij}\right) \tag{51}$$

where the indexed g is the covariant metric tensor. The Christoffel symbols of the second kind are obtained by raising the third index of the Christoffel symbols of the first kind, and hence they are given by:

$$\Gamma^k_{ij} = g^{kl}[ij,l] = \frac{g^{kl}}{2}\left(\partial_j g_{il} + \partial_i g_{jl} - \partial_l g_{ij}\right) \tag{52}$$

where g^{kl} is the contravariant metric tensor. This process is reversible and hence the Christoffel symbols of the first kind can be obtained from the Christoffel symbols of the second kind by lowering the upper index, that is:

$$g_{km}\Gamma^k_{ij} = g_{km}g^{kl}[ij,l] = \delta^l_m [ij,l] = [ij,m] \tag{53}$$

The Christoffel symbols of the first and second kind are symmetric in their paired indices, that is:

$$[ij,k] = [ji,k] \tag{54}$$
$$\Gamma^k_{ij} = \Gamma^k_{ji} \tag{55}$$

These properties can be verified by shifting the indices in the mathematical expressions of the Christoffel symbols, as given by Eqs. 51 and 52, noting that the metric tensor is symmetric (see § 2.5).[79]

In any coordinate system, all the Christoffel symbols of the first and second kind vanish identically *iff* all the components of the metric tensor in that system are constants (see exercise 7). Now, in affine coordinate systems all the components of the metric tensor are constants and hence all the Christoffel symbols of both kinds vanish identically. The prominent example is the orthonormal Cartesian coordinate systems where all the Christoffel symbols of the first and second kind are identically zero. In an nD space, the number of the Christoffel symbols of either kind is n^3 which represents all the possible permutations of the 3 indices including the repetitive ones. However, because of the symmetry property in their paired indices the number of independent Christoffel symbols of each kind in general coordinate systems is reduced to:[80]

$$N = \frac{n^2(n+1)}{2} \tag{56}$$

Problems

1. What are the other names given to the Christoffel symbols of the second kind in the literature of general relativity?
   Answer: The Christoffel symbols of the second kind are known as affine connections or connection coefficients or metric connections, as well as other names. In fact, Christoffel symbols and those terms (such as connection coefficients) are not the same conceptually although for our purpose in this book their difference can be ignored and hence we treat them as identical. In fact, the Christoffel symbols should be considered a special type of connection coefficients.

---

[79] This is demonstrated in question 5.10 of B3X. Also see Problem 4.
[80] This is proved in question 5.28 of B3X.

2.6 Christoffel Symbols

2. Justify the fact that the Christoffel symbols are an intrinsic property of the space.
   Answer: This can be easily concluded from Eqs. 51 and 52 since the Christoffel symbols of both kinds can be expressed solely in terms of the metric tensor and its derivatives, and the metric tensor is obviously intrinsic.
3. Develop the concept and fundamental definition of the Christoffel symbols.
   Answer: The partial derivatives of the basis vectors of a given coordinate system (whether rectilinear or curvilinear) of a given space (whether flat or curved) are vectors that belong to the space and hence they can be expressed as linear combination of the basis vectors of the space.[81] To cast this fact into a rigorous and fully informative mathematical form we write the following equation: $\partial_j \mathbf{E}_i = \Gamma_{ij}^k \mathbf{E}_k$ which can be considered as the fundamental definition of the Christoffel symbol Γ_{ij}^k. As we see, Γ_{ij}^k symbolizes the component of the partial derivative of the basis vector $\mathbf{E}_i$ with respect to the jth coordinate x^j in the direction of the basis vector $\mathbf{E}_k$ and hence it is fully informative with no ambiguity although we still need to develop a specific mathematical expression for this symbol.

Now, we should develop a similar fundamental definition for the Christoffel symbol in terms of the contravariant basis vectors, that is:

$$\begin{aligned}
\mathbf{E}_k \cdot \mathbf{E}^i &= \delta_k^i \\
\partial_j \left(\mathbf{E}_k \cdot \mathbf{E}^i \right) &= \partial_j \delta_k^i \\
(\partial_j \mathbf{E}_k) \cdot \mathbf{E}^i + \mathbf{E}_k \cdot \left(\partial_j \mathbf{E}^i \right) &= 0 \\
\Gamma_{kj}^l \mathbf{E}_l \cdot \mathbf{E}^i + \mathbf{E}_k \cdot \left(\partial_j \mathbf{E}^i \right) &= 0 \\
\Gamma_{kj}^l \delta_l^i + \mathbf{E}_k \cdot \left(\partial_j \mathbf{E}^i \right) &= 0 \\
\Gamma_{kj}^i + \mathbf{E}_k \cdot \left(\partial_j \mathbf{E}^i \right) &= 0 \\
\mathbf{E}_k \cdot \left(\partial_j \mathbf{E}^i \right) &= -\Gamma_{kj}^i \\
\partial_j \mathbf{E}^i &= -\Gamma_{kj}^i \mathbf{E}^k
\end{aligned}$$

where line 1 is the reciprocity relation between covariant and contravariant basis vectors, line 3 is the product rule plus the fact that the components of the unity tensor are constants, in line 4 we use the above fundamental definition of the Christoffel symbols in terms of the covariant basis vectors (i.e. $\partial_j \mathbf{E}_i = \Gamma_{ij}^k \mathbf{E}_k$), line 5 is the reciprocity relation, in line 6 we use Kronecker delta as an index replacement operator, and line 8 is another form of line 7 (noting that line 7 can be obtained from line 8 by inner product of both sides with $\mathbf{E}_k$ after changing the dummy index k).[82] Hence, the fundamental definition of the Christoffel symbols is given by the following basic relations:

$$\partial_j \mathbf{E}_i = \Gamma_{ij}^k \mathbf{E}_k \qquad \text{and} \qquad \partial_j \mathbf{E}^i = -\Gamma_{kj}^i \mathbf{E}^k$$

The next thing we need to do is to obtain a completely "covariant" version of the above Christoffel symbols. This is done by lowering the upper index using the covariant metric tensor, that is:

$$[ij, k] = g_{mk} \Gamma_{ij}^m$$

We note that in the above developments we are assuming that Γ_{ij}^k has certain tensor-like properties such as shifting indices by the metric tensor. This could be justified by the fact that these indices refer to covariant and contravariant basis vectors and hence they should be treated like tensor indices. We

---
[81] We are considering a space of sufficient dimensionality from an intrinsic perspective with no regard to embedding in a space of higher dimensionality. Also, there are theoretical subtleties about how the vectors belong to the space and how they can be expressed as linear combination of the basis vectors (where the projection onto the tangent space at the point of application is considered in addressing these subtleties). However, we follow an intuitive approach and hence we do not go through these details. In brief, we can take the above premise (i.e. the partial derivatives of the basis vectors of a given space are vectors that belong to the space) as assumption in the development of our definition of the Christoffel symbols.

[82] In fact, this in essence is the same as the relation $\mathbf{v} \cdot \mathbf{E}_i = v_i$ (or $\mathbf{V} \cdot \mathbf{E}_i = V_i^{\ k} \mathbf{E}_k$) which we proved in exercise 12 of § 2.5. Also see exercise 10.

2.6 Christoffel Symbols

should also note that Γ_{ij}^k is affine tensor which will become clear when we find specific mathematical expressions for the Christoffel symbols.

Finally, we need to find specific mathematical expressions for the Christoffel symbols of both versions (or kinds), i.e. $[ij,k]$ and Γ_{ij}^k. This will be dealt with in the next problem.

4. Using the fundamental definition of Γ_{ij}^k (i.e. $\partial_j \mathbf{E}_i = \Gamma_{ij}^k \mathbf{E}_k$) and the definition of $[ij,l]$ (i.e. $[ij,l] = g_{ml}\Gamma_{ij}^m$), show that:

$$[ij,l] = \frac{1}{2}(\partial_j g_{il} + \partial_i g_{jl} - \partial_l g_{ij})$$

$$\Gamma_{ij}^k = \frac{g^{kl}}{2}(\partial_j g_{il} + \partial_i g_{jl} - \partial_l g_{ij})$$

Answer: First, we should note that:

$$\Gamma_{ij}^k \mathbf{E}_k \equiv \partial_j \mathbf{E}_i = \partial_j \partial_i \mathbf{r} = \partial_i \partial_j \mathbf{r} = \partial_i \mathbf{E}_j \equiv \Gamma_{ji}^k \mathbf{E}_k$$

and hence the Christoffel symbols are symmetric in their paired indices, i.e. $\Gamma_{ij}^k = \Gamma_{ji}^k$. Moreover, since Γ_{ij}^k is symmetric in its paired indices then $[ij,k]$ (which is obtained from Γ_{ij}^k by lowering the non-paired index) should also be symmetric in its paired indices (see the upcoming note).

Now, we have:

$$\begin{aligned}
g_{il} &= \mathbf{E}_i \cdot \mathbf{E}_l \\
\partial_j g_{il} &= \partial_j (\mathbf{E}_i \cdot \mathbf{E}_l) \\
\partial_j g_{il} &= (\partial_j \mathbf{E}_i) \cdot \mathbf{E}_l + \mathbf{E}_i \cdot (\partial_j \mathbf{E}_l) \\
\partial_j g_{il} &= (\Gamma_{ij}^k \mathbf{E}_k) \cdot \mathbf{E}_l + \mathbf{E}_i \cdot (\Gamma_{lj}^k \mathbf{E}_k) \\
\partial_j g_{il} &= \Gamma_{ij}^k g_{kl} + \Gamma_{lj}^k g_{ik} \\
\partial_j g_{il} &= [ij,l] + [lj,i]
\end{aligned}$$

where line 1 is the relation between the metric coefficients and the basis vectors (see Eq. 45), line 3 is the product rule, line 4 is the fundamental definition of the Christoffel symbols, and line 6 is using the metric tensor as an index lowering operator (in conjunction with the definition of the symbols $[ij,l]$ and $[lj,i]$). On relabeling the indices in the last equation, we get two other relations:

$$\begin{aligned}
\partial_i g_{jl} &= [ji,l] + [li,j] \\
\partial_l g_{ij} &= [il,j] + [jl,i]
\end{aligned}$$

On adding these three relations algebraically we get:

$$\begin{aligned}
\partial_j g_{il} + \partial_i g_{jl} - \partial_l g_{ij} &= [ij,l] + [lj,i] + [ji,l] + [li,j] - [il,j] - [jl,i] \\
\partial_j g_{il} + \partial_i g_{jl} - \partial_l g_{ij} &= [ij,l] + [lj,i] + [ij,l] + [li,j] - [li,j] - [lj,i] \\
\partial_j g_{il} + \partial_i g_{jl} - \partial_l g_{ij} &= 2[ij,l] \\
[ij,l] &= \frac{1}{2}(\partial_j g_{il} + \partial_i g_{jl} - \partial_l g_{ij})
\end{aligned}$$

which completes the proof of the first part of the problem. We note that in line 2 we use the aforementioned symmetry property of the Christoffel symbols in their paired indices.

Now, if we multiply the last equation with g^{kl} we get:

$$\begin{aligned}
g^{kl}[ij,l] &= \frac{g^{kl}}{2}(\partial_j g_{il} + \partial_i g_{jl} - \partial_l g_{ij}) \\
\Gamma_{ij}^k &= \frac{g^{kl}}{2}(\partial_j g_{il} + \partial_i g_{jl} - \partial_l g_{ij})
\end{aligned}$$

2.6 Christoffel Symbols

which completes the proof of the second part of the problem.

Note: we should note that deviation from symmetry in the paired indices of the Christoffel symbols (or rather connection coefficients to be accurate) is considered in the literature (although from different perspective and starting point) where this deviation is quantified by the so-called torsion tensor which is the difference between Γ_{ij}^k and Γ_{ji}^k (i.e. $\Gamma_{ij}^k - \Gamma_{ji}^k$). However, we do not go through these details. In brief, we assume that the torsion tensor is identically zero (i.e. the space is torsion-free) and the covariant derivative of the metric tensor is also identically zero, and hence we have "metric connections".

5. Show that in orthogonal coordinate systems, the Christoffel symbols of the first kind are given by:

$$[ij,k] = 0 \qquad (i \neq j \neq k)$$

$$[ii,k] = -\frac{1}{2}\partial_k g_{ii} \qquad (i \neq k, \text{ no sum on } i)$$

$$[ij,i] = [ji,i] = \frac{1}{2}\partial_j g_{ii} \qquad (i \neq j, \text{ no sum on } i)$$

$$[ii,i] = \frac{1}{2}\partial_i g_{ii} \qquad (\text{no sum on } i)$$

Answer: The Christoffel symbols of the first kind are given by:

$$[ij,k] = \frac{1}{2}\left(\partial_j g_{ik} + \partial_i g_{jk} - \partial_k g_{ij}\right)$$

In orthogonal coordinate systems the metric tensor is diagonal (see exercise 15 of § 2.5) and hence $g_{ab} = 0$ when $a \neq b$. Accordingly:

- When all the indices are different (i.e. $i \neq j \neq k$) then $g_{ik} = g_{jk} = g_{ij} = 0$ identically and hence $[ij,k] = 0$.
- When the paired indices are identical (i.e. $i = j$ and $i \neq k$) then:

$$[ii,k] = \frac{1}{2}\left(\partial_i g_{ik} + \partial_i g_{ik} - \partial_k g_{ii}\right) = \frac{1}{2}\left(0 + 0 - \partial_k g_{ii}\right) = -\frac{1}{2}\partial_k g_{ii} \qquad (\text{no sum on } i)$$

- When two unpaired indices are identical (i.e. $i = k$ and $i \neq j$) then:

$$[ij,i] = \frac{1}{2}\left(\partial_j g_{ii} + \partial_i g_{ji} - \partial_i g_{ij}\right) = \frac{1}{2}\left(\partial_j g_{ii} + 0 - 0\right) = \frac{1}{2}\partial_j g_{ii} \qquad (\text{no sum on } i)$$

- When all the indices are identical (i.e. $i = j = k$) then:

$$[ii,i] = \frac{1}{2}\left(\partial_i g_{ii} + \partial_i g_{ii} - \partial_i g_{ii}\right) = \frac{1}{2}\partial_i g_{ii} \qquad (\text{no sum on } i)$$

We note that the above 4 cases are inclusive to all the possibilities of the Christoffel symbol with regard to the identicality and difference of its indices (where in the case of two unpaired indices the symmetry in the paired indices should be considered in the above demonstration).

6. Using the given expressions in the previous problem, derive the expressions of the Christoffel symbols of the second kind in orthogonal coordinate systems.
Answer: The Christoffel symbols of the second kind are given by:

$$\Gamma_{ij}^k = g^{kl}\,[ij,l]$$

In orthogonal coordinate systems we have $g_{ab} = g^{ab} = 0$ ($a \neq b$), and hence the Christoffel symbols of the second kind are given by:

$$\Gamma_{ij}^k = g^{kk}\,[ij,k] = \frac{[ij,k]}{g_{kk}} \qquad (\text{no sum on } k)$$

where $g^{kk} = 1/g_{kk}$ is justified by the diagonality of the metric tensor plus the fact that the covariant and contravariant metric tensors are inverses of each other. Accordingly:

- When all the indices are different (i.e. $i \neq j \neq k$) then $[ij, k] = 0$ (according to the previous problem) and hence:
$$\Gamma_{ij}^k = \frac{[ij,k]}{g_{kk}} = \frac{0}{g_{kk}} = 0$$

- When the paired indices are identical (i.e. $i = j$ and $i \neq k$) then:
$$\Gamma_{ii}^k = \frac{[ii,k]}{g_{kk}} = -\frac{1}{2g_{kk}}\partial_k g_{ii} \qquad \text{(no sum on } i \text{ or } k\text{)}$$

- When two unpaired indices are identical (i.e. $i = k$ and $i \neq j$) then:
$$\Gamma_{ij}^i = \Gamma_{ji}^i = \frac{[ij,i]}{g_{ii}} = \frac{1}{2g_{ii}}\partial_j g_{ii} \qquad \text{(no sum on } i\text{)}$$

- When all the indices are identical (i.e. $i = j = k$) then:
$$\Gamma_{ii}^i = \frac{[ii,i]}{g_{ii}} = \frac{1}{2g_{ii}}\partial_i g_{ii} \qquad \text{(no sum on } i\text{)}$$

Exercises
1. The Christoffel symbols are not tensors although they have indices that refer to coordinates and basis vectors like tensors. Justify.
2. Explain "affine tensors" within the context of Christoffel symbols.
3. What are the number of Christoffel symbols and the number of independent Christoffel symbols of each kind in the spacetime of general relativity?
4. What are the physical dimensions of the Christoffel symbols of the first and second kind?
5. Prove the following relation: $\partial_j g_{il} = [ij, l] + [lj, i]$.
6. Show that the Christoffel symbols vanish identically *iff* the coordinate system is rectilinear (and Cartesian in particular).
7. Prove that in any coordinate system, all the Christoffel symbols of either kind vanish identically *iff* all the components of the metric tensor in the given coordinate system are constants.[83]
8. Show that the vanishing of the Christoffel symbols identically is a sufficient but not necessary condition for space flatness.
9. Summarize the main properties of the Christoffel symbols.
10. Show that $\mathbf{E}^k \cdot \partial_j \mathbf{E}_i = \Gamma_{ij}^k$ and $\mathbf{E}_k \cdot \partial_j \mathbf{E}^i = -\Gamma_{kj}^i$.
11. Show that:
$$\Gamma_{ij}^k = \frac{\partial x^k}{\partial x_a}\frac{\partial^2 x_a}{\partial x^i \partial x^j}$$
where the lower-indexed x are Cartesian coordinates and the upper-indexed x are general coordinates.
12. Show that the Christoffel symbols transform as:
$$\Gamma'^k_{ij} = \frac{\partial x'^k}{\partial x^b}\frac{\partial x^d}{\partial x'^i}\frac{\partial x^c}{\partial x'^j}\Gamma^b_{dc} + \frac{\partial x'^k}{\partial x^c}\frac{\partial^2 x^c}{\partial x'^i \partial x'^j}$$

Hence, conclude that the Christoffel symbols are affine tensors but not general tensors.

2.7 Tensor Differentiation

We mean by tensor differentiation covariant and intrinsic (or absolute) differentiation.[84] In a sense, the former is the equivalent in curvilinear systems to the ordinary partial differentiation in rectilinear

---
[83] "Identically" should indicate the validity over the entire space. However, the statement may also apply in a given region.
[84] In fact, this is from a technical tensorial perspective (which is our main concern); otherwise we will investigate in the questions even ordinary types of differential operations on tensors (i.e. partial and total) and hence "tensor differentiation" is more general.

2.7 Tensor Differentiation

systems while the latter is the equivalent in curvilinear systems to the ordinary total differentiation in rectilinear systems. The objective of tensor differentiation is to ensure the invariance of derivative (i.e. being a tensor) in general coordinate systems, and this results in applying more sophisticated rules using Christoffel symbols. As we will see, using Christoffel symbols is based on differentiating the basis vectors due to the fact that the basis vectors in curvilinear systems are variable functions of coordinates and hence they should be differentiated like the components of the tensor (see § 2.6). The covariant derivative of a tensor is a tensor that is one covariant rank higher than the differentiated tensor while the intrinsic derivative of a tensor is a tensor that is the same rank and variance type as the differentiated tensor. However, we should note that the differentiation index in covariant derivative may be raised by the metric tensor and hence we get "contravariant differentiation" and therefore the resulting tensor is one contravariant rank higher than the differentiated tensor.

In brief, the covariant derivative is a partial derivative of the tensor that includes differentiating the basis vectors as well as differentiating the components of the tensor. Accordingly, the covariant derivative of a general tensor $\mathbf{A}$ is given generically by:

$$\mathbf{A}_{;k} \equiv \partial_k \left(A^{i\cdots m}_{j\cdots n} \mathbf{E}_i \cdots \mathbf{E}_m \mathbf{E}^j \cdots \mathbf{E}^n \right) = A^{i\cdots m}_{j\cdots n;k} \mathbf{E}_i \cdots \mathbf{E}_m \mathbf{E}^j \cdots \mathbf{E}^n \tag{57}$$

where the semicolon is a conventional notation for the covariant derivative of the tensor with respect to the coordinate indexed by the following index (i.e. the kth coordinate in this case). For example, the covariant derivative of a rank-2 mixed tensor, $\mathbf{A} = A_i{}^j \mathbf{E}^i \mathbf{E}_j$ is obtained as follows:

$$\begin{aligned}
\mathbf{A}_{;k} &\equiv \partial_k \left(A_i{}^j \mathbf{E}^i \mathbf{E}_j \right) \\
&= \left(\partial_k A_i{}^j \right) \mathbf{E}^i \mathbf{E}_j + A_i{}^j \left(\partial_k \mathbf{E}^i \right) \mathbf{E}_j + A_i{}^j \mathbf{E}^i \left(\partial_k \mathbf{E}_j \right) \\
&= \left(\partial_k A_i{}^j \right) \mathbf{E}^i \mathbf{E}_j + A_i{}^j \left(-\Gamma^i_{ak} \mathbf{E}^a \right) \mathbf{E}_j + A_i{}^j \mathbf{E}^i \left(\Gamma^a_{jk} \mathbf{E}_a \right) \\
&= \left(\partial_k A_i{}^j \right) \mathbf{E}^i \mathbf{E}_j - A_i{}^j \Gamma^i_{ak} \mathbf{E}^a \mathbf{E}_j + A_i{}^j \Gamma^a_{jk} \mathbf{E}^i \mathbf{E}_a \\
&= \left(\partial_k A_i{}^j \right) \mathbf{E}^i \mathbf{E}_j - A_a{}^j \Gamma^a_{ik} \mathbf{E}^i \mathbf{E}_j + A_i{}^a \Gamma^j_{ak} \mathbf{E}^i \mathbf{E}_j \\
&= \left(\partial_k A_i{}^j - A_a{}^j \Gamma^a_{ik} + A_i{}^a \Gamma^j_{ak} \right) \mathbf{E}^i \mathbf{E}_j \\
&= A_i{}^j{}_{;k} \mathbf{E}^i \mathbf{E}_j
\end{aligned}$$

where line 1 is a definition, line 2 is the product rule of differentiation, line 3 is based on the fundamental definition of Christoffel symbols (i.e. $\partial_j \mathbf{E}_i = \Gamma^k_{ij} \mathbf{E}_k$ and $\partial_j \mathbf{E}^i = -\Gamma^i_{kj} \mathbf{E}^k$), line 5 is relabeling of dummy indices, line 6 is taking a common factor, and line 7 is a conventional notation for the covariant derivative of the tensor with respect to the kth coordinate x^k.

Following this pattern, we can see that the covariant derivative of a differentiable tensor $\mathbf{A}$ of type (m,n) with respect to the kth coordinate is given by:

$$\begin{aligned}
A^{i_1\ldots i_m}_{j_1\ldots j_n;k} &= \partial_k A^{i_1\ldots i_m}_{j_1\ldots j_n} + \Gamma^{i_1}_{lk} A^{l\ldots i_m}_{j_1\ldots j_n} + \cdots + \Gamma^{i_m}_{lk} A^{i_1\ldots l}_{j_1\ldots j_n} \\
&\quad - \Gamma^{l}_{j_1 k} A^{i_1\ldots i_m}_{l\ldots j_n} - \cdots - \Gamma^{l}_{j_n k} A^{i_1\ldots i_m}_{j_1\ldots l}
\end{aligned} \tag{58}$$

The intrinsic derivative of a tensor along a t-parameterized curve $x^i(t)$ with respect to the parameter t is the inner product of the covariant derivative of the tensor and the tangent vector dx^i/dt to the curve.[85] In fact, the intrinsic derivative of a general tensor is a total derivative of the tensor that includes differentiating the basis vectors as well as differentiating the components of the tensor. Accordingly, the

---

[85] For this to be strictly valid, the tensor should be defined over the region enclosing the curve and not only over the curve. However, we can always imagine (possibly hypothetically) that this is the case to justify this definition from a formal (or superficial) viewpoint. It is worth noting that in some old textbooks the intrinsic derivative is labeled as "covariant derivative along a curve".

2.7 Tensor Differentiation

intrinsic derivative of a general tensor **A** is given generically by:

$$\frac{d}{dt}\left(A^{i\cdots m}_{j\cdots n}\mathbf{E}_i\cdots\mathbf{E}_m\mathbf{E}^j\cdots\mathbf{E}^n\right) = \frac{\partial\left(A^{i\cdots m}_{j\cdots n}\mathbf{E}_i\cdots\mathbf{E}_m\mathbf{E}^j\cdots\mathbf{E}^n\right)}{\partial x^k}\frac{dx^k}{dt} \quad (59)$$

$$= A^{i\cdots m}_{j\cdots n;k}\frac{dx^k}{dt}\mathbf{E}_i\cdots\mathbf{E}_m\mathbf{E}^j\cdots\mathbf{E}^n$$

$$\equiv \frac{\delta A^{i\cdots m}_{j\cdots n}}{\delta t}\mathbf{E}_i\cdots\mathbf{E}_m\mathbf{E}^j\cdots\mathbf{E}^n$$

where the chain rule is used in line 1 and Eq. 57 is used in line 2 while $\delta/\delta t$ in the last line is a conventional notation for the intrinsic derivative of the tensor with respect to the parameter t. For example, the intrinsic derivative of a contravariant vector $\mathbf{A} = A^i\mathbf{E}_i$ along a t-parameterized curve is obtained as follows:

$$\frac{d\mathbf{A}}{dt} = \frac{d}{dt}\left(A^i\mathbf{E}_i\right) \quad (60)$$

$$= \mathbf{E}_i\frac{dA^i}{dt} + A^i\frac{d\mathbf{E}_i}{dt}$$

$$= \mathbf{E}_i\frac{dA^i}{dt} + A^i\frac{\partial\mathbf{E}_i}{\partial x^j}\frac{dx^j}{dt}$$

$$= \mathbf{E}_i\frac{dA^i}{dt} + A^i\Gamma^k_{ij}\mathbf{E}_k\frac{dx^j}{dt}$$

$$= \mathbf{E}_i\frac{dA^i}{dt} + \mathbf{E}_iA^k\Gamma^i_{kj}\frac{dx^j}{dt}$$

$$= \left(\frac{dA^i}{dt} + A^k\Gamma^i_{kj}\frac{dx^j}{dt}\right)\mathbf{E}_i$$

$$= \frac{\delta A^i}{\delta t}\mathbf{E}_i$$

where line 2 is the product rule of differentiation, line 3 is the chain rule of differentiation, line 4 is the fundamental definition of Christoffel symbols (i.e. $\partial_j\mathbf{E}_i = \Gamma^k_{ij}\mathbf{E}_k$), line 5 is exchanging the dummy indices i and k, line 6 is taking a common factor, and line 7 is a conventional notation for the intrinsic derivative of the vector with respect to the parameter t. This pattern can be easily generalized to tensors of any rank and variance type. In fact, the pattern of intrinsic differentiation can be easily obtained from the pattern of covariant differentiation by noting that the intrinsic derivative is an inner product of the covariant derivative and the tangent vector (as seen in Eq. 59).

Problems

1. We started this section with the statement: "We mean by tensor differentiation covariant and intrinsic (or absolute) differentiation". Comment on this.
   Answer: In fact, "tensor differentiation" in its general sense should also include ordinary differentiation (i.e. partial and total differentiation) of tensors but this has no particular relation to tensors, moreover it does not preserve the tensorial nature of tensors in general. Therefore, it is not of prime interest to us apart from determining its status within the calculus of tensors (as will be investigated in the questions).

2. Make a list of the main properties and rules of tensor differentiation.
   Answer: We note the following:[86]
   • Tensor differentiation is the same as ordinary differentiation (i.e. partial and total differentiation) but with the application of the differentiation process on both the tensor components and its basis vectors using the product rule of differentiation.
   • The covariant and intrinsic derivatives of tensors are tensors.

---

[86] Some of the following properties and rules are demonstrated and proved in B3 and B3X (which the interested readers are referred to for details). These properties and rules are also investigated further in the Exercises.

2.7 Tensor Differentiation

- The covariant derivative is 1 covariant rank higher than the rank of the differentiated tensor while the rank and variance type of the intrinsic derivative is the same as the rank and variance type of the differentiated tensor. Hence, a tensor of type (m, n) will have a covariant derivative of type $(m, n + 1)$ and an intrinsic derivative of type (m, n).
- The sum and product rules of differentiation apply to tensor differentiation like ordinary differentiation.
- The covariant and intrinsic derivatives of scalars and affine tensors[87] of higher ranks are the same as the ordinary derivatives (i.e. partial and total derivatives).
- The covariant and intrinsic derivatives of the metric, Kronecker and permutation tensors as well as the basis vectors vanish identically in any coordinate system.
- Unlike ordinary differential operators, tensor differential operators do not commute with each other in general.
- Tensor differential operators commute with the index replacement operator and index shifting operators.
- Tensor differential operators commute with the contraction of indices.
- The covariant derivative is generally defined over a space or a region of space while the intrinsic derivative is defined over a curve.
- The intrinsic derivative of a tensor may be seen as the inner product of the covariant derivative of the tensor and the tangent vector to the curve (by contracting the differentiation index of the covariant derivative with the index of the tangent vector).

Exercises
1. What is the essence of tensor differentiation?
2. Show that the covariant and intrinsic derivatives of scalars and affine tensors of higher ranks are the same as the ordinary derivatives (i.e. partial and total derivatives).
3. Show that the partial derivative of a scalar is a tensor (i.e. rank-1 tensor which is vector).
4. Show that the total derivative of a scalar along a t-parameterized curve is a tensor (i.e. scalar or rank-0 tensor).
5. Demonstrate that the partial derivative of a non-scalar tensor is not a tensor (in general).
6. Demonstrate that the total derivative of a non-scalar tensor is not a tensor.
7. Show that the covariant derivative of a tensor is a tensor.
8. Show that the absolute derivative of a tensor is a tensor.
9. Derive, from first principles, the mathematical expression for the intrinsic derivative of a covariant vector $\mathbf{A} = A_i \mathbf{E}^i$ with full justification of each step.
10. Using the expressions for the intrinsic derivative of the tensors A_i and B^i along a curve $x^j(t)$, show that the intrinsic derivative is an inner product of the covariant derivative of the tensors and the tangent vector dx^j/dt to the curve.
11. Make a comparison between covariant derivative and absolute derivative.
12. Show that the covariant derivative of the basis vectors vanishes identically.
13. Show that the covariant and intrinsic derivatives of the metric tensor and the Kronecker delta tensor vanish identically.
14. Elucidate the fact that tensor differential operators do not commute with each other in general.
15. Show that tensor differential operators commute with the index shifting operators and the index replacement operator.
16. Show that tensor differential operators commute with the contraction of indices.
17. Show that the sum rule of differentiation applies to tensor differentiation like ordinary differentiation.
18. Show that the product rule of differentiation applies to tensor differentiation like ordinary differentiation.
19. What is the link between the following premises: "*the intrinsic derivative is a total derivative of the tensor that includes differentiating the basis vectors*" and "*the intrinsic derivative is the inner product of the covariant derivative of the tensor and the tangent vector*"?

---
[87] Affine tensors here means tensors in affine coordinate systems.

20. The rank and variance type of the intrinsic derivative is the same as the rank and variance type of the differentiated tensor. Why?

2.8 Parallel Transport

We discuss in this section parallel transport[88] which was used for characterizing the curvature of space (see § 2.3) and will be used for identifying geodesic paths (see § 2.9.4). The essence of the concept of "parallelism" is "having the same direction". In flat space, parallelism is a global and absolute property as it is defined without reference to a particular position or path where the absolute concept of straight line with a fixed and global direction is used. This is because "direction" in flat space is independent of position. However, in curved Riemannian space we need a generalization or adaptation to the concept of parallelism to suit the curved nature of the space because in curved space we do not have fixed direction in the global and absolute sense that we have in flat space. In other words, "direction" in curved space is dependent on position. This adaptation and generalization is made through the concept of parallel transport whose essence is the characterization of parallelism by the local direction along a given curve in the space. So, parallel transport means moving (or transporting) a vector along a curve in a series of infinitesimal steps while keeping the vector constant during each step so that the vector in the end of the step is identical to the vector in the start of the step.[89] In other words, keeping the vector infinitesimally parallel to itself during its journey along the curve. Hence, in Riemannian space the idea of parallelism is defined with respect to a local direction and with reference to a prescribed curve and therefore it is different from the idea of parallelism in the Euclidean sense.

To be more clear let have a vector at a given position P_1 in an nD space (whether the space is flat or curved) and we want to transport this vector to another position P_2 in the space such that the magnitude and direction of the vector at P_2 are the same as the magnitude and direction of the vector at P_1. If the space is flat then we just move the vector along any path connecting P_1 to P_2 while keeping it parallel to its direction at P_1 because in flat space parallelism is global and absolute and hence it is independent of the position of the vector and the path of transportation. However, if the space is curved then it is not obvious how to do this because the direction in curved space intuitively depends on the position and hence on the path. So, we need to generalize the idea of parallel transport from flat space to curved space in a sensible way. To do this we need to find the mathematical condition that represents the process of parallel transport in flat space and generalize this mathematical condition to curved space. Now, if we parallel-transport a vector $\mathbf{v}$ along a t-parameterized curve in a flat space[90] then we should keep the magnitude and direction of the vector constant along the path and this condition (i.e. keeping the magnitude and direction constant) is given mathematically by $\frac{d\mathbf{v}}{dt} = \mathbf{0}$, i.e. the total derivative of the vector with respect to t is identically zero along the path of transport since vanishing total derivative is equivalent to having a constant vector along the path. As we know from tensor differentiation (see § 2.7), the total derivative along a curve in flat space becomes an absolute derivative in curved space.[91] Therefore, the generalization of the condition $\frac{d\mathbf{v}}{dt} = \mathbf{0}$ is simply done by replacing the total derivative with the absolute derivative, that is $\frac{\delta \mathbf{v}}{\delta t} = \mathbf{0}$, i.e. the absolute derivative of the vector with respect to t is identically zero along the path of transport.[92]

In the following points we list a number of properties of parallel transport:[93]

---

[88] Parallel transport may also be called parallel propagation or parallel translation.
[89] In fact, the movement in infinitesimal steps is like the movement along straight segments since the curve is locally (or infinitesimally) straight. This should be supported by the fact that the space is supposed to be Riemannian and hence it is locally flat.
[90] We are assuming that we are using a Cartesian system which should be possible because the space is flat.
[91] In fact, becoming an absolute derivative in curved space is a requirement of the curvilinear nature of the coordinate system (since curved space must be coordinated by curvilinear system) and hence this "becoming" is also required in flat space if it is coordinated by a curvilinear system (as will be clarified later).
[92] Referring to the previous footnote, this generalization (i.e. replacing total derivative with absolute derivative) is needed for curvilinear systems in general even if the space is flat. The above description is essentially based, for the sake of simplicity, on employing a Cartesian (or affine) system to coordinate the flat space (as indicated in footnote [90]).
[93] These properties are general considering the inclusion of curved spaces.

2.8 Parallel Transport

1. The curve along which parallel transport takes place should be differentiable. However, it is sufficient to be piecewise differentiable.
2. Parallel transport is path dependent. Hence, given two points P_1 and P_2 in a space, the vector obtained at P_2 by parallel-transporting a vector from P_1 along a given curve C connecting P_1 to P_2 depends on the curve C.
3. Because parallel transport is path dependent then starting from a given point P on a closed curve C, parallel transport of a vector around C does not necessarily result in the same vector when arriving back at P. In fact, in curved space the final vector is generally different from the initial vector.
4. Parallel transport of a vector field along a curve is unique. Hence, knowing a vector field at a given point on a given curve determines its parallel transport at all other points on the curve.
5. The unit tangent vector of a geodesic curve is a parallel-transported field along the curve. In fact, this may be used as a definition of geodesic as "a curve that parallel-transports its tangent vector" or "a curve with a tangent vector field that is parallel-transported along the curve".
6. The magnitude of a parallel-transported vector field along a curve is constant.
7. The angle between two parallel-transported vector fields along a given curve is constant.
8. The inner product of two parallel-transported vectors along a given curve is constant.
9. Parallel transport is field dependent. Hence, a parallel-transported vector field along a given curve is different (at each point on the curve) from another parallel-transported vector field along that curve.
10. Parallel transport provides an indicator and measure of the curvature of space. As indicated earlier, in flat space a parallel-transported vector around a closed path returns to its starting point in the same direction at the starting point, while in curved space it generally (but not necessarily) returns to its starting point in a different direction to that at the starting point. In fact, the difference between its initial direction and its final direction can be used to quantify the intrinsic local curvature of the space. We note that for 2D space the angle between the initial and final vectors in this situation is a measure of the Gaussian curvature.
11. Although parallel transport is initially defined for vectors (i.e. rank-1 tensors), it can be easily generalized to include tensors of higher rank and hence we can talk about parallel transport of a rank-n tensor along a given curve.

Problems

1. Justify the fact that the curve along which parallel transport takes place should be differentiable.
   Answer: This condition is required for parallel transport to be defined sensibly because the mathematical condition for parallel transport is $\frac{\delta \mathbf{v}}{\delta t} = \mathbf{0}$ where $\frac{\delta \mathbf{v}}{\delta t} = \mathbf{v}_{;i} \frac{dx^i}{dt}$ with $\mathbf{v}_{;i}$ being the covariant derivative of the vector while $\frac{dx^i}{dt}$ (which is the tangent vector to the curve) is the derivative of the curve $x^i(t)$. Hence, the curve should be differentiable.
2. Justify the fact that parallel transport of a vector field along a given curve is unique.
   Answer: Since a parallel-transported vector is constant in an intrinsic sense (i.e. constant in magnitude and direction considering the intrinsic curvature of the space along the path of transport), then determining the vector at any point on the curve is sufficient for fully determining the vector at any other point on the curve because it is constant along the curve.

Exercises

1. Give a brief definition of parallel transport considering both flat space and curved space.
2. Explain how the definition of parallel transport of a vector along a curve is formally implemented.
3. Justify the fact that parallel transport is path dependent.
4. Justify the fact that starting from a given point P on a closed curve C, parallel-transporting a vector field around C does not necessarily result in the same vector field when arriving back at P.
5. Justify, qualitatively, the fact that the unit tangent vector of a geodesic curve is a parallel-transported field along the curve.
6. Show, formally, that the magnitude of a parallel-transported vector field along a curve is constant.
7. Show that the angle between two parallel-transported vector fields along a given curve is constant.
8. What you conclude from the last two exercises?

9. Show that the inner product of two parallel-transported vectors along a given curve is constant.
10. Justify the fact that parallel transport is field dependent.[94]
11. Justify the fact that parallel transport provides an indicator of the curvature of space (i.e. in flat space a parallel-transported vector around a closed path returns to its starting point in the same direction as that at the starting point, while in curved space it generally returns to its starting point in a different direction to that at the starting point).
12. Although parallel transport is primarily defined in relation to vectors, it can be extended to higher rank tensors. How?

2.9 Geodesic Path

In simple terms, geodesic path in curved space is the equivalent of straight line in flat space. Hence, geodesic path is commonly described as the most straight or most direct curve that connects two points in the space. This, of course, is not a technical definition and hence we still need to find a mathematically rigorous definition for geodesic path which will be developed later. However, as we will see the technical definition fundamentally rests on this generic definition, i.e. "the most straight curve".

Intrinsically, the geodesic paths are straight lines in the sense that an "intrinsic inhabitant" of the space to which they belong will see them straight since he cannot detect their curvature although they may be extrinsically curved as seen from an embedding space of higher dimensionality. This is due to the fact that only the intrinsic part of the curvature can be detected by the intrinsic inhabitant. Therefore, if this part vanishes the intrinsic inhabitant will fail to detect any curvature to the curve which is equivalent for him to having a straight line. Any deviation from such "straight lines" within the space is therefore an intrinsic curvature and hence it can be detected and measured by the intrinsic inhabitant.

Although geodesic curve is frequently the curve of the shortest distance between two given points in the space (and hence geodesic may be defined as "the path of shortest distance") it is not necessarily so. For instance, the largest of the two arcs forming a great circle on a sphere is a geodesic curve because it has no intrinsic (or geodesic) curvature but it is not the curve of the shortest distance on the sphere between its two end points. Accordingly, being the shortest path is a sufficient but not necessary condition for being a geodesic. A constraint may be imposed to make the criterion of minimal length apply to all geodesics by stating that geodesics minimize distance locally but not necessarily globally where an infinitesimal element of arc is considered in this constraint.[95] In fact, we may also generalize the above definition of geodesic by replacing "shortest" with "optimal" or "extremal" (where being optimal or extremal is related to a given characteristic parameter of the curve not necessarily the length). It is noteworthy that being geodesic is independent of the choice of the coordinate system and hence it is invariant under permissible coordinate transformations. It is also independent of the type of representation and parameterization and hence it is invariant in this sense. This is because being geodesic is a real geometric property and hence it should not depend on conventional attributes such as coordinate system or type of representation and parameterization.

To put the issue of geodesic paths in a proper context and to clarify many of its subtle aspects, we dedicate the next subsection to the issue of geodesic paths in 2D spaces or surfaces. These spaces are tangible and easy to visualize and analyze (by embedding them in the 3D Euclidean space that we live in) and hence we can get a beneficial insight into this issue with no need for highly abstract concepts and very sophisticated formal arguments that we usually need if we start this discussion from general nD spaces. However, this will serve as a case study and starting point for developing mathematically rigorous formal criteria for geodesic paths in general nD spaces.

Problems
1. List the different definitions of geodesic curve.
    Answer: There are three main definitions: (**a**) the most straight curve, (**b**) the curve of shortest

---

[94] This may sound trivial but the purpose of it is that two distinct vector fields that are parallel-transported along a given path will remain distinct at each point of the path.
[95] Accordingly, a geodesic is made of infinitesimal segments each of which is of minimal length.

2.9 Geodesic Path

distance (or optimal length) and (**c**) the curve that parallel-transports its tangent.

2. Give examples for geodesic paths in 2D, 3D and 4D spaces.
   Answer: The following are some examples:
   • 2D: arcs of great circles on sphere and meridians of surface of revolution.
   • 3D: straight lines in 3D Euclidean space.
   • 4D: trajectories (or world lines) of free particles and light rays in the Minkowskian spacetime of special relativity and in the Riemannian spacetime of general relativity.

3. Give a physical interpretation for geodesic curves within the context of relativity theories (i.e. special and general).
   Answer: From the answer of the previous problem we can see that a physical interpretation for the geodesic curves is that they are the trajectories (or world lines) of free particles (whether massive or massless) in the spacetime (whether the spacetime is the flat Minkowskian space of special relativity or the curved Riemannian space of general relativity).[96]

4. What are the main types of geodesic in the spacetime of special and general relativity and how are they characterized formally?
   Answer: There are three main types:
   • Timelike geodesics which are possible world lines of free massive objects. They are characterized by positive spacetime quadratic form, i.e. $(d\sigma)^2 > 0$.
   • Lightlike (or null) geodesics which are possible world lines of free massless objects. They are characterized by vanishing spacetime quadratic form, i.e. $(d\sigma)^2 = 0$.
   • Spacelike geodesics which cannot be world lines of any physical objects (due to the restrictions on the speed). They are characterized by negative spacetime quadratic form, i.e. $(d\sigma)^2 < 0$.
   We should note that the above classification of geodesics is based on the fact that the sign of the quadratic form $(ds)^2$ along any geodesic is constant (see next problem).

5. Show that the nature of the geodesic path (i.e. being timelike, lightlike or spacelike) is preserved along the entire path, i.e. if a geodesic path is timelike/lightlike/spacelike at one of its points then it remains so along the entire path.[97]
   Answer: Referring to § 1.3, a geodesic path with a tangent vector $\mathbf{T}$ is timelike/lightlike/spacelike according to whether $\mathbf{T}$ is timelike/lightlike/spacelike (i.e. $\mathbf{T} \cdot \mathbf{T} > 0$, $\mathbf{T} \cdot \mathbf{T} = 0$, $\mathbf{T} \cdot \mathbf{T} < 0$). Now, since geodesic path parallel-transports its tangent (see problem 1) and the inner product of parallel-transported vectors along a given curve is constant (see § 2.8) then $\mathbf{T} \cdot \mathbf{T}$ is constant and hence the nature of the geodesic path is preserved (i.e. constant) along the entire path.
   Note: in fact, the preservation of the nature of paths in spacetime is general for trajectories (or world lines) of real physical objects and is not restricted to geodesics. The reason is that according to the relativity theories massive objects always follow timelike trajectories and massless objects always follow lightlike trajectories while no physical object (whether massive or massless) can follow (even partly) spacelike trajectories. In fact, this can be inferred from the light cone structure where the trajectory of massive object is confined to inside the cone (which should be timelike) while the trajectory of massless object is confined to the cone itself (which should be lightlike).
   Accordingly, the above formal argument can be replaced by a simple physical argument, i.e. free massive/massless objects should always follow timelike/lightlike geodesics because they cannot change their nature (from massive to massless or from massless to massive),[98] moreover they cannot follow a spacelike trajectory in any part of their geodesic trajectories.

Exercises
1. Discuss the idea of being optimal or extremal (as a criterion for geodesics) in the relativity theories.
2. Compare between timelike geodesics and null geodesics within the framework of relativity theories.

---

[96] In fact, this applies to the timelike and null geodesics (see next problem).
[97] This question may be phrased more simply as: show that the sign of the quadratic form $(ds)^2$ along any geodesic is constant.
[98] This is not inconsistent with the mass-energy conversion because in this case we have two different objects rather than a single object of two natures.

3. What is the difference between special relativity and general relativity with regard to geodesics and null geodesics?
4. Give a brief characterization of geodesic path in Euclidean and Riemannian spaces.
5. Why is the concept of gravitational force seen redundant in general relativity? What is the implication of this? How can this explain the correspondence between special relativity and general relativity?
6. Compare between classical gravity and general relativistic gravity considering the concepts of mass and geodesic paths.
7. Can the paradigm of "geodesic" replace the paradigm of "gravitational force" entirely?
8. Is the paradigm of geodesic in general relativity a novelty of this theory?

2.9.1 Geodesic Path in 2D Spaces

Let discuss the issue of geodesic path in 2D spaces (i.e. surfaces) which are easy to envisage and hence we can get intuitive appreciation of the idea of geodesic and how it is formalized and quantified.[99] The curvature of a surface curve at a given point is represented by the curvature vector $\mathbf{K}$ which is given by:

$$\mathbf{K} = \mathbf{K}_n + \mathbf{K}_g = \kappa_n \mathbf{n} + \kappa_g \mathbf{u} \tag{61}$$

where $\mathbf{K}_n$ and $\mathbf{K}_g$ are the normal and geodesic components of the curvature vector $\mathbf{K}$, $\mathbf{n}$ and $\mathbf{u}$ are the normal unit vector to the surface and the geodesic unit vector, and κ_n and κ_g are the normal and geodesic curvatures of the curve at the given point. As we know, the geodesic curvature is intrinsic to the surface while the normal curvature is extrinsic and hence it belongs to the embedding space.

Now, based on our basic definition (or criterion) of geodesic curve as "the most straight path" we need to minimize (or rather eliminate) the curvature of the curve. Ideally, we want to have zero curvature (i.e. zero normal curvature and zero geodesic curvature) but from within the surface (i.e. intrinsically) this is achievable only for geodesic curvature because normal curvature belongs to the embedding space as it originates from the curvature of the surface itself within the embedding space. In brief, as long as we have a given surface that is curved within an embedding space and we have no access or control over the surface curvature then all we hope to minimize the curvature of a curve embedded into this surface is to eliminate the geodesic curvature which is intrinsic to the surface and hence it does not require access or control over the surface and its state within the embedding space. So, to have most straight path (i.e. geodesic) the geodesic curvature should vanish and this is the condition for the geodesic path in 2D space (or surface). Accordingly, if we want to generalize the concept of geodesic to spaces of higher dimensionality we need to impose the same condition, i.e. the intrinsic curvature (which is equivalent to the geodesic curvature in 2D space) must vanish.

Now, let see how we can achieve this condition (i.e. vanishing of geodesic curvature identically) in 2D space. In this regard, we use our knowledge of calculus, specifically the fact that if something vanishes identically (i.e. not accidentally) then it can represent the rate of change (or derivative) of a constant entity. So, what is the constant entity that we should look for in this context (i.e. minimizing the curvature). Intuitively, curvature means twisting or changing direction, so the constant entity is ideally represented by the unit tangent vector $\mathbf{T}$ of the curve since this vector solely represents the direction of the curve with no involvement of other factors like magnitude. In other words, to have a geodesic path whose intrinsic curvature is zero because it does not twist (or curve) intrinsically, the derivative of the unit tangent vector $\mathbf{T}$, which represents the direction of the geodesic curve, should vanish because $\mathbf{T}$ (considering the inherent curvature of the 2D space itself) is constant. i.e. $\mathbf{T}$ exactly follows the inherent curvature of the 2D space with no twisting or deviation from this inherent curvature.[100]

The last thing that we need to address is: what kind of derivative (e.g. partial, total, covariant, absolute, etc.) should we use to achieve our objective? The answer is that: as long as we are dealing with an intrinsic issue (i.e. intrinsic curvature) within a potentially curved space then we need an intrinsic derivative that detects the rate of change of something in the space intrinsically and only within the space itself. Now,

---

[99] The reader is advised to consult B2 and B2X for more details.
[100] The analog of this is straight line in a plane since this line follows exactly the inherent flatness of its space.

2.9.1 Geodesic Path in 2D Spaces

from our knowledge of tensor differentiation (see § 2.7) we can easily identify the absolute derivative (which is also known as *intrinsic* derivative for this reason) as the appropriate derivative for achieving our goal. As we know from the formulae of absolute derivative (e.g. $\frac{\delta A^i}{\delta t} = \frac{dA^i}{dt} + A^k \Gamma^i_{kj} \frac{dx^j}{dt}$), this derivative depends only on the surface coordinates and the Christoffel symbols and hence it is purely intrinsic to the surface. We should also note that the absolute derivative is defined along a curve (or path in space) and hence it is the pertinent derivative from this perspective.

To sum up, to have a "most straight path" in a given curved surface the curvature of the path should be eliminated. Since the curvature of the path consists of intrinsic part and extrinsic part and because the extrinsic part belongs to the surface itself (considering its existence in the embedding 3D space) which we cannot change, then all we can (and should) do is to eliminate the intrinsic part. So, the characteristic feature of geodesic path is that its intrinsic (or geodesic) curvature is identically zero. Now, the intrinsic curvature is demonstrated by the twisting (say "right" and "left") of the curve within the surface and this twisting is measured by the direction of the curve which is represented by the unit tangent vector **T**. Accordingly, the elimination of this intrinsic twisting (or curving) is equivalent to making the rate of change of **T** vanish, and because we are dealing with an intrinsic issue in a presumably curved space this rate of change should be intrinsic and of invariant (or tensorial) nature, i.e. absolute derivative (which is also the right type of derivative from the perspective of being defined over a curve). So, the required criterion for determining the geodesic path is that the intrinsic derivative of its unit tangent vector should vanish identically, that is:[101]

$$\frac{\delta}{\delta \lambda}\left(\frac{dx^\alpha}{d\lambda}\right) \equiv \frac{d^2 x^\alpha}{d\lambda^2} + \Gamma^\alpha_{\beta\gamma} \frac{dx^\beta}{d\lambda} \frac{dx^\gamma}{d\lambda} = 0 \qquad (62)$$

where λ is a natural (or affine) parameter of the curve, the indexed x are the coordinates of the curve, $\Gamma^\alpha_{\beta\gamma}$ symbolizes the Christoffel symbols of the surface, and the standard notation of absolute derivative is in use (see § 2.7). The demand for the parameter to be natural arises from the fact that $\mathbf{T} = d\mathbf{r}/d\lambda$ with $\mathbf{r}(\lambda)$ being the spatial representation of the curve and λ is a natural parameter.[102]

The above discussion and arguments about the nature and criterion of geodesic paths in 2D spaces (or surfaces) can be easily and intuitively generalized to geodesic paths in spaces of higher dimensionality and hence we obtain the same formal criterion (as given by Eq. 62) for characterizing and identifying geodesic paths in any space. In the following three subsections we will exploit the insight that we gained from the 2D case study to rigorously characterize geodesic curves and find the required formal condition for a curve to be geodesic in nD space using its different fundamental definitions (i.e. "straightest", "shortest" and "parallel-transporting its tangent"). As we will see, all these conditions are essentially based on the same fundamental principles and arguments that we discussed in this subsection and hence they can be seen as different demonstrations for a single geometrical principle on which the concept of geodesic rests, and this should explain why all these criteria are equivalent[103] as they lead to the same mathematical geodesic condition.

Problems

1. What natural and affine parameters mean?
   Answer: Natural parameter means that the parameter is curve length s while affine parameter means that the parameter is proportional to curve length, i.e. $\lambda = as$ where a is a non-zero constant.[104] Hence, affine parameter is more general than natural parameter. We should also note that "natural parameter" may be used to mean affine parameter (and vice versa) where the context should be

---
[101] We follow in indexing the coordinates with Greek letters our convention in B2.
[102] The relation $\mathbf{T} = d\mathbf{r}/d\lambda$ is fully justified in the textbooks of differential geometry (see for example B2 and B2X; also see Exercises). We note that in B2 and B2X we use s instead of λ. The shift in the present book from s to λ is because λ is commonly used in the books of general relativity to represent affine parameter (which is more general than natural parameter; see Problems) and hence it is more appropriate to use λ in this book.
[103] In fact, this equivalence applies in torsion-free spaces; otherwise there are details that should be sought in the extended literature of this subject.
[104] In fact, to be more general we should have $\lambda = as + b$ where b is a constant. However, b is rather arbitrary and hence it can be set to zero and discarded with no loss of generality.

consulted to determine the appropriate meaning. We remark that in the relativity theories s is replaced by σ (where for non-timelike paths its meaning is generalized).

Exercises
1. Investigate potential difficulties in the definitions of natural and affine parameters as given in the above Problem especially in the context of general relativity.
2. Show that the magnitude of the tangent vector of a curve is constant when the curve is parameterized with natural or affine parameter.

2.9.2 Geodesic Equation from Straightness

The purpose of this subsection is to obtain the geodesic condition from the basic definition of geodesic as the most straight path. In fact, this was thoroughly investigated in § 2.9.1 within the context of 2D spaces and we concluded there that the generalization to spaces of higher dimensionality is straightforward since the dimensionality of the space is irrelevant to the fundamental geometric principles which the obtained geodesic condition rests upon. Accordingly, the geodesic condition from the basic definition of geodesic as the most straight path is given by Eq. 62 where $\alpha, \beta, \gamma = 1, \cdots, n$ (with n being the space dimension).

Exercises
1. Write the geodesic equation in a compact tensorial form.

2.9.3 Geodesic Equation from Variational Principle

The purpose of this subsection is to obtain the geodesic condition from the basic definition of geodesic as the shortest (or rather the optimal) path using the variational principle. In brief, a sufficient and necessary condition for a curve to be geodesic is that the first variation of its length is zero. Figure 1 is an illustration of how the length of the geodesic curve between two given points is subject to the variational principle. In fact, this condition may be taken as the basis for the definition of geodesic as the curve that connects two fixed points, P_1 and P_2, whose length possesses a stationary value with regard to small variations in its neighborhood, that is:[105]

$$\delta \int_{P_1}^{P_2} ds = 0 \tag{63}$$

It is shown in the literature of calculus of variations that this condition is equivalent to the condition of being a solution to the Euler-Lagrange equation (see § 2.4) and hence being a solution to the Euler-Lagrange equation is a necessary and sufficient condition for extremizing the arc length. Accordingly, all we need for obtaining the geodesic condition from the variational principle is to show that the geodesic condition can be obtained from the Euler-Lagrange equation, and that is what we will do in the Problems where we show that the geodesic equation is a solution to the Euler-Lagrange equation (i.e. Eq. 42).

Problems
1. Derive the geodesic equation from the Euler-Lagrange equation.
   Answer: According to the variational definition of geodesic curve (i.e. the curve of shortest or optimal length), the geodesic curve should extremize the length which is given by the integral $\int ds$. So, the variational principle is given by:

   $$\delta \int ds = \delta \int \frac{ds}{d\lambda} d\lambda = 0 \tag{64}$$

   where λ is a parameter for the geodesic curve. Now, the Euler-Lagrange equation is given in its general form by Eq. 42 where in our case:

   $$x \equiv \lambda$$

---
[105] In other words, the geodesic is the optimal (in length) of all curves (connecting P_1 and P_2) that are obtained by small perturbations in that neighborhood.

2.9.3 Geodesic Equation from Variational Principle

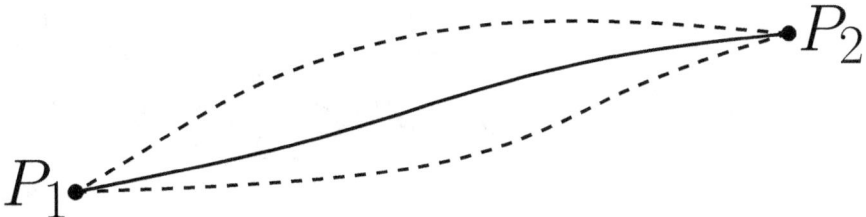

Figure 1: The length of a geodesic curve (solid) connecting two points, P_1 and P_2, as an extremum in comparison to the length of other curves (dashed) connecting these points where these curves are the result of small perturbations in its neighborhood.

$$y \equiv x^k$$
$$f \equiv \frac{ds}{d\lambda} = \sqrt{g_{ij}\frac{dx^i}{d\lambda}\frac{dx^j}{d\lambda}}$$

Hence, the Euler-Lagrange equation becomes:

$$\frac{\partial\sqrt{g_{ij}\dot{x}^i\dot{x}^j}}{\partial x^k} - \frac{d}{d\lambda}\left(\frac{\partial\sqrt{g_{ij}\dot{x}^i\dot{x}^j}}{\partial \dot{x}^k}\right) = 0 \qquad (65)$$

where the overdot represents derivative with respect to λ (i.e. $\frac{d}{d\lambda}$).
Now, we have:

$$\frac{\partial\sqrt{g_{ij}\dot{x}^i\dot{x}^j}}{\partial x^k} = \frac{(\partial g_{ij}/\partial x^k)\,\dot{x}^i\dot{x}^j}{2\sqrt{g_{ij}\dot{x}^i\dot{x}^j}} \qquad (66)$$

where the power and chain rules (or composite rule) of differentiation (as well as the product rule) are used noting that $\dot{x}^i$ and $\dot{x}^j$ (which represent the tangent) are independent of x^k and hence $\partial_k\dot{x}^i = \partial_k\dot{x}^j = 0$ (see exercise 2).
Also:

$$\begin{aligned}\frac{\partial\sqrt{g_{ij}\dot{x}^i\dot{x}^j}}{\partial \dot{x}^k} &= \frac{g_{ij}\left(\partial\dot{x}^i/\partial\dot{x}^k\right)\dot{x}^j + g_{ij}\dot{x}^i\left(\partial\dot{x}^j/\partial\dot{x}^k\right)}{2\sqrt{g_{ij}\dot{x}^i\dot{x}^j}}\\ &= \frac{g_{ij}\delta^i_k\dot{x}^j + g_{ij}\dot{x}^i\delta^j_k}{2\sqrt{g_{ij}\dot{x}^i\dot{x}^j}}\\ &= \frac{g_{kj}\dot{x}^j + g_{ik}\dot{x}^i}{2\sqrt{g_{ij}\dot{x}^i\dot{x}^j}}\\ &= \frac{g_{kj}\dot{x}^j + g_{kj}\dot{x}^j}{2\sqrt{g_{ij}\dot{x}^i\dot{x}^j}}\\ &= \frac{g_{kj}\dot{x}^j}{\sqrt{g_{ij}\dot{x}^i\dot{x}^j}}\end{aligned} \qquad (67)$$

where in line 1 we use the power, chain and product rules of differentiation noting that g_{ij} is independent of $\dot{x}^k$ (see exercise 3), in line 2 we use the fact that the coordinates are mutually independent and hence $\partial\dot{x}^i/\partial\dot{x}^k = \delta^i_k$ and $\partial\dot{x}^j/\partial\dot{x}^k = \delta^j_k$,[106] in line 3 we use the Kronecker delta as an index replacement operator, and in line 4 we relabel a dummy index and use the symmetry of the metric tensor. Hence:

$$\frac{d}{d\lambda}\left(\frac{\partial\sqrt{g_{ij}\dot{x}^i\dot{x}^j}}{\partial \dot{x}^k}\right) = \frac{d}{d\lambda}\left(\frac{g_{kj}\dot{x}^j}{\sqrt{g_{ij}\dot{x}^i\dot{x}^j}}\right)$$

---
[106] This is like the well-known identity: $\partial x^i/\partial x^j = \delta^i_j$ (see exercise 4).

2.9.3 Geodesic Equation from Variational Principle

$$= \frac{\sqrt{g_{ij}\dot{x}^i\dot{x}^j}\frac{d(g_{kj}\dot{x}^j)}{d\lambda} - (g_{kj}\dot{x}^j)\frac{d(\sqrt{g_{ij}\dot{x}^i\dot{x}^j})}{d\lambda}}{g_{ij}\dot{x}^i\dot{x}^j}$$

$$= \frac{\sqrt{g_{ij}\dot{x}^i\dot{x}^j}\frac{d(g_{kj}\dot{x}^j)}{d\lambda} - \frac{g_{kj}\dot{x}^j}{2\sqrt{g_{ij}\dot{x}^i\dot{x}^j}}\frac{d(g_{ij}\dot{x}^i\dot{x}^j)}{d\lambda}}{g_{ij}\dot{x}^i\dot{x}^j} \quad (68)$$

where in line 1 we substitute from Eq. 67, in line 2 we use the quotient rule of differentiation, and in line 3 we carry out differentiation in the second term (using the composite rule).
On substituting from Eqs. 66 and 68 into Eq. 65 we get:

$$\frac{(\partial g_{ij}/\partial x^k)\,\dot{x}^i\dot{x}^j}{2\sqrt{g_{ij}\dot{x}^i\dot{x}^j}} - \frac{\sqrt{g_{ij}\dot{x}^i\dot{x}^j}\frac{d(g_{kj}\dot{x}^j)}{d\lambda} - \frac{g_{kj}\dot{x}^j}{2\sqrt{g_{ij}\dot{x}^i\dot{x}^j}}\frac{d(g_{ij}\dot{x}^i\dot{x}^j)}{d\lambda}}{g_{ij}\dot{x}^i\dot{x}^j} = 0$$

$$\frac{\partial g_{ij}}{\partial x^k}\dot{x}^i\dot{x}^j - \frac{2g_{ij}\dot{x}^i\dot{x}^j\frac{d(g_{kj}\dot{x}^j)}{d\lambda} - g_{kj}\dot{x}^j\frac{d(g_{ij}\dot{x}^i\dot{x}^j)}{d\lambda}}{g_{ij}\dot{x}^i\dot{x}^j} = 0$$

$$\frac{\partial g_{ij}}{\partial x^k}\dot{x}^i\dot{x}^j - 2\frac{d(g_{kj}\dot{x}^j)}{d\lambda} + \frac{g_{kj}\dot{x}^j}{g_{ij}\dot{x}^i\dot{x}^j}\frac{d(g_{ij}\dot{x}^i\dot{x}^j)}{d\lambda} = 0$$

$$2\frac{d(g_{kj}\dot{x}^j)}{d\lambda} - \frac{\partial g_{ij}}{\partial x^k}\dot{x}^i\dot{x}^j - \frac{g_{kj}\dot{x}^j}{g_{ij}\dot{x}^i\dot{x}^j}\frac{d(g_{ij}\dot{x}^i\dot{x}^j)}{d\lambda} = 0 \quad (69)$$

where in line 2 we multiply with $2\sqrt{g_{ij}\dot{x}^i\dot{x}^j}$ and in lines 3 and 4 we simplify and carry out basic algebraic operations. Now:

$$2\frac{d(g_{kj}\dot{x}^j)}{d\lambda} = 2\frac{dg_{kj}}{d\lambda}\dot{x}^j + 2g_{kj}\frac{d\dot{x}^j}{d\lambda}$$

$$= 2\frac{\partial g_{kj}}{\partial x^i}\dot{x}^i\dot{x}^j + 2g_{kj}\frac{d\dot{x}^j}{d\lambda}$$

$$= \frac{\partial g_{kj}}{\partial x^i}\dot{x}^i\dot{x}^j + \frac{\partial g_{kj}}{\partial x^i}\dot{x}^i\dot{x}^j + 2g_{kj}\frac{d\dot{x}^j}{d\lambda}$$

$$= \frac{\partial g_{kj}}{\partial x^i}\dot{x}^i\dot{x}^j + \frac{\partial g_{ik}}{\partial x^j}\dot{x}^i\dot{x}^j + 2g_{kj}\frac{d\dot{x}^j}{d\lambda}$$

where in line 1 we use the product rule, in line 2 we use the chain rule in the first term, and in line 4 we exchange the dummy indices ij and use the symmetry of the metric tensor in the second term. Accordingly, Eq. 69 becomes:

$$\frac{\partial g_{kj}}{\partial x^i}\dot{x}^i\dot{x}^j + \frac{\partial g_{ik}}{\partial x^j}\dot{x}^i\dot{x}^j + 2g_{kj}\frac{d\dot{x}^j}{d\lambda} - \frac{\partial g_{ij}}{\partial x^k}\dot{x}^i\dot{x}^j - \frac{g_{kj}\dot{x}^j}{g_{ij}\dot{x}^i\dot{x}^j}\frac{d(g_{ij}\dot{x}^i\dot{x}^j)}{d\lambda} = 0$$

$$\left(\frac{\partial g_{kj}}{\partial x^i} + \frac{\partial g_{ik}}{\partial x^j} - \frac{\partial g_{ij}}{\partial x^k}\right)\dot{x}^i\dot{x}^j + 2g_{kj}\frac{d\dot{x}^j}{d\lambda} - \frac{g_{kj}\dot{x}^j}{g_{ij}\dot{x}^i\dot{x}^j}\frac{d(g_{ij}\dot{x}^i\dot{x}^j)}{d\lambda} = 0$$

Now, if we impose the condition that λ is a natural (or affine) parameter then $g_{ij}\dot{x}^i\dot{x}^j$ (which is the inner product of the tangent vector to the geodesic curve by itself and hence it represents the square of the magnitude of the tangent vector) is constant (see § 2.9.1) and hence the last term is zero, that is:

$$\left(\frac{\partial g_{kj}}{\partial x^i} + \frac{\partial g_{ik}}{\partial x^j} - \frac{\partial g_{ij}}{\partial x^k}\right)\dot{x}^i\dot{x}^j + 2g_{kj}\frac{d\dot{x}^j}{d\lambda} = 0$$

$$\frac{1}{2}\left(\frac{\partial g_{kj}}{\partial x^i} + \frac{\partial g_{ik}}{\partial x^j} - \frac{\partial g_{ij}}{\partial x^k}\right)\dot{x}^i\dot{x}^j + g_{kj}\frac{d\dot{x}^j}{d\lambda} = 0$$

$$\frac{g^{mk}}{2}\left(\frac{\partial g_{kj}}{\partial x^i}+\frac{\partial g_{ik}}{\partial x^j}-\frac{\partial g_{ij}}{\partial x^k}\right)\dot{x}^i\dot{x}^j + g^{mk}g_{kj}\frac{d\dot{x}^j}{d\lambda} = 0$$

$$\frac{g^{mk}}{2}\left(\frac{\partial g_{kj}}{\partial x^i}+\frac{\partial g_{ik}}{\partial x^j}-\frac{\partial g_{ij}}{\partial x^k}\right)\dot{x}^i\dot{x}^j + \delta^m_j\frac{d\dot{x}^j}{d\lambda} = 0$$

$$\frac{g^{mk}}{2}\left(\frac{\partial g_{kj}}{\partial x^i}+\frac{\partial g_{ik}}{\partial x^j}-\frac{\partial g_{ij}}{\partial x^k}\right)\dot{x}^i\dot{x}^j + \frac{d\dot{x}^m}{d\lambda} = 0$$

$$\Gamma^m_{ij}\dot{x}^i\dot{x}^j + \frac{d\dot{x}^m}{d\lambda} = 0$$

$$\Gamma^m_{ij}\frac{dx^i}{d\lambda}\frac{dx^j}{d\lambda} + \frac{d^2 x^m}{d\lambda^2} = 0$$

where in line 2 we divide by 2, in line 3 we multiply by g^{mk}, in line 4 we use the fact that g^{mk} and g_{kj} are inverses, in line 5 we use the Kronecker delta for index replacement, in line 6 we use the expression of Christoffel symbol, and in line 7 we use our convention that the overdot represents derivative with respect to λ. The last equation is the geodesic equation as given earlier (with some trivial notational differences).

Exercises
1. State the mathematical condition for the variational principle in the spacetime of general relativity and interpret this condition in the case of timelike trajectory.
2. Justify the following relation: $\partial_k \dot{x}^i = 0$ (as well as $\partial_k \dot{x}^j = 0$).
3. Justify the claim that g_{ij} is independent of $\dot{x}^k$.
4. Justify the following relation: $\partial \dot{x}^i / \partial \dot{x}^k = \delta^i_k$.

2.9.4 Geodesic Equation from Parallel Transport

In this subsection we discuss how to obtain the geodesic equation from the basic definition of geodesic as the curve that parallel-transports its own tangent. According to our previous investigation, parallel transport in curved space is the generalization of parallelism in flat space, while geodesic curve in curved space is the generalization of straight line in flat space. So, we need to make the analogy between the ideas of parallel transport and geodesic curve in curved space and the corresponding ideas of parallelism and straight line in flat space very clear. Now, in flat space straight line is characterized by having the same direction at any point, i.e. its tangent vector is constant and hence it remains parallel to itself as it moves along the line. So, to generalize the concept of "straight in flat space" to "geodesic in curved space" we can define geodesic as the curve whose tangential direction is "constant" with regard to parallel transport. This means that the tangent to a geodesic curve remains tangent to the curve when this tangent is parallel-transported along the curve and hence the geodesic curve in curved space has "fixed" or "constant" direction like straight line in flat space.

Accordingly, the required condition for a curve to "parallel-transport its own tangent" is that the absolute derivative of the tangent vector is identically zero along the curve, that is:

$$\frac{\delta T^i}{\delta t} \equiv \frac{dT^i}{dt} + T^k \Gamma^i_{kj}\frac{dx^j}{dt} = 0 \tag{70}$$

where the indexed T is the unit tangent vector to the curve while t is a parameter for the curve. Now, if t represents a natural parameter λ of the curve (i.e. $t \equiv \lambda$) then the unit tangent vector is given by $T^i = \frac{dx^i}{d\lambda}$.[107] On substituting this into the last equation we get:

$$\frac{d}{d\lambda}\left(\frac{dx^i}{d\lambda}\right) + \frac{dx^k}{d\lambda}\Gamma^i_{kj}\frac{dx^j}{d\lambda} = 0 \tag{71}$$

---

[107] Refer to B2 and B2X for details. Also see § 2.9.1. We note that the same geodesic equation will be obtained if the parameter is affine (and hence **T** represents a tangent vector of constant magnitude).

$$\frac{d^2 x^i}{d\lambda^2} + \Gamma^i_{kj} \frac{dx^k}{d\lambda} \frac{dx^j}{d\lambda} = 0 \qquad (72)$$

which is the geodesic equation that we obtained earlier from other criteria and methods. So, the geodesic equation is a mathematical expression of the fact that the absolute derivative of the unit tangent vector of a geodesic curve vanishes identically where the "vanishes identically" condition comes from the condition of parallel transport.

We note that in pseudo-Riemannian spaces (like the spacetime of general relativity) the line element $d\sigma$ can be zero or imaginary and hence the definition of geodesic as "the path of shortest distance" could be questioned.[108] However, this may be avoided by the definition of geodesic that is based on the concept of parallel transport[109] although this can also be questioned. Anyway, we consider these cases as generalizations of the concept of geodesic, regardless of the adopted definition of geodesic, and hence there should be no problem in extending the concept and criteria of geodesic to pseudo-Riemannian spaces.

Problems

1. Outline the essence and objective of the last three subsections (i.e. § 2.9.2, § 2.9.3 and § 2.9.4).
   Answer: The essence of the last three subsections is the investigation of three conditions (based on the three definitions of geodesic) for characterizing the geodesic curve in general nD space. The objective of this investigation is to show that these three conditions are equivalent since they lead to the same mathematical condition for the geodesic path, i.e. the condition given by the equation: $\frac{d^2 x^i}{d\lambda^2} + \Gamma^i_{jk} \frac{dx^j}{d\lambda} \frac{dx^k}{d\lambda} = 0$.

Exercises

1. State the mathematical condition for a vector field A^i to be parallel-transported along a given curve.
2. Briefly discuss the qualification of geodesic curve as "parallel-transporting its own tangent".
3. Show, qualitatively, that the geodesic criterion of "most straight" and the geodesic criterion of "parallel-transporting its tangent" are equivalent.
4. Justify the fact that in parallel transport of the tangent vector along a geodesic curve the parameter of the curve should be natural (or affine).
5. Show that the geodesic equation can be written as $\frac{dT_i}{d\lambda} = \frac{1}{2} \partial_i g_{jk} T^j T^k$ (where the indexed T is the unit tangent vector and λ is a natural parameter).

2.9.5 Geodesic Equation

Based on our investigation in the previous subsections, we conclude that the necessary and sufficient condition for a naturally parameterized curve in an nD space to be geodesic is given by the following tensorial second order non-linear differential equation:

$$\frac{d^2 x^\alpha}{d\lambda^2} + \Gamma^\alpha_{\beta\gamma} \frac{dx^\beta}{d\lambda} \frac{dx^\gamma}{d\lambda} = 0 \qquad (73)$$

where λ is a natural (or affine) parameter of the curve, the indexed x are the coordinates of the curve, $\Gamma^\alpha_{\beta\gamma}$ represents the Christoffel symbols of the space, and α, β, γ range over space dimensions (which for the spacetime of relativity theories become $\alpha, \beta, \gamma = 0, 1, 2, 3$ according to our indexing convention). Accordingly, all the λ-parameterized curves $x^\alpha(\lambda)$ that satisfy the above equation are geodesic paths in the space. So, what we need in our search for geodesic paths is to find the solutions of the above equation. However, because of its non-linear and tensorial nature this equation in most cases has no closed-form explicit solutions.

From Eq. 73, it can be seen that being a geodesic is an intrinsic property because this equation depends exclusively on the Christoffel symbols which depend only on the metric tensor and its partial derivatives (see § 2.6). Hence, geodesic curves can be detected and identified by an intrinsic inhabitant of the space.

---

[108] In fact, this questioning may even include the definition of geodesic as "the most straight path".
[109] This may be justified by the claim that parallel transport is based on an abstract analytical condition although it is originally obtained from an intuitive generalization of parallelism in Euclidean space.

2.9.5 Geodesic Equation

From Eq. 73, it can also be seen that for any flat space coordinated by a Cartesian system the geodesics are straight lines since in this case the Christoffel symbols vanish identically and Eq. 73 becomes $\frac{d^2 x^\alpha}{d\lambda^2} = 0$ which has a straight line solution. This should apply even if the flat space is coordinated by other types of coordinate system due to the invariance of geodesic (since it is a real geometric property) and hence geodesics in flat space are straight lines in general regardless of the employed coordinate system. We should finally note that the geodesic equation has only one free index and hence it is a vector equation. Therefore, in an nD space the geodesic equation represents n component equations (which in the case of the spacetime of relativity theories is 4).

Problems

1. Show that the affine parameter λ in the geodesic equation can be replaced by the proper time parameter τ (i.e. the proper time parameter is an affine parameter).
   Answer: First, we should remark that in the relativity theories this should be restricted to timelike geodesics where $\tau > 0$ (i.e. geodesics of massive objects). The reason is that for lightlike geodesics (i.e. geodesics of massless objects) we have $\tau = 0$ and hence τ is not a useful parameter, while for spacelike geodesics the proper time parameter is physically meaningless.
   Now, the geodesic equation is given by:
   $$\frac{d^2 x^\alpha}{d\lambda^2} + \Gamma^\alpha_{\beta\gamma} \frac{dx^\beta}{d\lambda} \frac{dx^\gamma}{d\lambda} = 0$$
   where λ is an affine parameter. Using the chain rule of differentiation, the above equation can be written as (see the upcoming note 1):
   $$\frac{d^2 x^\alpha}{d\tau^2} \left(\frac{d\tau}{d\lambda}\right)^2 + \Gamma^\alpha_{\beta\gamma} \frac{dx^\beta}{d\tau} \frac{dx^\gamma}{d\tau} \left(\frac{d\tau}{d\lambda}\right)^2 = 0$$
   $$\left[\frac{d^2 x^\alpha}{d\tau^2} + \Gamma^\alpha_{\beta\gamma} \frac{dx^\beta}{d\tau} \frac{dx^\gamma}{d\tau}\right] \left(\frac{d\tau}{d\lambda}\right)^2 = 0$$
   where τ is the proper time parameter. Now, since:
   $$\frac{d\tau}{d\lambda} = \frac{d(\sigma/c)}{d(a\sigma)} = \frac{1}{ac} \frac{d\sigma}{d\sigma} = \frac{1}{ac} \neq 0 \tag{74}$$
   then:
   $$\frac{d^2 x^\alpha}{d\tau^2} + \Gamma^\alpha_{\beta\gamma} \frac{dx^\beta}{d\tau} \frac{dx^\gamma}{d\tau} = 0 \tag{75}$$
   which is the geodesic equation with the affine parameter λ being replaced by the proper time parameter τ.
   Note 1: we have:
   $$\frac{d^2 x^\alpha}{d\lambda^2} = \frac{d}{d\lambda} \left(\frac{dx^\alpha}{d\lambda}\right) = \frac{d}{d\tau} \left(\frac{dx^\alpha}{d\tau} \frac{d\tau}{d\lambda}\right) \frac{d\tau}{d\lambda} = \left(\frac{d^2 x^\alpha}{d\tau^2} \frac{d\tau}{d\lambda} + \frac{dx^\alpha}{d\tau} \frac{d}{d\tau}\left[\frac{d\tau}{d\lambda}\right]\right) \frac{d\tau}{d\lambda}$$
   Now, since $\frac{d\tau}{d\lambda}$ is constant (according to Eq. 74 since a and c are constants) then $\frac{d}{d\tau}\left[\frac{d\tau}{d\lambda}\right] = 0$ and hence:
   $$\frac{d^2 x^\alpha}{d\lambda^2} = \frac{d^2 x^\alpha}{d\tau^2} \left(\frac{d\tau}{d\lambda}\right)^2$$
   We should remark that since we are dealing with the spacetime of general relativity then we replace s (which is the space interval) with σ (which is the spacetime interval) and hence the relation $\lambda = as$ becomes $\lambda = a\sigma$. We also use the relation between τ and σ, i.e. $\tau = \sigma/c$.
   Note 2: because $U^\alpha = dx^\alpha/d\tau$ (where U^α is the velocity 4-vector), Eq. 75 (which is the geodesic equation in terms of the proper time parameter) may be written as:
   $$\frac{d}{d\tau}\left(\frac{dx^\alpha}{d\tau}\right) + \Gamma^\alpha_{\beta\gamma} \frac{dx^\beta}{d\tau} \frac{dx^\gamma}{d\tau} = 0$$

$$\frac{dU^\alpha}{d\tau} + \Gamma^\alpha_{\beta\gamma} U^\beta U^\gamma = 0$$

The last equation may be written compactly as $\frac{\delta \mathbf{U}}{\delta \tau} = \mathbf{0}$.

Exercises

1. List the necessary and sufficient conditions for a curve to be geodesic.
2. What is the difference between the geodesics of special relativity and the geodesics of general relativity? Try to justify your answer technically.
3. In the last four sections (i.e. § 2.6, § 2.7, § 2.8 and § 2.9) we investigated the topics of Christoffel symbols, absolute derivative, parallel transport and geodesic path. Try to investigate the relation between these topics and how they are linked together. Also, outline how this leads to the derivation of the geodesic equation.
4. Show that in the Minkowski spacetime the geodesics of massive objects are described by the conditions: $\frac{d^2 t}{d\tau^2} = 0$ and $\frac{d^2 \mathbf{r}}{d\tau^2} = \mathbf{0}$. What this means?
5. Why the tensorial nature of the geodesic equation complicates its solution?
6. Justify the invariance of geodesic paths across various coordinate systems and its independence of the type of representation and parameterization of the space and curve.
7. What is the relation between the tensorial geodesic equation and the geodesic trajectory of an object (massive or massless) in spacetime?
8. The motion of free objects (i.e. free of forces other than gravity) is subject to the geodesic equation. How will this change when the objects are not free (i.e. they are subject to non-gravitational forces)?

2.10 Riemann-Christoffel Curvature Tensor

This is a rank-4 tensor that characterizes the intrinsic curvature of space and hence it plays an important role in non-Euclidean geometries and their applications like general relativity. The tensor is used, for instance, to test for the intrinsic curvature of space and as a building block for the Einstein tensor which represents the geometric component of the Field Equation (see § 2.13 and § 3.2). As indicated before, the Riemann-Christoffel curvature tensor vanishes identically *iff* the space is globally flat and hence the Riemann-Christoffel curvature tensor is zero in Euclidean spaces and quasi-Euclidean spaces like the Minkowski spacetime of special relativity. The Riemann-Christoffel curvature tensor depends only on the metric which, in general coordinate systems, is a function of position and hence the Riemann-Christoffel curvature tensor follows this dependency on position. Yes, for affine coordinate systems the metric tensor is constant and hence the Riemann-Christoffel curvature tensor vanishes identically throughout the space, i.e. it is independent of position.

There are two kinds of Riemann-Christoffel curvature tensor: first and second, where the first kind is a type $(0,4)$ tensor while the second kind is a type $(1,3)$ tensor.[110] The Riemann-Christoffel curvature tensor of the first kind, which may also be called the covariant (or totally covariant) Riemann-Christoffel curvature tensor, is given by:

$$\begin{aligned} R_{ijkl} &= \partial_k [jl, i] - \partial_l [jk, i] + [il, r] \Gamma^r_{jk} - [ik, r] \Gamma^r_{jl} \\ &= \frac{1}{2} \left(\partial_k \partial_j g_{li} + \partial_l \partial_i g_{jk} - \partial_k \partial_i g_{jl} - \partial_l \partial_j g_{ki} \right) + [il, r] \Gamma^r_{jk} - [ik, r] \Gamma^r_{jl} \\ &= \frac{1}{2} \left(\partial_k \partial_j g_{li} + \partial_l \partial_i g_{jk} - \partial_k \partial_i g_{jl} - \partial_l \partial_j g_{ki} \right) + g^{rs} \left([il, r][jk, s] - [ik, r][jl, s] \right) \end{aligned} \qquad (76)$$

where line 2 is based on Eq. 51,[111] while line 3 is based on Eq. 52. Similarly, the Riemann-Christoffel curvature tensor of the second kind, which may also be called the mixed Riemann-Christoffel curvature

---

[110] In fact, these are the types that are in common use in the literature (due to their special significance in the formulation); otherwise the indices of the Riemann-Christoffel curvature tensor can in principle be shifted up and down (using the index shifting operators) like the indices of any tensor.

[111] That is $[ij, l] = \frac{1}{2} \left(\partial_j g_{il} + \partial_i g_{jl} - \partial_l g_{ij} \right)$ and hence $\partial_k [ij, l] = \frac{1}{2} \left(\partial_k \partial_j g_{il} + \partial_k \partial_i g_{jl} - \partial_k \partial_l g_{ij} \right)$.

2.10 Riemann-Christoffel Curvature Tensor

tensor, is given by:

$$R^i_{jkl} = \partial_k \Gamma^i_{jl} - \partial_l \Gamma^i_{jk} + \Gamma^r_{jl}\Gamma^i_{rk} - \Gamma^r_{jk}\Gamma^i_{rl} \tag{77}$$

The Riemann-Christoffel curvature tensor of the first and second kinds can be obtained from each other by the index shifting operator, that is:

$$R^i{}_{jkl} = g^{ia} R_{ajkl} \qquad \text{and} \qquad R_{ijkl} = g_{ia} R^a{}_{jkl} \tag{78}$$

The covariant Riemann-Christoffel curvature tensor satisfies the following symmetric and anti-symmetric relations in its four indices:

$$\begin{aligned} R_{ijkl} &= +R_{klij} && \text{(block symmetry)} \\ &= -R_{jikl} && \text{(anti-symmetry in the first two indices)} \\ &= -R_{ijlk} && \text{(anti-symmetry in the last two indices)} \end{aligned} \tag{79}$$

The anti-symmetry property of the covariant Riemann-Christoffel curvature tensor with respect to the last two indices also applies to the mixed Riemann-Christoffel curvature tensor, that is:

$$R^i{}_{jkl} = -R^i{}_{jlk} \tag{80}$$

As a consequence of the two anti-symmetry properties of the covariant Riemann-Christoffel curvature tensor, the components of the tensor with identical values of the first two indices (e.g. R_{iijk}) or/and the last two indices (e.g. R_{ijkk}) are zero. We remark that all the above symmetry and anti-symmetry properties of the Riemann-Christoffel curvature tensor of the first and second kinds can be proved by using the above mathematical expressions of this tensor.[112]

In an nD space, the Riemann-Christoffel curvature tensor has n^4 components because it is a rank-4 tensor. However, due to its symmetry and anti-symmetry properties (as well as the first Bianchi identity), the tensor has only $\frac{n^2(n^2-1)}{12}$ independent non-vanishing components.[113] For example, in 2D space the tensor has $2^4 = 16$ components with only one independent non-vanishing component, while in 3D space the tensor has $3^4 = 81$ components but only 6 of these are independent non-vanishing components. Similarly, in 4D space (such as the spacetime of general relativity) the tensor has $4^4 = 256$ components but only 20 of these are independent non-vanishing components.

An important set of identities related to the Riemann-Christoffel curvature tensor and frequently used in the formulation of general relativity are the Bianchi identities. These identities, which are based on combinations of this tensor or its covariant derivative, are given by:

$$\begin{aligned} R_{ijkl} + R_{iljk} + R_{iklj} &= 0 && \text{(first Bianchi identity)} \end{aligned} \tag{81}$$

$$\begin{aligned} R_{ijkl;m} + R_{ijmk;l} + R_{ijlm;k} &= 0 && \text{(second Bianchi identity)} \end{aligned} \tag{82}$$

These are called the Bianchi identities of the first kind. By raising the first index of the tensors in these identities the Bianchi identities of the second kind are obtained, that is:

$$R^i{}_{jkl} + R^i{}_{ljk} + R^i{}_{klj} = 0 \tag{83}$$

$$R^i{}_{jkl;m} + R^i{}_{jmk;l} + R^i{}_{jlm;k} = 0 \tag{84}$$

The first Bianchi identity as stated above is an instance of the pattern that by fixing the position of one of the four indices and permuting the other three indices cyclically, the sum of these three permuting forms is zero, that is:[114]

$$R_{ijkl} + R_{iljk} + R_{iklj} = 0 \qquad (i \text{ fixed}) \tag{85}$$

---

[112] See question 7.20 of B3X.
[113] See question 7.22 of B3X. We should note that the number of independent non-vanishing components (as given by the above formula) represents what is required by the symmetry and anti-symmetry properties; otherwise the actual number can be less than this (e.g. in flat spaces all the components are zero; also in certain locations in the space the number can be less than this).
[114] The first form of the first Bianchi identity is proved in § 7.2.2 of B3 while the fourth form is proved in question 7.28 of B3X. The other two forms can be easily proved following similar method.

2.10 Riemann-Christoffel Curvature Tensor

$$R_{ijkl} + R_{ljik} + R_{kjli} = 0 \qquad (j \text{ fixed}) \qquad (86)$$
$$R_{ijkl} + R_{likj} + R_{jlki} = 0 \qquad (k \text{ fixed}) \qquad (87)$$
$$R_{ijkl} + R_{kijl} + R_{jkil} = 0 \qquad (l \text{ fixed}) \qquad (88)$$

The pattern of the second Bianchi identity in its both kinds is also simple, that is the first two indices are fixed while the last three indices are cyclically permuted in the three terms.

Problems

1. Justify the fact that the Riemann-Christoffel curvature tensor is an intrinsic property of the space.
   Answer: This can be easily concluded from Eq. 76, for example, since the Riemann-Christoffel curvature tensor can be expressed in terms of the Christoffel symbols (and their derivatives) which are intrinsic attributes of the space (see § 2.6).
2. Give an example of the involvement of the Riemann-Christoffel curvature tensor in tensor differentiation.
   Answer: The covariant differential operators in mixed derivatives are not commutative in general and hence for a covariant vector **A** and a contravariant vector **B** we have:[115]

   $$A_{j;kl} - A_{j;lk} = R^i{}_{jkl} A_i \qquad \text{and} \qquad B^j{}_{;kl} - B^j{}_{;lk} = R^j{}_{ilk} B^i$$

   These equations show that the covariant differential operators are commutative *iff* the Riemann-Christoffel curvature tensor vanishes identically which means that the space is flat.[116]
3. Outline a proof for the statement: the space is flat *iff* the Riemann-Christoffel curvature tensor vanishes identically.
   Answer: First, we need to show that the Riemann-Christoffel curvature "tensor" is really a tensor. From the equation $A_{j;kl} - A_{j;lk} = R^i{}_{jkl} A_i$ (which we stated in the previous problem) we see that $A_{j;kl} - A_{j;lk}$ is a tensor (because it is the difference between two tensors)[117] and A_i is a tensor (by assumption). Hence, by the quotient rule of tensors $R^i{}_{jkl}$ is also a tensor.
   To prove the above statement, we have two parts:
   • If the space is flat then the Riemann-Christoffel curvature tensor vanishes identically: this is obvious because if the space is flat then it can be coordinated by a Cartesian (or pseudo-Cartesian) system where the Christoffel symbols vanish identically (see exercise 6 of § 2.6) and hence the Riemann-Christoffel curvature tensor also vanishes identically according to its definition (see Eqs. 76 and 77). Now, because it is a tensor then it should vanish in any other coordinate system (whether Cartesian or not) due to the value-invariance of the zero tensor (which was established in § 2.2).
   • If the Riemann-Christoffel curvature tensor vanishes identically then the space is flat: it is shown in the literature that if the Riemann-Christoffel curvature tensor vanishes identically then the space can be coordinated by a Cartesian (or pseudo-Cartesian) system and hence according to the definition and criterion of flat space (see § 1.3 and § 2.3) the space should be flat.
   Note: there are several approaches in the literature (such as the use of parallel transport or the use of commutativity of covariant derivatives) to prove the second part. However, in our view it may be more straightforward to prove this by using the geodesic deviation formula (i.e. $\frac{\delta^2 D^\alpha}{\delta \lambda^2} + R^\alpha{}_{\beta\mu\gamma} D^\mu \frac{dx^\beta}{d\lambda} \frac{dx^\gamma}{d\lambda} = 0$) which is derived in § 7.13 where the vanishing of $R^\alpha{}_{\beta\mu\gamma}$ leads to the vanishing of the second order absolute derivative of the vector D^α (i.e. $\frac{\delta^2 D^\alpha}{\delta \lambda^2} = 0$) which should imply space flatness.
4. Show that in rectilinear coordinate systems $R_{ijkl} = 0$.
   Answer: In rectilinear coordinate systems the components of the metric tensor are constants because the basis vectors are constants and hence all the Christoffel symbols of both kinds vanish identically (see exercises 6 and 7 of § 2.6). Hence, from Eq. 76 we get $R_{ijkl} = 0$.

Exercises

---

[115] For more details, refer to § 5.2 of B3.
[116] The commutativity should also apply in flat regions of space (though not identically over the entire space).
[117] We note that the covariant derivative of a tensor is a tensor (see § 2.7) and hence the covariant derivative (of any order) of a tensor is a tensor (e.g. $A_{j;k}$ is a tensor and hence its covariant derivative $A_{j;kl}$ is also a tensor).

1. What are the physical dimensions of the Riemann-Christoffel curvature tensor?
2. Make a list of all the main properties of the Riemann-Christoffel curvature tensor (i.e. rank, type, symmetry, etc.).
3. When the Riemann-Christoffel curvature tensor vanishes the covariant differential operators become commutative. Justify.
4. Show that $R_{ijkl} = 0$ if the Christoffel symbols vanish identically. What is the implication of this?
5. Give a proof for the second Bianchi identity.
6. The first Bianchi identity $R_{ijkl} + R_{iljk} + R_{iklj} = 0$ may be expressed as $R_{i[jkl]} = 0$ where the square brackets [ ] mean anti-symmetrization in the enclosed indices. Justify.

2.11 Ricci Curvature Tensor

The Ricci curvature tensor, which is a rank-2 tensor, is derived from the Riemann-Christoffel curvature tensor and hence it plays a similar role in characterizing the space and describing its curvature (although partly). There are two kinds of Ricci curvature tensor: first and second, where the first kind is a type $(0, 2)$ tensor while the second kind is a type $(1, 1)$ tensor.[118] The first kind is obtained by contracting the contravariant index of the Riemann-Christoffel curvature tensor of the second kind with its last covariant index, that is:

$$R_{ij} = R^a_{\ ija} = \partial_j \Gamma^a_{ia} - \partial_a \Gamma^a_{ij} + \Gamma^b_{ia}\Gamma^a_{bj} - \Gamma^b_{ij}\Gamma^a_{ba} \tag{89}$$

where Eq. 77 is used to obtain this expression. As we see, the Ricci curvature tensor can be defined directly by the Christoffel symbols although it is originally based on the Riemann-Christoffel curvature tensor.

Because $\Gamma^j_{ij} = \partial_i \left(\ln \sqrt{g}\right)$,[119] the Ricci curvature tensor can also be written in the following forms as well as several other forms:

$$\begin{aligned}
R_{ij} &= \partial_j \partial_i \left(\ln \sqrt{g}\right) - \partial_a \Gamma^a_{ij} + \Gamma^a_{bj}\Gamma^b_{ia} - \Gamma^b_{ij}\partial_b \left(\ln \sqrt{g}\right) \\
&= \partial_j \partial_i \left(\ln \sqrt{g}\right) + \Gamma^a_{bj}\Gamma^b_{ia} - \frac{1}{\sqrt{g}}\partial_a \left(\sqrt{g}\,\Gamma^a_{ij}\right)
\end{aligned} \tag{90}$$

where g is the determinant of the covariant metric tensor. The Ricci curvature tensor of the first kind is symmetric, that is:

$$R_{ij} = R_{ji} \tag{91}$$

This can be easily verified by exchanging the i and j indices in the last line of Eq. 90 taking account of the commutativity of partial differential operators and the fact that the Christoffel symbols are symmetric in their paired indices (Eq. 55).[120]

The Ricci curvature tensor of the first and second kinds can be obtained from each other by the index shifting operators, that is:

$$R^i_{\ j} = g^{ik} R_{kj} \qquad \text{and} \qquad R_{ij} = g_{ik} R^k_{\ j} \tag{92}$$

In an nD space, the Ricci curvature tensor has n^2 components but due to its symmetry only $\frac{n(n+1)}{2}$ are independent (see § 2.2).

Problems

---

[118] In this we follow the literature where these two kinds may be considered as the main types; otherwise we will see in § 2.13 that for the contravariant Einstein tensor $G^{\mu\nu}$ we need to define a contravariant Ricci curvature tensor $R^{\mu\nu}$. Anyway, the Ricci curvature tensor is a tensor and hence in principle it can have contravariant form (as well as covariant and mixed forms) like other tensors.

[119] See § 5.1 of B3.

[120] See question 7.32 of B3X.

1. Obtain a mathematical expression for the mixed Ricci curvature tensor R^i_j in terms of the Christoffel symbols and the metric tensor.
   Answer: We have:
   $$\begin{aligned} R^i_j &= g^{ik} R_{kj} \\ &= g^{ik} \left(\partial_j \Gamma^a_{ka} - \partial_a \Gamma^a_{kj} + \Gamma^b_{ka} \Gamma^a_{bj} - \Gamma^b_{kj} \Gamma^a_{ba} \right) \end{aligned}$$
   where in line 1 we use Eq. 92 and in line 2 we use Eq. 89.

Exercises
1. What are the physical dimensions of the Ricci curvature tensor?
2. What are the number of components and the number of independent components of the Ricci curvature tensor in the spacetime of general relativity?
3. Make a list of the main properties of the Ricci curvature tensor (rank, type, symmetry, etc.).
4. The first kind of the Ricci curvature tensor may also be obtained by contracting the contravariant index of the Riemann-Christoffel curvature tensor of the second kind with its covariant index before the last. Comment on this.
5. Show that $R_{ij} = 0$ identically is a necessary but not sufficient condition for a space to be flat.

2.12 Ricci Curvature Scalar

The Ricci curvature scalar R, which is also known as the curvature scalar and the curvature invariant, is the result of contracting the indices of the Ricci curvature tensor of the second kind, that is:

$$R = g^{ij} R_{ij} = R^j_j \tag{93}$$

Problems
1. Obtain a mathematical expression for the Ricci curvature scalar R in terms of the Christoffel symbols and metric tensor.
   Answer: We have:
   $$\begin{aligned} R &= g^{ij} R_{ij} \\ &= g^{ij} \left(\partial_j \Gamma^a_{ia} - \partial_a \Gamma^a_{ij} + \Gamma^b_{ia} \Gamma^a_{bj} - \Gamma^b_{ij} \Gamma^a_{ba} \right) \end{aligned}$$
   where in line 1 we use Eq. 93 and in line 2 we use Eq. 89.
2. How is the Ricci curvature scalar obtained from the covariant Riemann-Christoffel curvature tensor?
   Answer: It is obtained as follows:
   • The mixed Riemann-Christoffel curvature tensor is obtained by raising the first index of the covariant Riemann-Christoffel curvature tensor, that is:
   $$R^a{}_{ijk} = g^{ab} R_{bijk}$$
   • The covariant Ricci curvature tensor is obtained by contracting the contravariant index with the last covariant index of the mixed Riemann-Christoffel curvature tensor, that is:
   $$R_{ij} = \delta^k_a R^a{}_{ijk} = R^a{}_{ija}$$
   • The mixed Ricci curvature tensor is obtained by raising the first index of the covariant Ricci curvature tensor, that is:
   $$R^k{}_j = g^{ki} R_{ij}$$
   • The Ricci curvature scalar R is obtained by contracting the indices of the mixed Ricci curvature tensor, that is:
   $$R = \delta^j_k R^k{}_j = R^j{}_j$$

Exercises
1. What are the physical dimensions of the Ricci curvature scalar?
2. Make a list of the main properties of the Ricci curvature scalar.
3. Show that $R = 0$ identically is a necessary but not sufficient condition for a space to be flat.
4. Compare the curvature information contained in the Riemann-Christoffel curvature tensor, the Ricci curvature tensor and the Ricci curvature scalar.

2.13 Einstein Tensor

The Einstein tensor $G_{\mu\nu}$ is a rank-2 tensor defined in terms of the Ricci curvature tensor $R_{\mu\nu}$, the Ricci curvature scalar R and the metric tensor $g_{\mu\nu}$ as follows:

$$G_{\mu\nu} = R_{\mu\nu} - \frac{1}{2}g_{\mu\nu}R \tag{94}$$

The tensor may also be given in contravariant and mixed forms, that is:

$$G^{\mu\nu} = R^{\mu\nu} - \frac{1}{2}g^{\mu\nu}R \qquad \text{and} \qquad G^{\mu}_{\nu} = R^{\mu}_{\nu} - \frac{1}{2}\delta^{\mu}_{\nu}R \tag{95}$$

As we will see in § 3, the Einstein tensor represents the geometric part of the Field Equation while the energy-momentum tensor represents its physical part.

Since both the Ricci curvature tensor and the metric tensor are symmetric, the Einstein tensor is also symmetric (see § 2.2). The divergence[121] of the Einstein tensor vanishes identically for any Riemannian metric, that is:

$$G^{\mu\nu}_{;\nu} = 0 \tag{96}$$

This can be easily proved by contracting the second Bianchi identity twice with the use of the anti-symmetry properties of the Riemann-Christoffel curvature tensor (see exercise 4).

Problems
1. Make a list of the main properties of the Einstein tensor.
   Answer: The main properties are:
   • It is synthesized from the Ricci curvature tensor, the Ricci curvature scalar and the metric tensor.
   • It is absolute rank-2 symmetric tensor and hence in nD space it has n^2 components and $\frac{n(n+1)}{2}$ independent components. Therefore, in the 4D spacetime of general relativity it has 16 components 10 of which are independent.
   • It depends only on the metric tensor and hence it is intrinsic to the space.
   • It characterizes the geometry of space and expresses its curvature (partly). In fact, it represents the geometric part of the Field Equation of general relativity (see § 3.2).
   • It has the physical dimensions of reciprocal area.
   • Its divergence is zero identically.
   • It vanishes in flat space.
2. Explain how to build the Einstein tensor from the metric tensor.
   Answer: We build it as follows:
   (a) We obtain the Christoffel symbols from the metric tensor (see Eqs. 51 and 52).
   (b) We obtain the Riemann-Christoffel curvature tensor from the Christoffel symbols (see Eq. 76).
   (c) We obtain the Ricci curvature tensor $R_{\mu\nu}$ and the Ricci curvature scalar R from the Riemann-Christoffel curvature tensor and the metric tensor (see problem 2 of § 2.12).
   (d) We obtain the Einstein tensor from the Ricci curvature tensor, the Ricci curvature scalar and the

---

[121] We note that the divergence in the textbooks of general relativity may be called covariant divergence or 4-divergence (or terms like these) to indicate its tensorial 4D nature and distinguish it from the ordinary divergence in 3D space coordinated by Cartesian system. However, we think this distinction is redundant because we are already assuming 4D spacetime coordinated by general coordinate system and hence divergence in this context is just a generalization of the ordinary divergence.

metric tensor using Eq. 94.

Note: we may replace step (b) and step (c) with using Eqs. 89 and 93 without building the Riemann-Christoffel curvature tensor.

Exercises

1. Show that the physical dimensions of the Einstein tensor are reciprocal area.
2. In flat space the Einstein tensor is zero. Explain why.
3. What the identity $G^{\mu\nu}_{;\nu} = 0$ means?
4. Prove that the divergence of Einstein tensor is identically zero (i.e. $G^{\mu\nu}_{;\nu} = 0$) using the second Bianchi identity.
5. Show that $g^{\nu\alpha}G_{\alpha\mu;\nu} = 0$.
6. Show that $G = -R$ where G is the trace of the Einstein tensor and R is the Ricci curvature scalar.
7. Show that $G_{\mu\nu} = 0$ iff $R_{\mu\nu} = 0$.
8. Show the following: (a) $G = 0$ iff $R = 0$ and (b) if $G_{\mu\nu} = 0$ then $R = 0$.
9. Compare the curvature information contained in the Riemann-Christoffel curvature tensor and the Einstein tensor. What is the implication of this?

2.14 Energy-Momentum Tensor

The energy-momentum tensor $T^{\mu\nu}$, which expresses the distribution and flow of energy and momentum in spacetime, is a rank-2 symmetric tensor and hence it has 16 components with only 10 of these being independent. These components are generally functions of spacetime coordinates and hence each point in spacetime is characterized by a specific value of the energy-momentum tensor. As we will see in § 3.2, the energy-momentum tensor represents the physical part of the Field Equation while the Einstein tensor represents its geometric part.[122] We note that the energy-momentum tensor may also be called the stress-energy-momentum tensor or the stress-energy tensor. We should also remark that due to the mass-energy equivalence, according to the Poincare relation $E = mc^2$, energy here includes mass since it is a form of energy. In the following discussion we generally assume an orthonormal 4D coordinate system whose all coordinates have physical dimensions of length. This should be valid at least locally in the Riemannian spacetime of general relativity.

To determine the physical significance of the components of the energy-momentum tensor, let define this tensor in simple terms as the tensor whose $T^{\mu\nu}$ component represents the flux of the μth component of the momentum 4-vector across a surface perpendicular to the νth coordinate.[123] Now, "flux" in simple terms means "per unit area per unit time". So, all we need to find the $T^{\mu\nu}$ component is to take the μth component of the momentum 4-vector and divide it by the area of a surface element perpendicular to the νth coordinate and by the time. Now, the surface element perpendicular to the νth coordinate is defined by the product of two line elements of other two coordinates (i.e. other than the νth coordinate). So, if we divide and multiply by a line element along the direction of the νth coordinate then the $T^{\mu\nu}$ component can be seen as the μth component of the momentum 4-vector divided by volume (i.e. the volume passing through the surface in a unit time) and multiplied by speed (i.e. the velocity component in the νth direction).[124] So in brief, the component $T^{\mu\nu}$ of the energy-momentum tensor represents the density (i.e. per volume) of the μth component of the momentum 4-vector times the νth component of

---

[122] As we will see, the energy-momentum tensor may involve the metric and hence it may not be seen as purely physical.

[123] We remind the reader that the momentum 4-vector is given by $\mathbf{P} = m\gamma\left(c, u^1, u^2, u^3\right)$ where m is the invariant mass, γ is the Lorentz factor, c is the characteristic speed of light and $\left(u^1, u^2, u^3\right)$ is the velocity 3-vector. We also note that all the components of the coordinate (or position) 4-vector have the dimension of length (assuming certain type of spatial coordinates) since $x^0 = ct$ (see B4). The reader should also be aware that expressions like "across a surface" are more general than being perpendicular or parallel to the surface since the surface can be tangential to the component of the momentum 4-vector (i.e. when the μth and ν^{th} coordinates are perpendicular). Also, "surface perpendicular to the νth coordinate" means coordinate surface of constant ν^{th} coordinate.

[124] The reader should note that this is a recipe (or pedagogical device) for synthesizing and analyzing the energy-momentum tensor rather than a depiction and reflection of the actual physical situation and process (in fact we mainly consider the case in which the μth and ν^{th} coordinates are identical). The validity of this approach will be verified in the discussion of the examples of the energy-momentum tensor in the questions.

2.14 Energy-Momentum Tensor

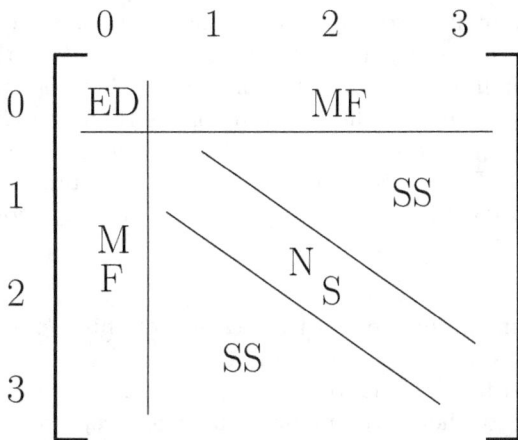

Figure 2: The general structure of the energy-momentum tensor showing the positions of the components of energy density (ED), momentum flow (MF), spatial shear stress (SS) and spatial normal stress (NS).

the velocity 4-vector (but without the γ factor).[125] Accordingly, the physical dimensions of this tensor are energy per volume (e.g. J/m^3).

To be more clear, let state the above description in a more practical and definite form considering a simple case. The component T^{00} represents the energy density at the point of spacetime represented by the tensor while the components $T^{0i} = T^{i0}$ ($i = 1, 2, 3$) represent the temporal rate of energy flow per unit area perpendicular to the ith direction divided by the speed of light c (the division by c is because the zeroth component of the 4-momentum is E/c). The components T^{ij} ($i, j = 1, 2, 3$) represent the temporal rate of flow of the ith component of momentum per unit area perpendicular to the jth direction. In more simple (and rather different) terms, in orthonormal systems T^{00} represents energy density, $T^{0i} = T^{i0}$ represents momentum flow,[126] and T^{ij} represents spatial stress (see Figure 2).

As indicated earlier and will be detailed later (refer to § 3.2), the energy-momentum tensor represents the physical part in the Field Equation (in contrast to the geometrical part that is represented by the Einstein tensor which is based on the space metric) since it mathematically portrays the distribution and flow of matter and energy in the "physical" spacetime. Accordingly, it is the source term in the gravitational field formulation of general relativity like mass density ρ in the gravitational field formulation of the classical theory (see Eq. 14 and refer to § 3.2). A characteristic property of the energy-momentum tensor is that its divergence is identically zero, that is:

$$T^{\mu\nu}_{;\nu} = 0 \qquad (97)$$

The vanishing of divergence is a requirement for the conservation of energy-momentum as represented by the momentum 4-vector (see exercises 8, 9 and 11).

The computation of the energy-momentum tensor in real physical situations is generally a complicated process if it is viable at all. However, in the literature of general relativity there are several simple examples that are used as prototypes to illustrate the structure, physical significance and method of computation of this tensor (in fact these prototypes are commonly used in real physical models such as modeling the Universe in cosmology). These examples include dust of non-interacting particles, ideal fluid and electromagnetic fields. We will discuss these examples in the questions of this section.

In a spacetime region that is void of matter and energy, the energy-momentum tensor vanishes (i.e. $T_{\mu\nu} = 0$). As we will see in § 3.2, the Field Equation can be written as $R_{\mu\nu} = \kappa \left(T_{\mu\nu} - \frac{1}{2} g_{\mu\nu} T \right)$ and hence

---

[125] The velocity 4-vector is given by $\mathbf{U} = \gamma \left(c, u^1, u^2, u^3 \right)$ where γ is the Lorentz factor, c is the characteristic speed of light and (u^1, u^2, u^3) is the velocity 3-vector (see B4).

[126] To be more consistent (regarding the physical dimensions) we may call it "momentum density flow" or "momentum flow density".

the Field Equation in such case reduces to $R_{\mu\nu} = 0$ and the solutions are called vacuum solutions.[127] In this case only the Ricci curvature tensor is needed for the solution rather than the entire Einstein tensor. It should be noted that the condition $R_{\mu\nu} = 0$ does not mean that the spacetime is flat in the vacuum region because the condition for flatness is the vanishing of the Riemann-Christoffel curvature tensor (i.e. $R_{\mu\nu\sigma\omega} = 0$), and the condition $R_{\mu\nu} = 0$ is not sufficient for achieving flatness. This is a direct result of the fact that $R_{\mu\nu}$ is a contraction of $R_{\mu\nu\sigma\omega}$ and hence $R_{\mu\nu}$ does not contain the entire curvature information of the spacetime (see exercise 4 of § 2.12). This similarly applies to $G_{\mu\nu}$ (noting that $G_{\mu\nu} = 0$ iff $R_{\mu\nu} = 0$; see exercises 7 and 9 of § 2.13).

Problems

1. Justify, in detail, the statements in the third paragraph of this section, i.e. the paragraph starting with: "To be more clear, let state the above description etc.".
    Answer: We use in our justification the following statement (which is in the second paragraph): "So in brief, the component $T^{\mu\nu}$ of the energy-momentum tensor represents the density (i.e. per volume) of the μth component of the momentum 4-vector times the νth component of the velocity 4-vector (but without the γ factor)."
    Accordingly, the components of $T^{\mu\nu}$ are as follows (where we use loose symbolism in some cases):
    - T^{00} is the density (i.e. per volume V) of the 0th component of the momentum 4-vector (which is $m\gamma c$) times the 0th component of the velocity 4-vector without the γ factor (which is c), i.e.
    $$T^{00} = \frac{m\gamma c}{V} \times c = \frac{m\gamma c^2}{V} = \frac{E}{V}$$
    which is the energy density, as stated in the text.
    - T^{0i} is the density (i.e. per V) of the 0th component of the momentum 4-vector (which is $m\gamma c$) times the ith component of the velocity 4-vector without the γ factor (which is u^i), that is:
    $$T^{0i} = \frac{m\gamma c}{V} \times u^i = \frac{m\gamma c u^i}{V} = \frac{m\gamma c^2 u^i}{Vc} = \frac{E u^i}{Vc} = \frac{E\left(dx^i/dt\right)}{\left(dx^i dx^j dx^k\right)c} = \frac{E/dt}{\left(dx^j dx^k\right)c}$$
    which is the temporal rate of energy flow per unit area perpendicular to the ith direction divided by the speed of light c, as stated in the text.
    - T^{i0} is the density (i.e. per V) of the ith component of the momentum 4-vector (which is $m\gamma u^i$) times the 0th component of the velocity 4-vector without the γ factor (which is c), that is:
    $$T^{i0} = \frac{m\gamma u^i}{V} \times c = \frac{m\gamma c u^i}{V} = \frac{m\gamma c^2 u^i}{Vc} = \frac{E u^i}{Vc} = \frac{E\left(dx^i/dt\right)}{\left(dx^i dx^j dx^k\right)c} = \frac{E/dt}{\left(dx^j dx^k\right)c}$$
    which is the temporal rate of energy flow per unit area perpendicular to the ith direction divided by the speed of light c, as stated in the text.
    - T^{ii} (no summation over i) is the density (i.e. per V) of the ith component of the momentum 4-vector (which is $m\gamma u^i$) times the ith component of the velocity 4-vector without the γ factor (which is u^i), that is:
    $$T^{ii} = \frac{m\gamma u^i}{V} \times u^i = \frac{p^i u^i}{V} = \frac{p^i\left(dx^i/dt\right)}{dx^i dx^j dx^k} = \frac{p^i/dt}{dx^j dx^k}$$
    which is the temporal rate of flow of the ith component of momentum p^i per unit area perpendicular to the ith direction.
    - T^{ij} ($i \neq j$) is the density (i.e. per V) of the ith component of the momentum 4-vector (which is $m\gamma u^i$) times the jth component of the velocity 4-vector without the γ factor (which is u^j), that is:
    $$T^{ij} = \frac{m\gamma u^i}{V} \times u^j = \frac{p^i u^j}{V} = \frac{p^i\left(dx^j/dt\right)}{dx^i dx^j dx^k} = \frac{p^i/dt}{dx^i dx^k}$$

---

[127] We refer the readers to exercise 13 of § 3.2 where the meaning and significance of local and global vanishing of the energy-momentum tensor is investigated and analyzed. Also see § 2.13 where it is shown that $G_{\mu\nu} = 0$ iff $R_{\mu\nu} = 0$, i.e. we have $G_{\mu\nu} = \kappa T_{\mu\nu}$ (according to the Field Equation) and since $T_{\mu\nu} = 0$ then $G_{\mu\nu} = 0$ and hence $R_{\mu\nu} = 0$ (according to exercise 7 of § 2.13).

2.14 Energy-Momentum Tensor

which is the temporal rate of flow of the ith component of momentum per unit area perpendicular to the jth direction.

- T^{ji} ($i \neq j$) is the density (i.e. per V) of the jth component of the momentum 4-vector (which is $m\gamma u^j$) times the ith component of the velocity 4-vector without the γ factor (which is u^i), that is:

$$T^{ji} = \frac{m\gamma u^j}{V} \times u^i = \frac{p^j u^i}{V} = \frac{p^j \left(dx^i/dt\right)}{dx^i dx^j dx^k} = \frac{p^j/dt}{dx^j dx^k}$$

which is the temporal rate of flow of the jth component of momentum per unit area perpendicular to the ith direction. We note that we may also write:

$$T^{ji} = \frac{m\gamma u^j}{V} \times u^i = \frac{m\gamma u^i}{V} \times u^j = \frac{p^i u^j}{V} = \frac{p^i \left(dx^j/dt\right)}{dx^i dx^j dx^k} = \frac{p^i/dt}{dx^i dx^k}$$

which is the temporal rate of flow of the ith component of momentum per unit area perpendicular to the jth direction. This (added to the above results regarding the components T^{0i}, T^{i0} and T^{ij}) should ensure the symmetry of the energy-momentum tensor.

Note 1: we do not need to obtain the expression of T^{ji} since it is contained in the expression of T^{ij} considering that $i, j = 1, 2, 3$ (and $i \neq j$ since T^{ii} was considered separately). However, we did this for pedagogical purpose and to show the symmetry.

Note 2: as we are supposed to use the energy-momentum tensor for general relativistic calculations, the use of the special relativistic formalism (like 4-momentum) in the above determination of the formalism of this tensor may be questioned. However, since the energy-momentum tensor is a function of coordinates and hence it is identified and determined point by point in the spacetime (where the spacetime at each point is supposedly flat by the locality condition) then the use of special relativity is justified (although this justification may also be questioned).

2. Show that the energy-momentum tensor is symmetric.
   Answer: The answer of the previous question is sufficiently general to demonstrate the symmetry of the energy-momentum tensor in general.
3. Find the energy-momentum tensor of a homogeneous cloud of dust made of identical non-interacting particles that move with a common velocity.
   Answer: This is one of the simplest examples of the energy-momentum tensor. Let the mass of each particle be m (since they are identical) and the common 3-velocity of these particles be $\mathbf{u} \equiv (u^1, u^2, u^3)$.[128] Accordingly, the energy of each particle is γmc^2 and its 3-momentum is $\gamma m\mathbf{u}$ where γ is the Lorentz factor which is a function of $u \equiv |\mathbf{u}|$. Now, if the number density (i.e. the number of particles per unit volume) is N then we have:
   - The energy density of the cloud (which is represented by the component T^{00}) is:

$$T^{00} = N\gamma mc^2 = \rho\gamma c^2$$

where ρ is the mass density.
   - The temporal rate of energy flow per unit area perpendicular to the ith direction divided by the speed of light c (which is represented by the components $T^{0i} = T^{i0}$ where $i = 1, 2, 3$) is:

$$T^{0i} = T^{i0} = \frac{\left(Nu^i At\right)\left(\gamma mc^2\right)}{(Atc)} = N\gamma mcu^i = \rho\gamma cu^i$$

where $\left(Nu^i At\right)$ is the number of particles crossing an area A (perpendicular to the ith direction) in time t, $\left(\gamma mc^2\right)$ is the energy of each one of these crossing particles, and (Atc) represents "per unit area" (i.e. A), "temporal rate" (i.e. t) and "divided by the speed of light c".
   - The temporal rate of flow of the ith component of momentum per unit area perpendicular to the jth direction (which is represented by the components T^{ij} where $i, j = 1, 2, 3$) is:

$$T^{ij} = \frac{\left(Nu^j At\right)\left(\gamma mu^i\right)}{(At)} = N\gamma mu^i u^j = \rho\gamma u^i u^j$$

---

[128] We use m here as invariant, i.e. there is no difference between rest and non-rest mass.

where $\left(Nu^j At\right)$ is the number of particles crossing an area A (perpendicular to the jth direction) in time t, $\left(\gamma m u^i\right)$ is the ith component of momentum of each one of these crossing particles, and (At) represents "per unit area" (i.e. A), and "temporal rate" (i.e. t).

So, the energy-momentum tensor for this problem is given by:

$$[T^{\mu\nu}] \equiv \begin{bmatrix} T^{00} & T^{01} & T^{02} & T^{03} \\ T^{10} & T^{11} & T^{12} & T^{13} \\ T^{20} & T^{21} & T^{22} & T^{23} \\ T^{30} & T^{31} & T^{32} & T^{33} \end{bmatrix} = \begin{bmatrix} N\gamma m c^2 & N\gamma m c u^1 & N\gamma m c u^2 & N\gamma m c u^3 \\ N\gamma m c u^1 & N\gamma m u^1 u^1 & N\gamma m u^1 u^2 & N\gamma m u^1 u^3 \\ N\gamma m c u^2 & N\gamma m u^2 u^1 & N\gamma m u^2 u^2 & N\gamma m u^2 u^3 \\ N\gamma m c u^3 & N\gamma m u^3 u^1 & N\gamma m u^3 u^2 & N\gamma m u^3 u^3 \end{bmatrix}$$

$$= N\gamma m \begin{bmatrix} c^2 & cu^1 & cu^2 & cu^3 \\ cu^1 & u^1 u^1 & u^1 u^2 & u^1 u^3 \\ cu^2 & u^2 u^1 & u^2 u^2 & u^2 u^3 \\ cu^3 & u^3 u^1 & u^3 u^2 & u^3 u^3 \end{bmatrix} = \rho\gamma \begin{bmatrix} c^2 & cu^1 & cu^2 & cu^3 \\ cu^1 & u^1 u^1 & u^1 u^2 & u^1 u^3 \\ cu^2 & u^2 u^1 & u^2 u^2 & u^2 u^3 \\ cu^3 & u^3 u^1 & u^3 u^2 & u^3 u^3 \end{bmatrix}$$

Note 1: we may also use the definition of the energy-momentum tensor (i.e. the flux of 4-momentum) and hence:

- $$T^{00} = \frac{(E/c)}{V} \times c = \frac{E}{V} = \frac{nmc^2\gamma}{V} = Nmc^2\gamma = \rho\gamma c^2$$

where (E/c) is the 0th component of the momentum 4-vector, V is volume, c (which we multiply with) is the 0th component of the velocity 4-vector (without γ), and n is the number of particles inside volume V.

- $$T^{0i} = \frac{(E/c)}{V} \times u^i = \frac{nmc\gamma}{V} u^i = Nmc\gamma u^i = \rho\gamma c u^i$$
$$T^{i0} = \frac{p^i}{V} \times c = \frac{nm\gamma u^i c}{V} = Nm\gamma u^i c = \rho\gamma c u^i$$

where the steps are similarly justified.

- $$T^{ii} = \frac{p^i}{V} \times u^i = \frac{nm\gamma u^i u^i}{V} = Nm\gamma u^i u^i = \rho\gamma u^i u^i$$
$$T^{ij} = \frac{p^i}{V} \times u^j = \frac{nm\gamma u^i u^j}{V} = Nm\gamma u^i u^j = \rho\gamma u^i u^j$$
$$T^{ji} = \frac{p^j}{V} \times u^i = \frac{nm\gamma u^j u^i}{V} = Nm\gamma u^j u^i = \rho\gamma u^i u^j$$

where the steps are similarly justified.

Note 2: it should be obvious that the above energy-momentum tensor can be written as:

$$[T^{\mu\nu}] = N\gamma m \begin{bmatrix} c^2 & cu^1 & cu^2 & cu^3 \\ cu^1 & u^1 u^1 & u^1 u^2 & u^1 u^3 \\ cu^2 & u^2 u^1 & u^2 u^2 & u^2 u^3 \\ cu^3 & u^3 u^1 & u^3 u^2 & u^3 u^3 \end{bmatrix} = \frac{N}{\gamma} \mathbf{P}^T \mathbf{U} = \frac{N}{\gamma} \mathbf{U}^T \mathbf{P} = \frac{N}{m\gamma} \mathbf{P}^T \mathbf{P} = \frac{Nm}{\gamma} \mathbf{U}^T \mathbf{U}$$

where $\mathbf{P} = m\gamma \left[c, u^1, u^2, u^3\right]$ is the momentum 4-vector of the individual particles, $\mathbf{U} = \gamma \left[c, u^1, u^2, u^3\right]$ is their velocity 4-vector, and the superscript T symbolizes transposition. It is noteworthy that we are using here matrix notation. If we use symbolic notation of tensors (where $\mathbf{T}$ represents the energy-momentum tensor and $\mathbf{UU}$ represents the outer product of the velocity 4-vector by itself which may also be written as $\mathbf{U} \otimes \mathbf{U}$) and note that $\frac{Nm}{\gamma} = \frac{N_0 \gamma m}{\gamma} = N_0 m = \rho_0$ (where the subscript 0 refers to the proper quantity) then the above relation can be written as $\mathbf{T} = \rho_0 \mathbf{UU}$ which is the form that is commonly used in the literature.

4. Obtain the energy-momentum tensor for the problem in the previous question according to a co-moving frame.[129]

   Answer: The physical dimensions of the energy-momentum tensor is energy density (i.e. energy per unit volume). Hence, the components of this tensor can be generically expressed as mass density times speed squared (since mass times speed squared is energy like mc^2). So, if we want to express this tensor in an invariant and simple form then we can use the proper mass density (which is invariant) and the 4-velocity of the cloud (which is also invariant), that is:[130]

   $$T^{\mu\nu} = \rho_0 U^\mu U^\nu$$

   where ρ_0 is the proper mass density (i.e. the density as seen from the co-moving frame) while U^μ and U^ν represent components of the 4-velocity of the cloud. Now, in a co-moving frame $\mathbf{u} = \mathbf{0}$ and $\gamma = 1$ and hence the velocity 4-vector becomes $\mathbf{U} = (c, 0, 0, 0)$. Therefore, all the components of the energy-momentum tensor are zero except the energy density of the cloud (i.e. the component T^{00}) which is $T^{00} = \rho_0 U^0 U^0 = \rho_0 cc = \rho_0 c^2$. Therefore, the energy-momentum tensor according to a co-moving frame is:

   $$[T^{\mu\nu}] = \begin{bmatrix} \rho_0 c^2 & 0 & 0 & 0 \\ 0 & 0 & 0 & 0 \\ 0 & 0 & 0 & 0 \\ 0 & 0 & 0 & 0 \end{bmatrix}$$

   In fact, this can be obtained directly from the solution of the previous problem, i.e.

   $$[T^{\mu\nu}] = \rho\gamma \begin{bmatrix} c^2 & cu^1 & cu^2 & cu^3 \\ cu^1 & u^1 u^1 & u^1 u^2 & u^1 u^3 \\ cu^2 & u^2 u^1 & u^2 u^2 & u^2 u^3 \\ cu^3 & u^3 u^1 & u^3 u^2 & u^3 u^3 \end{bmatrix}$$

   by setting $\rho \equiv \rho_0$, $\gamma = 1$ and $u^1 = u^2 = u^3 = 0$. So, the solutions of the two problems are consistent. In fact, the solution of this problem is a special case of the solution in the previous problem.

   Note: the easiest way to obtain the result of this question is to start from the general result that we obtained in the previous question (i.e. $\mathbf{T} = \rho_0 \mathbf{UU}$ in symbolic notation which becomes $T^{\mu\nu} = \rho_0 U^\mu U^\nu$ in indicial notation). The result of this question can then be obtained immediately by noting that in a co-moving frame $\mathbf{U} = (c, 0, 0, 0)$ as explained above.

5. Find the energy-momentum tensor of ideal fluid in equilibrium considering the case of co-moving frame and assuming a locally flat spacetime.[131]

   Answer: Ideal (or perfect) fluid is characterized by having no internal friction forces (or viscosity) and hence it may be seen from this perspective as the continuum analog of the discrete model (i.e. cloud of non-interacting particles) of the previous two problems. However, ideal fluid is distinguished by having internal pressure due to its continuum nature.

   Now, if we use the argument and result of the previous problem then the components of the energy-momentum tensor of ideal fluid should have a term like the term in the previous problem (i.e. $\rho_0 U^\mu U^\nu$) because the two systems are essentially identical (apart from being discrete and continuum models). Moreover, since ideal fluid has internal pressure (due to its continuum nature) then it should have two more terms: dynamic pressure term and static pressure term.[132] It can be shown that these terms are given by $(p/c^2) U^\mu U^\nu$ and $-p g^{\mu\nu}$ where p is the pressure (which is coordinate dependent) and $g^{\mu\nu}$

---

[129] Co-moving frame is a frame that moves with the cloud and hence the particles of the cloud are stationary relative to the frame. Co-moving frame may also be called rest frame.

[130] In fact, this form was obtained in a more rigorous way in the previous question (see note 2 of the previous question). So, the purpose of this rather loose approach in this question is to obtain the result independently of the result of the previous question. Also see the upcoming note.

[131] Local flatness can be justified in general by the Riemannian nature of the spacetime. In fact, flatness may be justified even globally by weak gravity.

[132] We are assuming negligible thermodynamic effects (due to net directional bulk flow of heat energy) although thermostatic effects (due to random motion of molecules) should be included in the total energy (or mass-energy) of the system. In

is the metric tensor. Accordingly, the energy-momentum tensor of ideal fluid in a general frame is given by:

$$T^{\mu\nu} = \rho_0 U^\mu U^\nu + \frac{p}{c^2} U^\mu U^\nu - p g^{\mu\nu} \tag{98}$$

We note that all the terms of this equation have the same physical dimensions (i.e. energy density) and hence it is dimensionally consistent.

Now, in a co-moving frame $\mathbf{u} = \mathbf{0}$ and $\gamma = 1$; moreover if the spacetime is locally flat then the metric tensor can be approximated by the Minkowski metric (i.e. $[g^{\mu\nu}] \simeq [\eta^{\mu\nu}] = \text{diag}\,[+1, -1, -1, -1]$) and hence the energy-momentum tensor in this case becomes:

$$[T^{\mu\nu}] = \begin{bmatrix} \rho_0 c^2 & 0 & 0 & 0 \\ 0 & p & 0 & 0 \\ 0 & 0 & p & 0 \\ 0 & 0 & 0 & p \end{bmatrix}$$

Note 1: the justification of the form of $[T^{\mu\nu}]$ as given in the last equation is that: due to the conditions $\mathbf{u} = \mathbf{0}$ and $[g^{\mu\nu}] \simeq [\eta^{\mu\nu}] = \text{diag}\,[+1, -1, -1, -1]$, all the off-diagonal components will vanish, i.e. the first two terms of Eq. 98 will vanish by the condition $\mathbf{u} = \mathbf{0}$ and the last term will vanish by the condition $[g^{\mu\nu}] \simeq [\eta^{\mu\nu}]$. Moreover, we have:

$$\begin{aligned} T^{00} &= \rho_0 U^0 U^0 + \frac{p}{c^2} U^0 U^0 - p g^{00} \\ &= \rho_0 (\gamma c)(\gamma c) + \frac{p}{c^2}(\gamma c)(\gamma c) - p \times 1 \\ &= \rho_0 c^2 + p - p \\ &= \rho_0 c^2 \end{aligned}$$

where line 3 is justified by $\gamma = 1$ since $u \equiv |\mathbf{u}| = 0$. We note that the subscript 0 in ρ_0 is a label (for proper) and not an index.
Also:

$$\begin{aligned} T^{ii} &= \rho_0 U^i U^i + \frac{p}{c^2} U^i U^i - p g^{ii} \\ &= 0 + 0 - p(-1) \\ &= p \end{aligned}$$

where $i = 1, 2, 3$ (with no summation over i) and line 2 is justified by $U^i = 0$ since $\mathbf{u} = \mathbf{0}$.

Note 2: referring to the aforementioned similarity between the ideal fluid system and the dust cloud system, we see that if we assume that the pressure of the ideal fluid is negligible (i.e. $p \simeq 0$) then the energy-momentum tensor of the ideal fluid system will converge to the energy-momentum tensor of the dust cloud system of the previous problem. On the other hand, if we assume that the particles in the dust cloud system are interacting (by non-viscous forces simulating the pressure) then the energy-momentum tensor of the dust cloud system should become similar to the energy-momentum tensor of the ideal fluid system.

Note 3: the physical dimensions of pressure are the same as the physical dimensions of energy density (i.e. $\text{kg}\,\text{m}^{-1}\,\text{s}^{-2}$) and hence the components of the above tensor are dimensionally consistent.

Note 4: Eq. 98 represents one of the most general forms of the energy-momentum tensor in curved spacetimes and hence it is used in many applications of general relativity such as relativistic cosmology. In fact, it is possibly the most commonly used form in the applications of general relativity (due to the simplicity of ideal fluid as a model for gravity source as well as being relatively realistic in comparison to other simpler models).

---

fact, some authors impose this in the definition of ideal fluid and hence ideal fluid is characterized by having no heat conduction as well as being inviscid. There are other definitions of ideal fluid (e.g. being isotropic in its rest frame or being totally characterized by density and pressure) some of which are equivalent or similar to others while others are questionable. However, most these definitions are incomplete or inaccurate since they focus on the properties of main concern in the given context.

2.14 Energy-Momentum Tensor

6. Referring to the result that we obtained in problem 3 (which is a typical example of energy-momentum tensor), justify the generality of the definition of the energy-momentum tensor as "the flux of the μth component of the momentum 4-vector across a surface perpendicular to the νth coordinate" with regard to the temporal coordinate.

Answer: For simplicity, consistency and clarity we use rather loose symbolism in the following answer (as we did in some previous answers).

- For T^{00} the 0th component of the momentum 4-vector is $n\gamma mc$ and the 0th coordinate is $x^0 \equiv ct$ and hence:

$$T^{00} = \frac{n\gamma mc}{At} = \frac{n\gamma mc^2}{Act} = \frac{n\gamma mc^2}{dx^i dx^j dx^0} = \gamma \frac{nm}{V}c^2 = \gamma\rho c^2$$

- For T^{0i} the 0th component of the momentum 4-vector is $n\gamma mc$ and the ith coordinate is x^i and hence:

$$T^{0i} = \frac{n\gamma mc}{At} = \frac{n\gamma mc}{(dx^j dx^k)t} = \frac{n\gamma mc\, dx^i}{(dx^i dx^j dx^k)t} = \gamma\frac{nm}{V}c\frac{dx^i}{t} = \gamma\rho cu^i$$

- For T^{i0} the ith component of the momentum 4-vector is $n\gamma mu^i$ and the 0th coordinate is $x^0 \equiv ct$ and hence:

$$T^{i0} = \frac{n\gamma mu^i}{At} = \frac{n\gamma mcu^i}{(dx^j dx^k)ct} = \frac{n\gamma mcu^i}{dx^j dx^k dx^0} = \gamma\frac{nm}{V}cu^i = \gamma\rho cu^i$$

Exercises

1. Why the energy-momentum tensor has 16 components with only 10 of these being independent?
2. What are the physical dimensions of the energy-momentum tensor? Provide detailed explanation.
3. Make a list of the main properties of the energy-momentum tensor.
4. Show that the divergence of the energy-momentum tensor is zero (i.e. $T^{\mu\nu}_{;\nu} = 0$) assuming that the Field Equation is given as a postulate.
5. What is the energy-momentum tensor of electromagnetic field? What is the significance of this?
6. Show that the trace of the energy-momentum tensor of electromagnetic field is zero.
7. What is the state of the Ricci curvature tensor $R_{\mu\nu}$ and the Ricci curvature scalar R in the vacuum problems[133] of general relativity? What is the state of the curvature of spacetime in these problems? Comment on the result.
8. Justify the premise that the vanishing divergence of the energy-momentum tensor (i.e. $T^{\mu\nu}_{;\nu} = 0$) is a requirement (or an implication) for the conservation of the momentum 4-vector.
9. Show that the divergence of the energy-momentum tensor is zero, i.e. $T^{\mu\nu}_{;\nu} = 0$.
10. Justify the existence of vacuum problems and vacuum solutions in general relativity, i.e. why we should have curved spacetime at locations where $T_{\mu\nu} = 0$.
11. What is the significance of the fact that the divergence of the energy-momentum tensor is zero? Also, link this to the vanishing of the divergence of the Einstein tensor.
12. Make a comparison between the Einstein tensor **G** and the energy-momentum tensor **T**.
13. Discuss briefly the historical origin of the energy-momentum tensor and if it is a novelty of general relativity.
14. Comment on the following:
    $R_{ij} = 0$ (as well as $R = 0$) globally does not imply flatness (globally) from a geometric perspective but it implies flatness (globally) from a general relativistic perspective (according to the Field Equation and as required by special relativity). However, $R_{ij} = 0$ (as well as $R = 0$) locally does not imply flatness (locally) from a geometric perspective and it does not imply flatness (locally) from a general relativistic perspective.

---

[133] "Vacuum problem" means gravitational problem related to a region of spacetime where there is no matter or energy. Accordingly, vacuum solutions are solutions of vacuum problems.

Chapter 3
Formalism of General Relativity

The focus of this chapter is the formalism of general relativity which is represented by the Field Equation. All the formal results of this theory (as a gravity theory and not as a "General Theory") should be derived directly or indirectly from this equation possibly with the aid of some known physical and mathematical facts as well as certain assumptions and approximations. In the first section of this chapter we outline the background of the Field Equation and the rationale that is followed in developing this equation and applying it in general. This will provide a prior insight into the physical and mathematical framework of general relativity. This will also help to understand and appreciate this equation and its formal consequences which are highly abstract and difficult to comprehend. The subsequent sections will then investigate the Field Equation in its full form (with and without the cosmological constant) and in its linearized form. The chapter will be culminated by a brief investigation of the two main aspects of general relativity and its formalism (i.e. as a gravity theory and as a "General Theory").

We should remark that in general relativity we actually have two formalisms (which are independent of each other): a general-relativity-specific formalism represented by the Field Equation (and its direct consequences) and a special-relativity-specific formalism represented by the Lorentz transformations (and their direct consequences). In fact, the former formalism represents the gravitational component of the theory of general relativity while the latter formalism represents the local application of Lorentz mechanics in the spacetime of general relativity (where this spacetime is described and formalized by the former formalism) due to the local flatness of the spacetime according to general relativity. In other words, the former formalism creates the spacetime (whose structure incorporates gravity and its effects) while the latter formalism sets the physical rules that should be followed locally in this spacetime.[134] So, our investigation in the present chapter is about the general-relativity-specific formalism since the special-relativity-specific formalism belongs to the theory of special relativity and its literature and textbooks (see for example B4).

In this context, it is important to appreciate the above-indicated fact that on one side general relativity is a gravity theory that is based on the Field Equation, and on another side it is a "General Theory" that is based on generalizations of the application and invariance of physical laws in various types of reference frame and the nature of this application and invariance where the principles of general relativity (i.e. the principles of equivalence and invariance associated with the claim of local application of special relativity in the spacetime of general relativity) play the major role in the latter side.[135] Accordingly, the main focus of our investigation in the present chapter is the gravity theory and not the "General Theory" although the latter aspect of the theory will be briefly investigated in the final section, as indicated above.

3.1 Rationale of the Field Equation

In the following points we outline the rationale that is followed (within the theoretical framework of general relativity) in developing and applying the Field Equation:[136]
• We start from the principle of invariance which general relativity rests upon. According to this principle the laws of physics in general should be independent of the employed coordinate systems and hence if a

---

[134] Although these rules essentially come from special relativity (due to the local flatness of spacetime) they could be modified (due to their dependency on the frames and the transformations between them). In fact, this modification could be seen as a link between the two formalisms (and even as a link between the gravity theory and the "General Theory").
[135] In fact, other principles (see § 1.8.4) may also play a role.
[136] It is worth noting that in the development of the Field Equation (according to the above rationale) elements and principles that mainly belong to the "General Theory" may be employed although the Field Equation in itself represents the gravity theory. This can also be seen as an example of the links between the two sides of the theory.

3.1 Rationale of the Field Equation

physical law is valid in one frame it must be valid in all other frames. So, the privileged frames of special relativity (i.e. inertial frames) are discarded and replaced by a more general type of frames (to include non-inertial frames), and consequently the Minkowskian spacetime and its flat metric are replaced by the more general Riemannian space and metric.[137] Accordingly, the spacetime interval $d\sigma$, which is given in special relativity by $(d\sigma)^2 = \eta_{\mu\nu} dx^\mu dx^\nu$, will be given by:

$$(d\sigma)^2 = g_{\mu\nu} dx^\mu dx^\nu \tag{99}$$

where $g_{\mu\nu}$ is the covariant metric tensor of the spacetime (which is curved in general) while dx^μ and dx^ν are differentials of general coordinates.

- The metric $g_{\mu\nu}$ of the spacetime should be determined by the distribution of matter and energy in the spacetime, and this can be seen as an implication of the principle of metric gravity (see § 1.8.4).[138] Accordingly, the geometry of the "mathematical" spacetime (as represented by $G_{\mu\nu}$ which is ultimately determined by $g_{\mu\nu}$) should be linked to the "physical" spacetime (as represented by $T_{\mu\nu}$).
- The distribution of matter and energy is defined by the energy-momentum tensor $T^{\mu\nu}$ which in the spacetime of special relativity and with the absence of forces satisfies the relation $T^{\mu\nu}_{,\nu} = 0$ which originates from the conservation principles of energy and momentum and expresses the divergence-free nature of this tensor.[139] Now, in the spacetime of general relativity we are dealing with gravitational effects only (i.e. excluding any other force) and the gravity is accounted for by the space metric (i.e. it is not a force). Hence, by the principle of invariance the energy-momentum tensor $T^{\mu\nu}$ should also be divergence-free (as in the spacetime of special relativity) and hence it should satisfy the relation $T^{\mu\nu}_{;\nu} = 0$ (which is the curved-space version of the relation $T^{\mu\nu}_{,\nu} = 0$ which is the flat-space version).
- The divergence-free Einstein tensor **G** suggests a proportionality with the energy-momentum tensor **T** which is also divergence-free (as seen in the previous point). This should lead to the Field Equation $\mathbf{G} = \kappa \mathbf{T}$ where κ is a proportionality constant (see § 3.2; we should also refer to § 3.3 where the cosmological term may be seen as generalization from proportionality to linearity).
- The available physical setting of the problem (plus appropriate simplifications and assumptions) should lead to the determination of **T**. Employing the Field Equation with the use of the determined **T** should lead to the determination of **G** and hence the determination of the metric tensor which depicts the geometry of the spacetime.
- The determination of the metric tensor will lead to the determination of the geodesic trajectories (or world lines) in the spacetime as well as other required physical consequences and results that depend on the metric, and hence the problem of gravitation as a tempo-spatial phenomenon is solved (with the extraction of all its natural conclusions).

Problems

1. Outline the theoretical approach which general relativity (as represented by the Field Equation) follows in addressing the kinematics of gravity.
   Answer: The theoretical approach may be outlined as follows:
   - Matter and energy in the spacetime (represented by the energy-momentum tensor) shape the geometry of the spacetime.
   - The geometry of the spacetime (represented by the metric tensor which is embedded in the Einstein tensor) determines the geodesic and null geodesic paths which are the trajectories (or world lines) of massive and massless objects in the spacetime (noting that these objects should be free because these trajectories presumably represent the effect of gravity alone).

---

[137] This should imply discarding the privileged Lorentz transformations and replacing them by more general coordinate (or frame) transformations. We should note that the generalization to Riemannian space and metric requires more than the invariance principle and the independence of laws from the employed coordinate systems (because this relates to the nature of the coordinated space and not to the nature of the coordinate system). In fact, the equivalence principle (with its locality condition) or some other principle (see § 1.8.4) may be used to justify this generalization.

[138] In fact, some may consider this as an implication of the principle of equivalence.

[139] As noted earlier (see § 2.14), the energy-momentum tensor can be defined in the spacetime of special relativity but not as a source of gravity.

- The kinematical effects of gravitation are then described as free fall motion of massive and massless objects along these geodesic and null geodesic paths.

Exercises
1. What we mean by the "Rationale of the Field Equation" which is the title of this section?
2. Outline the transition from classical gravity to general relativistic gravity.
3. In the literature of general relativity we can find several attempts and methods for deriving the Field Equation from more fundamental principles in a rigorous way. Discuss this issue.

3.2 The Field Equation of General Relativity

As seen in § 2.13 and § 2.14, the Einstein tensor represents the geometric component of gravity while the energy-momentum tensor represents its physical component. We have also seen that the mathematical properties of these two tensors are similar, e.g. both are rank-2 symmetric tensors in 4D spacetime. So, it is logical that if we have to correlate the physics of gravity to the geometry of spacetime (as demanded by the rationale of general relativity) in an invariant form then we should find a formal tensorial relation between these two tensors. We have also noted that the divergence of both these tensors is zero, and hence it is logical to assume that these two tensors are proportional, that is:

$$\mathbf{G} = \kappa \mathbf{T} \tag{100}$$

where $\mathbf{G}$ is the Einstein tensor, κ is a constant scalar (i.e. proportionality factor) and $\mathbf{T}$ is the energy-momentum tensor. Using the definition of the Einstein tensor and employing the indicial notation of tensors, the last equation can be written in covariant form as:

$$G_{\mu\nu} = \kappa T_{\mu\nu} \tag{101}$$

$$R_{\mu\nu} - \frac{1}{2} g_{\mu\nu} R = \kappa T_{\mu\nu} \tag{102}$$

$$R_{\mu\nu} - \frac{1}{2} g_{\mu\nu} R = -\frac{8\pi G}{c^4} T_{\mu\nu} \tag{103}$$

where line 2 is based on the definition of Einstein tensor (see Eq. 94) while line 3 is based on the expression of κ which is determined by the correspondence principle (i.e. convergence of general relativity to Newtonian gravity in the classical limit) as will be justified later (see § 5.1). It is important to note that G here is the gravitational constant and not the trace of Einstein tensor. We should also note that depending on the convention about the tensors in the Field Equation and the sign of κ, a minus sign may be added to the right hand side, i.e. $G_{\mu\nu} = -\kappa T_{\mu\nu}$.

As we will see, the main objective of formulating and solving the Field Equation in any particular physical situation is to find the metric of the spacetime which represents the geometric properties of the spacetime and hence it contains all the required information about gravitation according to the general relativistic doctrine. However, different strategies are followed in formulating and solving the Field Equation where these strategies mainly depend on the physical and mathematical considerations of the concerned problem as well as practical and procedural factors.

Problems
1. Compare the essence of the Newtonian gravity as a field theory represented by the equation $\nabla \cdot \mathbf{g} = -4\pi G \rho$ (see § 1.5) to the essence of the gravity of general relativity represented by the Field Equation.
   Answer: The Newtonian formulation correlates the gravitational field $\mathbf{g}$ to the mass density ρ as a source for the field. Similarly, the general relativistic formulation correlates the "gravitational field" represented by the spacetime geometry (as embedded in $\mathbf{G}$ and ultimately in the metric tensor) to the mass-energy source (as embedded in $\mathbf{T}$). Hence, the two theories (as represented by their field formulations) are similar in essence although this similarity is not seen when we compare the Newtonian primitive formulation (represented by the form $f = \frac{G m_1 m_2}{d^2}$) to the field formulation of general relativity. An important difference between the two formulations is that the Newtonian formulation is purely

spatial (in 3D Euclidean space) while the general relativistic formulation is tempo-spatial (in 4D Riemannian spacetime). Also, the source of gravitation in the Newtonian formulation is mass while the source of gravitation in the general relativistic formulation is mass-energy. We should also note that the classical formulation is scalar and hence it has a single component while the general relativistic formulation is tensorial and hence it has 16 components (10 of which are independent).

2. What are the physical dimensions of κ?
   Answer: We have $|\kappa| = \frac{8\pi G}{c^4}$ where the minus sign is dropped in evaluating the physical dimensions. Now, 8π is dimensionless, G (which is the gravitational constant) has the dimensions of force (which is mass times acceleration) times area per mass squared (e.g. $\mathrm{N\,m^2\,kg^{-2}} = \mathrm{kg^{-1}m^3\,s^{-2}}$) while c^4 has the dimensions of length^4 divided by time^4 (e.g. $\mathrm{m^4\,s^{-4}}$). Hence, κ has the dimensions of time squared per length per mass (e.g. $\mathrm{kg^{-1}\,m^{-1}\,s^2}$).

3. Verify that the Field Equation is dimensionally consistent and comment on the result.
   Answer: We have $\mathbf{G} = \kappa \mathbf{T}$. Now, if we use SI units to represent the physical dimensions then $\mathbf{G}$ has the dimensions of $\mathrm{m^{-2}}$ (see § 2.13) while $\kappa \mathbf{T}$ has the dimensions of $\mathrm{kg^{-1}\,m^{-1}\,s^2} \times \mathrm{kg\,m^{-1}\,s^{-2}} = \mathrm{m^{-2}}$ (see the previous problem and § 2.14). Hence, both sides of the Field Equation have the dimensions of $\mathrm{m^{-2}}$ and therefore the Field Equation is dimensionally consistent.
   Comment: the fact that the Field Equation has the physical dimensions of reciprocal area (e.g. $\mathrm{m^{-2}}$) which are the same as the physical dimensions of curvature reveals the nature of the Field Equation as a curvature equation. In fact, this also reveals the nature of general relativity as a geometric theory about the curvature of spacetime as a model for physical gravity.

4. In the 4D spacetime of general relativity the Riemann-Christoffel curvature tensor $R_{\mu\nu\rho\omega}$ has 20 independent components while there are only 10 Field Equations. What are the implications of this?
   Answer: We can claim that there are two main implications:
   • It is possible to have vacuum solutions (i.e. curved spacetime and gravitational field in empty regions) since in this case we have only 10 vacuum Field Equations $R_{\mu\nu} = 0$ (since $R_{\mu\nu}$ has only 10 independent components) while we have twice this number of $R_{\mu\nu\rho\omega}$ components. In other words, the vanishing of the 10 components (or constraints) on $R_{\mu\nu}$ (i.e. $R_{\mu\nu} = 0$) is consistent with the non-vanishing of the 20 components (or constraints) on $R_{\mu\nu\rho\omega}$ (noting that the non-vanishing of $R_{\mu\nu\rho\omega}$ implies curved spacetime and the existence of gravitational fields since a space is flat *iff* $R_{\mu\nu\rho\omega} = 0$).
   • It is possible that some of the spacetime curvature information are not included in the general relativistic gravitational formulation since 10 Field Equations are not sufficient in general to incorporate the full curvature information contained in $R_{\mu\nu\rho\omega}$. In other words, gravity in general relativity is correlated to only part of the curvature of spacetime and not to the entire curvature of spacetime.

Exercises

1. How many component equations the tensorial Field Equation represents, and how many of these component equations are independent?
2. What "solving the tensorial Field Equation" means mathematically and physically?
3. What is the classical equation of gravity that corresponds to the tensorial Field Equation of general relativity? How many component equations the classical equation of gravity represents?
4. Show that $G = \kappa T$ and $R = -\kappa T$ where G, T, R are the traces of $G^\mu_\nu, T^\mu_\nu, R^\mu_\nu$.
5. Show that $R_{\mu\nu} = \kappa \left(T_{\mu\nu} - \frac{1}{2} g_{\mu\nu} T \right)$.
6. Show that if a region of spacetime is void of matter and energy then the Field Equation in that region becomes: $R_{\mu\nu} = 0$. What is the significance of this?
7. Does the condition $R_{\mu\nu} = 0$ at a given region of spacetime means that the given region is flat? Justify your answer.
8. Show that $G_{\mu\nu} = 0$ iff $T_{\mu\nu} = 0$ (where these conditions apply locally or globally).
9. Show that $R_{\mu\nu} = 0$ iff $T_{\mu\nu} = 0$ (where these conditions apply locally or globally).
10. Compare the relations that we derived in § 2.13 (e.g. $G = -R$) with the relations that we obtained in this section, e.g. $R_{\mu\nu} = \kappa \left(T_{\mu\nu} - \frac{1}{2} g_{\mu\nu} T \right)$.
11. Can you see a potential problem in the Field Equation from the perspective of the fact that the Einstein tensor contains less curvature information than the Riemann-Christoffel curvature tensor?

12. Discuss the principle of geodesic motion in general relativity and its relation to the Field Equation.
13. Analyze the significance of the vanishing of the energy-momentum tensor locally and globally within the context of the Field Equation.
14. Consider obtaining the Field Equation as a generalization of the classical gravity with the help of special relativity and some general principles (e.g. principle of invariance, principle of equivalence, principle of metric gravity, etc.).

3.3 The Field Equation with Cosmological Constant

A cosmological term may be added to the Field Equation and hence it becomes:

$$G_{\mu\nu} + \Lambda g_{\mu\nu} = \kappa T_{\mu\nu} \qquad \text{or} \qquad R_{\mu\nu} - \left(\frac{1}{2}R - \Lambda\right)g_{\mu\nu} = \kappa T_{\mu\nu} \qquad (104)$$

where Λ is the cosmological constant. The cosmological term was introduced by Einstein using cosmological arguments to make the Universe static, i.e. not expanding or shrinking.[140] However, the cosmological term is still in use in modern cosmology (although for other purposes) even with non-static cosmological models. The cosmological term may be interpreted as representing the contribution of the so-called "dark energy" and hence the cosmological term may be considered as part of the right hand side, i.e. part of the gravity source as represented initially by the energy-momentum tensor.[141] Accordingly, the Field Equation will be written as:

$$R_{\mu\nu} - \frac{1}{2}g_{\mu\nu}R = \kappa T_{\mu\nu} - \Lambda g_{\mu\nu} \qquad \text{or} \qquad R_{\mu\nu} - \frac{1}{2}g_{\mu\nu}R = \kappa\left(T_{\mu\nu} - \frac{\Lambda}{\kappa}g_{\mu\nu}\right) \qquad (105)$$

In fact, the cosmological term may be absorbed in the $\kappa T_{\mu\nu}$ term and hence the term $\kappa T_{\mu\nu}$ is understood to incorporate the effect of the cosmological term, i.e. $T_{\mu\nu}$ includes the effects of the $-\frac{\Lambda}{\kappa}g_{\mu\nu}$ term in the last equation. However, it should be noted that in this case the cosmological term is still independent of other sources of gravitation and hence the vanishing of mass and energy in the spacetime does not imply the vanishing of cosmological term, i.e. $T_{\mu\nu}$ could stand for the energy density represented by the cosmological term alone.

We should remark that dark energy (which is supposedly represented by the cosmological term) may also be labeled as "vacuum energy" although vacuum energy may be seen as only one possible form or source of dark energy. The use of "vacuum energy" to refer to dark energy originates from the premise that the source of dark energy (or one of its sources) is the alleged "quantum mechanical vacuum fluctuations". In fact, the disastrous failure of the quantum mechanical calculations to provide a sensible quantitative estimate of this alleged vacuum energy should discredit (at least for the time being) any attempt to explain the alleged dark energy by the alleged vacuum fluctuations. There may also be another reason for the "vacuum" label that is it is associated with the ground state of the Universe, i.e. its vacuum state. We should also remark that because of its cosmological origin and significance (as its name suggests) and because we have no primary interest in cosmology in the present book, we do not go through the details of the cosmological term and constant. The interested reader should therefore consult the relevant literature of relativistic cosmology.

Finally, the reader should note that the use of the Field Equation with the cosmological constant term (i.e. Eq. 105) should be restricted to the cosmological problems and hence in all non-cosmological problems the original form of the Field Equation (i.e. Eq. 101) should be used. This is because the contribution of the cosmological term in non-cosmological problems is completely negligible due to the extreme smallness of the cosmological constant (assuming it is not zero). This should also be justified from a purely conceptual and theoretical perspective as can be easily inferred from the "cosmological" label that is attached to this term and this constant.

---

[140] It appeared later that this proposed universe is unstable (i.e. it deviates from its static state by any perturbation and hence it becomes expanding or shrinking).

[141] In fact, this move from the left hand side to the right hand side may be seen as converting the cosmological term from being a geometric factor to be a physical factor.

Problems

1. What are the physical dimensions of the cosmological constant?
   Answer: Due to the requirement of dimensional consistency, all the terms of physical equations should have the same physical dimensions. Now, if we remember that $g_{\mu\nu}$ is dimensionless (see § 2.5) and the dimensions of the other terms of the Field Equation are reciprocal area (see § 3.2) then from Eq. 104 we can conclude that the physical dimensions of the cosmological constant are reciprocal area (and hence it essentially represents curvature).

2. Verify that the dark energy density term (i.e. $-\frac{\Lambda}{\kappa}g_{\mu\nu}$) has the physical dimensions of energy density (like $T_{\mu\nu}$).
   Answer: $g_{\mu\nu}$ is dimensionless, Λ has the dimensions of reciprocal area (e.g. m^{-2}) according to the previous problem, and κ has the dimensions of time squared per length per mass (e.g. kg^{-1} m^{-1} s^2) according to problem 2 of § 3.2. Hence, the term $-\frac{\Lambda}{\kappa}g_{\mu\nu}$ has the physical dimensions of m$^{-2}/\left(\text{kg}^{-1}\,\text{m}^{-1}\,\text{s}^2\right) = $ kg m^{-1} s^{-2} = J/m^3, i.e. energy density.

3. Comment on the physical and epistemological significance of the cosmological term.
   Answer: The cosmological term represents the physical effects of the so-called dark energy (or vacuum energy where vacuum energy may be seen as an instance or potential source of dark energy). In our view, dark energy is a metaphysical concept that was introduced into modern physics to address certain limitations in its theoretical structure. In fact, it is just one example of the numerous metaphysical objects and concepts that infest modern physics and cosmology, and the main credit for this belongs to general relativity. These metaphysical entities usually originate from highly hypothetical considerations and contemplations that are totally detached from physical reality and hence they are not legitimate in physical science although they may be legitimate in other disciplines (such as religion or philosophy or pure mathematics).

Exercises

1. Using the Field Equation with the cosmological term, show that $G = \kappa T - 4\Lambda$ and $R = -\kappa T + 4\Lambda$.
2. Show that $R_{\mu\nu} = \kappa\left(T_{\mu\nu} - \frac{1}{2}g_{\mu\nu}T\right) + \Lambda g_{\mu\nu}$ and comment on its significance.
3. What is the relation between the cosmological term and dark energy?
4. Discuss dark energy and assess it as a physical or metaphysical concept. Also, discuss and assess dark matter from the same perspective.
5. Outline the characteristics of the cosmological term $\Lambda g_{\mu\nu}$.
6. Outline the characteristics of the cosmological constant Λ.
7. Discuss briefly some nonsensical consequences and implications that are commonly associated with the cosmological term and constant.
8. Discuss the implication of the presumed existence of dark energy on special relativity.

3.4 The Linearized Field Equation

In this section we briefly investigate the linearized form of the Field Equation which is used as a good approximation to the full (non-linear) Field Equation in the case of weak gravitational fields where the spacetime is pseudo- or quasi-Minkowskian and hence it deviates slightly from the flat geometry of the authentic Minkowskian spacetime.[142] The importance of the linearized form is not only because of its status as a practically-useful approximation (and hence it can save valuable effort when high accuracy is not required) but it is also because of its use in some applications of general relativity (e.g. gravitational waves) where formulations based on the full Field Equation are prohibitive or unavailable. In this regard, the linearized form serves like numerical methods and classical gravity in providing reasonable solutions. In fact, the linearized form can also be used in some theoretical arguments and contentions and hence it is useful to have regardless of any other reason.

In the following points we outline the derivation of the linearized form of the Field Equation. However, we should remark that we omitted some detailed technicalities and justifications (as well as background

---

[142] In fact, gravity in the weak field regime can be seen as a minor curvature in spacetime and as a Lorentzian field in a flat Minkowskian spacetime. From the latter perspective, it is like electromagnetism in special relativity.

3.4 The Linearized Field Equation

materials) to keep inline with the intended size, style and level of the book. In fact, this section is mainly included for the purpose of completeness and hence the reader can skip it with no loss of continuity (although some of the contents in this section are used casually as background materials later in the book; see for example § 5.1).

• We note first that all the approximations in the derivation of the linearized form are to the first order and hence any non-linear function will be linearized by discarding higher order dependencies. In fact, this is the essence of the linearization process which the linearized form is based on. However, we should remark that the linearization is solely related to the metric and the geometric entities that are based on the metric, i.e. it is related to the geometry of the spacetime exclusively.

• As indicated above, the spacetime in the linearized form is quasi-Minkowskian and hence in a first order approximation the covariant metric tensor $g_{\mu\nu}$ can be expressed as:

$$g_{\mu\nu} = \eta_{\mu\nu} + \varepsilon_{\mu\nu} \tag{106}$$

where $\eta_{\mu\nu}$ is the covariant Minkowski metric tensor and $\varepsilon_{\mu\nu}$ is a perturbation tensor representing first order deviation from flatness such that $|\varepsilon_{\mu\nu}| \ll 1$ and where for the sake of simplicity we used (and will continue to use) exact equality to represent approximate equality to first order.[143] The symmetry of $g_{\mu\nu}$ and $\eta_{\mu\nu}$ implies the symmetry of $\varepsilon_{\mu\nu}$ (see § 2.2).

• By a similar argument to the argument in the previous point we can conclude that in a first order approximation the contravariant metric tensor $g^{\mu\nu}$ can be expressed as:

$$g^{\mu\nu} = \eta^{\mu\nu} + \chi^{\mu\nu} \tag{107}$$

where $\eta^{\mu\nu}$ is the contravariant Minkowski metric tensor and $\chi^{\mu\nu}$ is a perturbation tensor representing first order deviation from flatness such that $|\chi^{\mu\nu}| \ll 1$. It can be shown (see exercise 2) that $\chi^{\mu\nu} = -\varepsilon^{\mu\nu}$ and hence:

$$g^{\mu\nu} = \eta^{\mu\nu} - \varepsilon^{\mu\nu} \tag{108}$$

The symmetry of $g^{\mu\nu}$ and $\eta^{\mu\nu}$ implies the symmetry of $\varepsilon^{\mu\nu}$ (see § 2.2).

• For the purpose of lowering and raising indices of minute tensorial quantities we can use $\eta_{\mu\nu}$ and $\eta^{\mu\nu}$ as index lowering and index raising operators instead of $g_{\mu\nu}$ and $g^{\mu\nu}$. The justification is that in the first order approximation the product of $\varepsilon_{\mu\nu}$ and $\varepsilon^{\mu\nu}$ with the minute tensorial quantity should be discarded because it is of higher order. For example, if A^ν is a minute tensorial quantity then we have:

$$g_{\mu\nu}A^\nu = (\eta_{\mu\nu} + \varepsilon_{\mu\nu})A^\nu = \eta_{\mu\nu}A^\nu + \varepsilon_{\mu\nu}A^\nu = \eta_{\mu\nu}A^\nu = A_\mu \tag{109}$$

where in the third equality we discarded $\varepsilon_{\mu\nu}A^\nu$ because it is of higher order. Similarly, if B_ν is a minute tensorial quantity then we have:

$$g^{\mu\nu}B_\nu = (\eta^{\mu\nu} - \varepsilon^{\mu\nu})B_\nu = \eta^{\mu\nu}B_\nu - \varepsilon^{\mu\nu}B_\nu = \eta^{\mu\nu}B_\nu = B^\mu \tag{110}$$

where in the third equality we discarded $\varepsilon^{\mu\nu}B_\nu$ because it is of higher order. Accordingly, for the purpose of index lowering and index raising of minute tensorial quantities we have:

$$g_{\mu\nu} = \eta_{\mu\nu} \quad \text{and} \quad g^{\mu\nu} = \eta^{\mu\nu} \quad \text{(for lowering and raising only)} \tag{111}$$

• The Field Equation of general relativity is given by:

$$R_{\mu\nu} - \frac{1}{2}g_{\mu\nu}R = \kappa T_{\mu\nu} \tag{112}$$

As indicated above, the linearization process is related to the metric and its companions. In other words, to obtain the linearized Field Equation we need to find the linearized forms of the geometric quantities in

---

[143] In fact, $\varepsilon_{\mu\nu}$ may be considered as Lorentzian tensor but not general tensor. We should also note that we need to assume the smallness of the derivatives of the components of $\varepsilon_{\mu\nu}$ as well (and hence their products are negligible according to the first order approximation) although this may be understood from our previous explanation.

3.4 The Linearized Field Equation

the Field Equation (which are $R_{\mu\nu}$ and R, noting that $g_{\mu\nu}$ is already linearized) and insert these linearized forms into Eq. 112.

• To obtain the linearized form of $R_{\mu\nu}$ and R we need first to obtain the linearized form of the Christoffel symbols, that is:

$$\begin{aligned}
\Gamma^{\omega}_{\mu\nu} &= \frac{g^{\omega\alpha}}{2}\left(\partial_{\nu}g_{\mu\alpha} + \partial_{\mu}g_{\nu\alpha} - \partial_{\alpha}g_{\mu\nu}\right) \\
&= \frac{g^{\omega\alpha}}{2}\left(\partial_{\nu}\varepsilon_{\mu\alpha} + \partial_{\mu}\varepsilon_{\nu\alpha} - \partial_{\alpha}\varepsilon_{\mu\nu}\right) \\
&= \frac{\eta^{\omega\alpha}}{2}\left(\partial_{\nu}\varepsilon_{\mu\alpha} + \partial_{\mu}\varepsilon_{\nu\alpha} - \partial_{\alpha}\varepsilon_{\mu\nu}\right) \\
&= \frac{1}{2}\left(\partial_{\nu}\varepsilon_{\mu}{}^{\omega} + \partial_{\mu}\varepsilon_{\nu}{}^{\omega} - \partial^{\omega}\varepsilon_{\mu\nu}\right)
\end{aligned} \quad (113)$$

where line 2 is because $\eta_{\mu\nu}$ is constant, in line 3 we use Eq. 111, and in line 4 we raise the index α by $\eta^{\omega\alpha}$.[144] We note that ∂^{ω} is the *contravariant* partial differential operator.

• $R_{\mu\nu}$ is given by (see Eq. 89):

$$R_{\mu\nu} = \partial_{\nu}\Gamma^{\alpha}_{\mu\alpha} - \partial_{\alpha}\Gamma^{\alpha}_{\mu\nu} + \Gamma^{\beta}_{\mu\alpha}\Gamma^{\alpha}_{\beta\nu} - \Gamma^{\beta}_{\mu\nu}\Gamma^{\alpha}_{\beta\alpha} \quad (114)$$

Now, the last two terms are second order in ε's (or their derivatives) and hence the linearized $R_{\mu\nu}$ is given by:

$$\begin{aligned}
R_{\mu\nu} &= \partial_{\nu}\Gamma^{\alpha}_{\mu\alpha} - \partial_{\alpha}\Gamma^{\alpha}_{\mu\nu} \\
&= \frac{1}{2}\left(\partial_{\nu}\partial_{\alpha}\varepsilon^{\alpha}_{\mu} + \partial_{\nu}\partial_{\mu}\varepsilon^{\alpha}_{\alpha} - \partial_{\nu}\partial^{\alpha}\varepsilon_{\mu\alpha}\right) - \frac{1}{2}\left(\partial_{\alpha}\partial_{\nu}\varepsilon^{\alpha}_{\mu} + \partial_{\alpha}\partial_{\mu}\varepsilon^{\alpha}_{\nu} - \partial_{\alpha}\partial^{\alpha}\varepsilon_{\mu\nu}\right) \\
&= \frac{1}{2}\left(\partial_{\nu}\partial_{\mu}\varepsilon^{\alpha}_{\alpha} - \partial_{\nu}\partial^{\alpha}\varepsilon_{\mu\alpha} - \partial_{\alpha}\partial_{\mu}\varepsilon^{\alpha}_{\nu} + \partial_{\alpha}\partial^{\alpha}\varepsilon_{\mu\nu}\right) \\
&= \frac{1}{2}\left(\partial_{\nu}\partial_{\mu}\varepsilon^{\alpha}_{\alpha} - \partial_{\nu}\partial^{\alpha}\varepsilon_{\mu\alpha} - \partial_{\alpha}\partial_{\mu}\varepsilon^{\alpha}_{\nu} + \Box^{2}\varepsilon_{\mu\nu}\right) \\
&= \frac{1}{2}\left(\partial_{\nu}\partial_{\mu}\varepsilon - \partial_{\nu}\partial^{\alpha}\varepsilon_{\mu\alpha} - \partial_{\alpha}\partial_{\mu}\varepsilon^{\alpha}_{\nu} + \Box^{2}\varepsilon_{\mu\nu}\right) \\
&= \frac{1}{2}\left(\partial_{\nu}\partial_{\mu}\varepsilon - \partial_{\nu}\partial_{\alpha}\varepsilon^{\alpha}_{\mu} - \partial_{\alpha}\partial_{\mu}\varepsilon^{\alpha}_{\nu} + \Box^{2}\varepsilon_{\mu\nu}\right)
\end{aligned} \quad (115)$$

where in line 2 we use Eq. 113, in line 3 we use the commutativity of the partial differential operators, in line 4 we use the definition of the d'Alembertian operator (i.e. $\Box^{2} \equiv \partial_{\alpha}\partial^{\alpha}$; see B4 and exercise 15 of § 2.2), and in line 5 ε represents the trace of ε^{ν}_{μ}. The second term in the last line is justified by the following:

$$\partial^{\alpha}\varepsilon_{\mu\alpha} = \left(\eta^{\alpha\beta}\partial_{\beta}\right)\varepsilon_{\mu\alpha} = \partial_{\beta}\left(\eta^{\alpha\beta}\varepsilon_{\mu\alpha}\right) = \partial_{\beta}\left(\varepsilon^{\beta}_{\mu}\right) = \partial_{\alpha}\varepsilon^{\alpha}_{\mu} \quad (116)$$

where the constancy of $\eta^{\alpha\beta}$ allows it to pass the partial differential operator (i.e. they commute).

• We obtain the linearized form of R by contracting the linearized form of $R_{\mu\nu}$ (after raising its first index), that is:

$$\begin{aligned}
R &= R^{\nu}_{\nu} \\
&= g^{\mu\nu}R_{\mu\nu} \\
&= \eta^{\mu\nu}R_{\mu\nu} \\
&= \frac{\eta^{\mu\nu}}{2}\left(\partial_{\nu}\partial_{\mu}\varepsilon - \partial_{\nu}\partial_{\alpha}\varepsilon^{\alpha}_{\mu} - \partial_{\alpha}\partial_{\mu}\varepsilon^{\alpha}_{\nu} + \Box^{2}\varepsilon_{\mu\nu}\right)
\end{aligned}$$

---

[144] The step in line 3 may seem unnecessary. However, it should be needed for instance for justifying the commutativity of the metric with the partial differential operators (see for example Eq. 116). Moreover, it is more appropriate considering that $\varepsilon_{\mu\nu}$ is essentially a perturbation tensor on a Minkowskian spacetime background.

$$
\begin{aligned}
&= \frac{1}{2}\left(\Box^2\varepsilon - \partial_\nu\partial_\alpha\varepsilon^{\nu\alpha} - \partial_\alpha\partial_\mu\varepsilon^{\mu\alpha} + \Box^2\varepsilon_\mu^\mu\right)\\
&= \frac{1}{2}\left(\Box^2\varepsilon - \partial_\alpha\partial_\mu\varepsilon^{\mu\alpha} - \partial_\alpha\partial_\mu\varepsilon^{\mu\alpha} + \Box^2\varepsilon\right)\\
&= \Box^2\varepsilon - \partial_\alpha\partial_\mu\varepsilon^{\mu\alpha}\\
&= \Box^2\varepsilon - \partial_\alpha\partial_\beta\varepsilon^{\beta\alpha}
\end{aligned}
\qquad (117)
$$

where in line 3 we use Eq. 111, in line 4 we substitute from Eq. 115, in line 5 we use $\eta^{\mu\nu}$ as an index raising operator (noting that $\eta^{\mu\nu}\partial_\nu\partial_\mu\varepsilon = \partial_\nu\partial^\nu\varepsilon \equiv \Box^2\varepsilon$), and in line 6 we relabel the dummy indices in the second term (noting the symmetry of $\varepsilon^{\nu\alpha}$ or the commutativity of the partial differential operators).[145]

- On substituting from Eqs. 115, 106 and 117 into Eq. 112 we get:

$$
\frac{1}{2}\left(\partial_\nu\partial_\mu\varepsilon - \partial_\nu\partial_\alpha\varepsilon_\mu^\alpha - \partial_\alpha\partial_\mu\varepsilon_\nu^\alpha + \Box^2\varepsilon_{\mu\nu}\right) - \frac{1}{2}\left(\eta_{\mu\nu} + \varepsilon_{\mu\nu}\right)\left(\Box^2\varepsilon - \partial_\alpha\partial_\beta\varepsilon^{\beta\alpha}\right) = \kappa T_{\mu\nu} \qquad (118)
$$

$$
\left(\partial_\nu\partial_\mu\varepsilon - \partial_\nu\partial_\alpha\varepsilon_\mu^\alpha - \partial_\alpha\partial_\mu\varepsilon_\nu^\alpha + \Box^2\varepsilon_{\mu\nu}\right) - \eta_{\mu\nu}\left(\Box^2\varepsilon - \partial_\alpha\partial_\beta\varepsilon^{\beta\alpha}\right) = 2\kappa T_{\mu\nu} \qquad (119)
$$

$$
\Box^2\varepsilon_{\mu\nu} + \partial_\nu\partial_\mu\varepsilon - \partial_\nu\partial_\alpha\varepsilon_\mu^\alpha - \partial_\alpha\partial_\mu\varepsilon_\nu^\alpha - \eta_{\mu\nu}\left(\Box^2\varepsilon - \partial_\alpha\partial_\beta\varepsilon^{\beta\alpha}\right) = 2\kappa T_{\mu\nu} \qquad (120)
$$

where in line 2 we discard the higher order terms, i.e. $\varepsilon_{\mu\nu}\left(\Box^2\varepsilon - \partial_\alpha\partial_\beta\varepsilon^{\beta\alpha}\right)$. The last equation is the linearized form of the Field Equation.

Exercises

1. The linearization process is solely related to the metric and the geometric entities that are based on the metric. Why?
2. Show that $g^{\mu\nu} = \eta^{\mu\nu} - \varepsilon^{\mu\nu}$.
3. Assess the reliability of the linearized form of the Field Equation.
4. What is the significance of linearizing the Field Equation?

3.5 General Relativity as Gravity Theory and as General Theory

As indicated already repeatedly, general relativity is a "General Theory" as well as a "Gravity Theory". In the following points we discuss this issue and highlight its main implications and significance:

- The formalism of the "Gravity Theory" is represented by the Field Equation and its consequences, while the formalism of the "General Theory" is represented by the combined formalism of the "Gravity Theory" plus the formalism of special relativity. Accordingly, the "General Theory" consists of two independent formalisms: gravitational formalism and special relativistic formalism. The role of the gravitational formalism is to set the spacetime and determine its geometric properties and consequences (i.e. metric, geodesics, etc.), while the role of the special relativistic formalism is to set the physical rules that should take place locally in the spacetime.
- The fundamental principle which the "Gravity Theory" is based on is the weak equivalence principle (plus the principle of metric gravity which claims that gravity is a geometric attribute of the spacetime whose geometry is determined by the presence of matter), while the fundamental principle which the "General Theory" is based on is the strong equivalence principle.
- The independence of the gravitational and special relativistic formalisms of each other makes the validity and invalidity of these formalisms (as well as the evidence for and against them) independent, i.e. the validity or invalidity of one of these formalisms does not imply the validity or invalidity of the other formalism.
- Although there is some evidence for the validity of the "Gravity Theory", there is no evidence for the validity of the "General Theory" (in fact, there could be evidence against the "General Theory" since the strong equivalence principle is not valid in general). This is inline with the previous point since the validity of the "Gravity Theory" does not imply the validity of the "General Theory".

Exercises

---

[145] Again, the step in line 3 may seem unnecessary (but it is needed as before).

3.5 General Relativity as Gravity Theory and as General Theory

1. Outline the essence of the theory of general relativity as a "General Theory" that has gravitational and non-gravitational formalisms.
2. Give examples of the features of general relativity that characterize its nature as a "General Theory".
3. How to generalize a special relativistic law to become a general relativistic law?
4. Analyze and assess the procedure proposed in the previous exercise.

Chapter 4
Solutions of the Field Equation

The tensorial Field Equation consists of terms of rank-2 symmetric tensors in 4D spacetime. Each one of these terms has 10 independent components and hence the tensorial Field Equation represents 10 independent equations where each one of these equations is a non-linear partial differential equation. Accordingly, the solution of the Field Equation means solving 10 coupled non-linear partial differential equations. So, it is no surprise that the Field Equation is notorious for its complexity and difficulty to solve especially analytically. Hence, there are very few analytical solutions to the Field Equation related to simple cases, and therefore most solutions (if they exist at all) of general relativity problems are obtained computationally with the aid of numerical methods and computers. Approximate solutions may also be obtained from other inexact methods such as linearization (see § 3.4) and following semi-classical approach (where part of the problem is formulated classically). In fact, even the exact analytical solutions are usually based on a number of approximations and compromising assumptions (at least in their application).

The main task in solving general relativistic problems is to find the appropriate metric tensor that corresponds to the particular energy-momentum tensor that represents the physical situation of the problem. Ideally, we start from a certain physical setting (mass, energy, momentum, etc.) and we are required to find the spacetime geometry that corresponds to that physical setting.[146] On obtaining the geometry of the spacetime, as represented by the metric tensor, the geodesic paths (which describe the trajectories of massive and massless free objects in spacetime) are then obtained from the geodesic equation (see § 2.9.5). Apart from obtaining the geodesic paths, many other consequences and results can also be obtained directly or indirectly from the derived metric, as we will see later (refer for example to § 6, § 7 and § 8).

In the next sections we investigate a number of analytical solutions to the Field Equation. In fact, these represent a sample of the metrics that can be obtained from solving the Field Equation in particular physical situations. The main purpose of presenting these solutions is to have a feeling of the theory and its formulation and application. Therefore, the description of these solutions is very brief except the most important one of these which is the Schwarzschild solution where we provide derivation and rather detailed investigation because of its simplicity and versatile applicability as well as being a typical case study for representing and featuring the solutions of general relativistic problems. The Schwarzschild solution will also be used for developing other formal aspects of the theory that depend on the metric of the spacetime (see for example § 5.2, § 6.3, § 6.4, § 8.1 and § 8.2).

We should remark that being a mathematically valid solution to the Field Equation does not imply being a physically valid solution. In other words, being a valid mathematical solution is a necessary but not sufficient condition for being a valid physical solution.[147] Accordingly, we may obtain a mathematically legitimate solution which is physically nonsensical or unrealistic. Therefore, any solution that is obtained from solving the Field Equation should be checked meticulously for physical validity and sensibility. In fact, the checks for physical validity and sensibility are more required in general relativity than in most other theories due (among other reasons) to its exceptional mathematical and theoretical sophistication (which makes mathematical artifacts and theoretical fantasies more common) and the use of curved spacetime paradigm (which introduces further complications such as the duality of metrical and coordinate quantities

---

[146] As we will see in the method that we follow in finding the solution of the Schwarzschild problem (see § 4.1.1) the above-described approach is a rather ideal recipe for solving the Field Equation and hence it is not followed in most cases. We should also refer the reader to § 5.1 for another simple (one component) application of the Field Equation that represents a different solution approach and strategy which can help the reader to understand and appreciate how the solutions of general relativistic problems are obtained.

[147] In fact, this is true in general although it should be more obvious in general relativity.

as we will see in § 6). We should also note that all the methods and strategies for solving the Field Equation (some of which will be outlined in the questions) have serious limitations, breakdowns and traps and this should put more emphasis on the need for verifications and tests for physical sensibility and validity (apart from the required ultimate verification and validation by experimental and observational evidence).

Problems
1. Why is it very difficult to obtain a solution for the Field Equations?
   Answer: There are several reasons such as:
   • The Field Equations are highly non-linear and hence exact analytical solutions are generally unavailable. Also, the principle of superposition cannot be used while the perturbation techniques become questionable.
   • The tensorial Field Equation represents 10 independent equations. Therefore, obtaining a solution means solving a system of 10 simultaneous second order partial differential equations which is very difficult (if not impossible).
   • The Field Equations are tensor equations in 4D spacetime and hence they involve many variables, components, indices, etc. and that makes them very complicated and confusing. Also, tensor calculus notations and techniques in such a space are highly sophisticated and messy.
   • In realistic physical situations it is difficult to formulate the physical ingredient of the Field Equations which is the energy-momentum tensor. Hence, a number of approximations, compromises and sophisticated tackling strategies are usually required to make the problem solvable.
2. Explain why the Field Equations of general relativity are non-linear.
   Answer: Because the Einstein tensor is a non-linear function of the metric (refer to § 2.13). Non-linearity may also affect the energy-momentum tensor.

Exercises
1. We may describe a solution of a scientific problem as "ideal solution". Discuss this issue and indicate the approximation methods that are generally used in science to obtain second best solutions.
2. Outline some of the approximation methods that are used to overcome the technical difficulties in solving general relativistic problems.
3. What "solving the Field Equations" means?
4. Outline the steps that should be followed in formulating and solving gravity problems according to general relativity.
5. Outline some alternative tackling strategies that are used in formulating and solving general relativistic problems.

4.1 Schwarzschild Solution

The Schwarzschild solution (or metric) is one of the simplest analytical solutions to the Field Equations and it is the first solution to these equations from a historical perspective. The solution is used to describe the geometry of spacetime around various types of astronomical objects such as stars and planets and hence it provides a basis for the solution of many astronomical problems. It also provides a basis for many predictions and results related for example to the gravitational time dilation (see § 6.3.3) or the gravitational light bending (see § 8.2) or the physics of static classic black holes (see § 8.7) and therefore it is one of the most important solutions in general relativity with many instantiations and applications. In fact, most of the investigations in general relativity are based on or related to the Schwarzschild metric.

The Schwarzschild solution is a spherically symmetrical and time independent metric that describes the geometry of spacetime exterior to a static, spherically symmetric and electrically uncharged gravitating body of mass M surrounded by empty space. It should be remarked that "static" here means that the object does not vary in space or/and time (e.g. by translation or rotation in space or by changing its mass density or distribution in time) while "spherically symmetric" means that its mass distribution has a spherical symmetry and hence its mass density is a function of the radial coordinate r alone (i.e. with no dependency on θ or ϕ).[148] Also, "empty space" means space void of energy (e.g. electromagnetic) as

---
[148] In fact, our use of "static" here is rather different from the common use of this term in the literature of general relativity

well as matter. We should also note that "spherically symmetric" condition suggests that the object has a finite size and hence it may exclude point-like objects (even as approximation) or black holes if they are considered real physical singularities. So, to avoid any potential challenges or disputes we can generalize this by claiming that spherical symmetry includes even geometric points (i.e. these points are spherically symmetric).

It should be obvious that "exterior" (associated with "surrounded by empty space") in the above description means that the metric is a vacuum solution, i.e. it corresponds to a region of spacetime where the energy-momentum tensor vanishes, that is $T_{\mu\nu} = 0$. Accordingly, the Field Equation reduces to $G_{\mu\nu} = 0$ and hence $R_{\mu\nu} = 0$ (noting that $G_{\mu\nu} = 0$ *iff* $R_{\mu\nu} = 0$ as demonstrated in exercise 7 of § 2.13; also see exercises 8 and 9 of § 3.2). In brief, the solution corresponds to the spacetime region where $R_{\mu\nu} = 0$. As will be stated later, we are assuming zero cosmological constant (see § 3.3). In fact, we state this for formality and clarity; otherwise the problem is not cosmological and hence the cosmological term is not supposed to exist in this context (or at least it is utterly negligible).

4.1.1 Derivation of Schwarzschild Metric

In this subsection we present a rather simple derivation of the Schwarzschild metric. However, before that we need to draw the attention to the following points about the solution and how it is obtained:

1. As indicated earlier, obtaining a solution means obtaining the metric of the spacetime. So, our objective in this subsection is to obtain the metric tensor (as well as the line element which is based on the metric tensor) of the spacetime of the Schwarzschild problem (i.e. static, spherically symmetric and electrically uncharged gravitating body etc.).
2. The procedure that is followed in obtaining the solution is to start with a guess about the generic form of the metric tensor (or line element). This guess is then used with some assumptions, insights and guidance by the classical gravitation theory and special relativity to obtain the final solution. As we will see, the main assumptions and insights are the spherical symmetry and time independence of the solution (which are essentially justified by the spherical symmetry and time independence of the source of gravity although we should take account of the upcoming Birkhoff theorem).[149]
3. From a physical perspective, the Schwarzschild object is characterized by a single physical parameter which is its mass.[150] Hence, from this perspective the Schwarzschild metric must be a function of the mass only. Also, from a geometric perspective, the metric tensor in general is a function of spacetime coordinates and hence in principle the Schwarzschild metric must be a function of these coordinates as well. So in brief, the Schwarzschild metric must be a function of the mass and coordinates.
4. We use a spherical coordinate system (centered on the center of the gravitating body) for coordinating the spatial part of the spacetime[151] and hence the coordinates of any point in the spacetime are given by (ct, r, θ, ϕ) which are commonly called the Schwarzschild coordinates.[152] The justification of this choice is the spherical symmetry of the problem which makes this choice intuitive. In brief, the spherical symmetry of the gravity source (i.e. gravitating body) implies the spherical symmetry of the metric tensor and this justifies the desirability of using a spherical coordinate system for coordinating the spatial part of the spacetime.

---

which means time independent (or stationary in time) and reversible in time. However, our meaning is not far from this. We should also note that r, θ, ϕ refer effectively to spherical coordinates centered on the gravitating object (as will be clarified further later on). It is also important to note that certain restrictions imposed by "static" condition will be lifted by the upcoming Birkhoff theorem.

[149] The essence of the Birkhoff theorem is that all spherically symmetric gravitational fields in vacuum are static (i.e. even if the source of gravity is not static) and they are uniquely given by the Schwarzschild metric. In fact, it may also be stated as: all gravitational fields in vacuum generated by spherically symmetric gravitating objects are static (even if the objects are not static). However, this statement is less general.

[150] In fact, mass here should include energy (due to the mass-energy equivalence) as long as the other conditions of the Schwarzschild problem are respected.

[151] In fact, this system applies strictly only in the regions of space that are very far from the gravitating object where the spacetime becomes flat. This will be clarified later.

[152] We note that the Schwarzschild metric can be obtained using other coordinate systems, e.g. Eddington-Finkelstein coordinates. However, in this book all the coordinate-specific investigations of the Schwarzschild metric are based on the use of Schwarzschild coordinates.

4.1.1 Derivation of Schwarzschild Metric

5. Because of the presumed spherical symmetry, the Schwarzschild metric should be independent of the spatial coordinates θ and ϕ (i.e. independent of orientation and direction relative to the center). For the same reason (i.e. spherical symmetry) the line element should be reversible in θ and ϕ and hence it should not vary by the transformations $d\theta \leftrightarrow -d\theta$ and $d\phi \leftrightarrow -d\phi$.
6. The physical setting of the Schwarzschild problem dictates that the Schwarzschild metric should be independent of time. This is justified by the fact that according to the presumed physical setting and the stated assumptions of this problem there is nothing that varies in time and hence the metric must not depend on time. We should also assume that the Schwarzschild line element (which is based on the Schwarzschild metric) does not vary by reversing the sign of time (i.e. replacing dt with $-dt$) because the line element should be reversible in time since the object is static and the sign of time in this case is rather arbitrary. In fact, the reversibility in time is mainly a demonstration of the fact that the object is not rotating (as compared for example to the upcoming Kerr solution; see § 4.2).
7. To sum up, the Schwarzschild metric tensor should be independent of the coordinates ct, θ, ϕ and hence it should be a function of r and M only (where M is the mass of the gravitating object). Moreover, the line element should be reversible in t, θ, ϕ and hence it should be invariant under the transformations $dt \leftrightarrow -dt$, $d\theta \leftrightarrow -d\theta$ and $d\phi \leftrightarrow -d\phi$.
8. As the distance from the gravitating body increases the curvature of the spacetime diminishes progressively, and hence at very large distance from the gravitating body (i.e. $r \to \infty$) the spacetime is effectively flat. Therefore, at regions very far away from the gravitating object the Schwarzschild metric should converge asymptotically to the Lorentz metric[153] of special relativity (see § 2.5). Similarly, as the mass of the gravitating object M decreases the curvature of the spacetime diminishes progressively, and hence as the mass tends to zero (i.e. $M \to 0$) the spacetime becomes effectively flat Minkowskian. Therefore, when the mass vanishes (i.e. $M = 0$) the Schwarzschild metric should converge to the Lorentz metric of special relativity.

   In brief, the expected metric should converge to the Lorentz metric of special relativity in certain limiting cases, i.e. the case of distant regions ($r \to \infty$) and the case of vanishing mass ($M = 0$). These limiting cases provide guiding conditions during our derivation of the Schwarzschild metric. Moreover, any metric that we obtain as a solution to the Schwarzschild problem must yield the Lorentz metric of special relativity in these limiting cases and hence the solution must be rejected if it does not pass these tests (see exercises 6 and 7).
9. We consider the Field Equation without the cosmological term since this term is not supposed to exist in this type of problems (i.e. non-cosmological). This can also be understood from the aforementioned fact that we are solving the following Field Equation: $R_{\mu\nu} = 0$.

To derive the Schwarzschild metric, we start from the Field Equation $R_{\mu\nu} = 0$ since the problem is a vacuum problem. Based on the insight that we gained from the above points we can guess that the Schwarzschild quadratic form in its generic and most general form must be given by the following expression:

$$\begin{aligned}(d\sigma)^2 &= g_{00}(cdt)^2 + g_{11}(dr)^2 + g_{22}(d\theta)^2 + g_{33}(d\phi)^2 \\ &= f_0(cdt)^2 - f_1(dr)^2 - r^2(d\theta)^2 - r^2\sin^2\theta(d\phi)^2\end{aligned} \quad (121)$$

where f_0 and f_1 are real positive functions of r and M only that should converge to 1 as $r \to \infty$ (due to the boundary condition at infinity, i.e. the convergence to the Lorentz metric of special relativity at infinity) and as $M \to 0$ (due to the demand that the Schwarzschild metric should reduce to the Lorentz metric when the spacetime is void).

Now, at this point we need to provide a number of justifications for the above form which looks restricted and not as general as we claimed:

(a) The justification for the $+ - - -$ sign pattern of the terms in line 2 is that we expect the metric to converge to the Lorentz metric of special relativity (which has the $+ - - -$ sign pattern) at certain limiting cases (i.e. distant regions $r \to \infty$ and vanishing mass $M = 0$) and hence the Schwarzschild metric

---

[153] We use "Lorentz metric" to be more general since "Minkowski metric" may be restricted to the form diag $[1, -1, -1, -1]$ and its variants (like diag $[-1, 1, 1, 1]$) which are based on using Cartesian system for coordinating the spatial part.

4.1.1 Derivation of Schwarzschild Metric

should resemble the Lorentz metric in this sign pattern (or signature).[154] This should also justify the requirement that f_0 and f_1 are real positive functions that converge to 1 as $r \to \infty$ and as $M \to 0$.

(**b**) The justification for f_0 and f_1 to be functions of r and M only has been outlined earlier, i.e. the time independence of the physical setting (and hence f_0 and f_1 are independent of t) and its spherical symmetry (and hence f_0 and f_1 are independent of θ and ϕ) as well as the sole physical characterization of the gravity source by its mass M.

(**c**) The justification for the form of g_{22} and g_{33} (i.e. $g_{22} = -r^2$ and $g_{33} = -r^2 \sin^2 \theta$) is the use of spherical coordinates due to the presumed spherical symmetry (as well as the convergence to the Lorentz metric of special relativity at the above limits).

(**d**) The justification for the absence of the mixed differential terms (i.e. terms involving $dx^\mu dx^\nu$ where $\mu \ne \nu$) is the fact that the mixed terms that involve time differential (i.e. dt) are zero because the line element is reversible in time and hence it should be independent of the sign of dt, while the mixed differential terms that do not involve time differential (i.e. those involving only space differentials such as $drd\theta$) are zero because the line element must be independent of the sign of $d\theta$ and $d\phi$ due to the spherical symmetry of the problem. In brief, the Schwarzschild line element has no mixed differential terms and hence the Schwarzschild metric should be diagonal.

So, what we need in order to have a completely definite form of the Schwarzschild metric is to obtain the form of f_0 and f_1. It can be argued that if we have to keep the above sign pattern and satisfy the boundary condition at infinity then we need to use exponential functions (which are non-negative), i.e. $f_0 = e^H$ and $f_1 = e^K$ where H and K are functions of r only[155] that should vanish at infinity (i.e. $H = K = 0$ as $r \to \infty$) as required by the indicated boundary condition at infinity (i.e. convergence to the Lorentz metric at this limit).[156] Accordingly, we should have (refer to Eq. 121):

$$[g_{\mu\nu}] = \text{diag}\left[e^H, -e^K, -r^2, -r^2 \sin^2 \theta\right] \tag{122}$$

$$[g^{\mu\nu}] = \text{diag}\left[e^{-H}, -e^{-K}, -r^{-2}, -r^{-2} \sin^{-2} \theta\right] \tag{123}$$

As stated before, we are supposed to solve the tensorial equation $R_{\mu\nu} = 0$ and hence we need to find the coefficients of the Ricci curvature tensor $R_{\mu\nu}$, set them to zero and solve the resulting system of equations $R_{\mu\nu} = 0$ simultaneously to obtain the Schwarzschild metric in its final and definite form (by determining the functions H and K). Now, the coefficients of the Ricci curvature tensor are given by (see Eq. 89):

$$R_{\mu\nu} = \partial_\nu \Gamma^\alpha_{\mu\alpha} - \partial_\alpha \Gamma^\alpha_{\mu\nu} + \Gamma^\beta_{\mu\alpha}\Gamma^\alpha_{\beta\nu} - \Gamma^\beta_{\mu\nu}\Gamma^\alpha_{\beta\alpha} \tag{124}$$

Hence, we need to find the Christoffel symbols first where these symbols are given by (see Eq. 52):

$$\Gamma^\xi_{\mu\nu} = \frac{g^{\xi\alpha}}{2}\left(\partial_\nu g_{\mu\alpha} + \partial_\mu g_{\nu\alpha} - \partial_\alpha g_{\mu\nu}\right) \tag{125}$$

Accordingly, our plan in order to obtain the definite form of the Schwarzschild metric is to obtain the Christoffel symbols using Eq. 125 with the aid of Eqs. 122 and 123. We then find the coefficients of the Ricci curvature tensor $R_{\mu\nu}$ using Eq. 124. We then form the system of equations $R_{\mu\nu} = 0$ and solve them simultaneously (subject to the aforementioned assumptions and boundary conditions) to obtain the Schwarzschild metric in its definite form by identifying the functions $H(r)$ and $K(r)$. In fact, there

---

[154] "Signature" in this context means the number of positive metric coefficients (or terms of quadratic form) minus the number of negative metric coefficients noting that the metric is diagonal. So, according to the sign pattern $+---$ the signature is $1 - 3 = -2$ while according to the sign pattern $-+++$ the signature is $3 - 1 = 2$.

[155] From now on, we ignore the dependency of f_0 and f_1 (and hence the dependency of H and K) on M because for a given Schwarzschild object the mass is fixed (i.e. $M = $ constant) and hence f_0 and f_1 are functions of r only although from the point of view of a general metric they are functions of both r and M. As we will see, M will be incorporated in the metric through a constant of integration with the help of the classical limit guidance, and hence M will be embedded into H and K (and hence f_0 and f_1) as a constant. So we may say: H and K depend on M as a constant and on r as a variable (noting that H and K should vanish as $M = 0$ and as $r \to \infty$).

[156] The choice of exponential functions may be additionally justified by being a convenient form to work with. Anyway, the exponential form is an acceptable form to use according to the imposed conditions and requirements and hence it does not require further justification.

4.1.1 Derivation of Schwarzschild Metric

are other plans for obtaining the definite form of the Schwarzschild metric such as solving the tensorial equation $G_{\mu\nu} = 0$ directly instead of solving the equation $R_{\mu\nu} = 0$ (which is equivalent to the equation $G_{\mu\nu} = 0$ as seen in exercise 7 of § 2.13). However, in our view the method that we follow is the clearest and most straightforward although the other methods may also have some advantages.

Regarding the Christoffel symbols, it can be shown (see exercise 11) that the non-vanishing Christoffel symbols are only:

$$\begin{aligned}
&\Gamma^0_{01} = \Gamma^0_{10} = \tfrac{1}{2}H' \\
&\Gamma^1_{00} = \tfrac{1}{2}e^{H-K}H' \quad \Gamma^1_{11} = \tfrac{1}{2}K' \quad \Gamma^1_{22} = -re^{-K} \quad \Gamma^1_{33} = -re^{-K}\sin^2\theta \\
&\Gamma^2_{12} = \Gamma^2_{21} = \tfrac{1}{r} \quad \Gamma^2_{33} = -\sin\theta\cos\theta \\
&\Gamma^3_{13} = \Gamma^3_{31} = \tfrac{1}{r} \quad \Gamma^3_{23} = \Gamma^3_{32} = \cot\theta
\end{aligned} \tag{126}$$

where the prime means total derivative with respect to r (i.e. d/dr). It can also be shown (see exercise 12) that the non-vanishing coefficients of the Ricci curvature tensor are only the diagonal elements, that is:

$$R_{00} = e^{H-K}\left(\frac{H'K'}{4} - \frac{(H')^2}{4} - \frac{H''}{2} - \frac{H'}{r}\right) \tag{127}$$

$$R_{11} = -\frac{H'K'}{4} + \frac{(H')^2}{4} + \frac{H''}{2} - \frac{K'}{r} \tag{128}$$

$$R_{22} = e^{-K}\left[1 + \frac{r(H' - K')}{2}\right] - 1 \tag{129}$$

$$R_{33} = \sin^2\theta\left\{e^{-K}\left[1 + \frac{r(H' - K')}{2}\right] - 1\right\} \tag{130}$$

Now, solving the system of equations $R_{\mu\nu} = 0$ simultaneously means solving the system of the last four coefficients simultaneously (since the other coefficients will produce the trivial equation $0 = 0$), that is:

$$e^{H-K}\left(\frac{H'K'}{4} - \frac{(H')^2}{4} - \frac{H''}{2} - \frac{H'}{r}\right) = 0 \tag{131}$$

$$-\frac{H'K'}{4} + \frac{(H')^2}{4} + \frac{H''}{2} - \frac{K'}{r} = 0 \tag{132}$$

$$e^{-K}\left[1 + \frac{r(H' - K')}{2}\right] - 1 = 0 \tag{133}$$

$$\sin^2\theta\left\{e^{-K}\left[1 + \frac{r(H' - K')}{2}\right] - 1\right\} = 0 \tag{134}$$

Now, since $e^{H-K} \neq 0$ and $\sin^2\theta$ is not identically zero[157] then the last four equations will reduce to:

$$\frac{H'K'}{4} - \frac{(H')^2}{4} - \frac{H''}{2} - \frac{H'}{r} = 0 \tag{135}$$

$$-\frac{H'K'}{4} + \frac{(H')^2}{4} + \frac{H''}{2} - \frac{K'}{r} = 0 \tag{136}$$

$$e^{-K}\left[1 + \frac{r(H' - K')}{2}\right] - 1 = 0 \tag{137}$$

$$e^{-K}\left[1 + \frac{r(H' - K')}{2}\right] - 1 = 0 \tag{138}$$

---

[157] In fact, $\sin^2\theta$ is zero only at the poles which should be excluded.

4.1.1 Derivation of Schwarzschild Metric

As we see, the last two equations are identical and hence the system will reduce to the following three equations:

$$\frac{H'K'}{4} - \frac{(H')^2}{4} - \frac{H''}{2} - \frac{H'}{r} = 0 \tag{139}$$

$$-\frac{H'K'}{4} + \frac{(H')^2}{4} + \frac{H''}{2} - \frac{K'}{r} = 0 \tag{140}$$

$$e^{-K}\left[1 + \frac{r(H' - K')}{2}\right] - 1 = 0 \tag{141}$$

Now, if we add the first two equations we get:[158]

$$-\frac{H'}{r} - \frac{K'}{r} = 0 \qquad \text{that is} \qquad H' = -K' \tag{142}$$

and hence on integrating both sides with respect to r (noting that H and K are functions of r only) we get:

$$H = -K + C \tag{143}$$

where C is the constant of integration. Now, since H and K should converge to zero at infinity (according to the aforementioned infinity boundary condition)[159] then $C = 0$ and we get:

$$H = -K \tag{144}$$

On substituting from Eqs. 142 and 144 into Eq. 141 we get:

$$e^{H}\left[1 + \frac{r(H' + H')}{2}\right] - 1 = 0 \tag{145}$$

$$e^{H}(1 + rH') = 1 \tag{146}$$

It can be shown (see exercise 13) that this differential equation has the following solution:

$$e^{H} = 1 - \frac{C_s}{r} \tag{147}$$

where C_s is a constant. By using the Newtonian gravity theory (see § 1.5) in conjunction with the correspondence principle (see § 1.8.3) at the classical gravity limit it can be shown (see the upcoming notes in the end of this subsection) that C_s is equal to the Schwarzschild radius[160] (i.e. $C_s = R_S \equiv \frac{2GM}{c^2}$), and hence:

$$e^{H} = 1 - \frac{2GM}{c^2 r} \tag{148}$$

$$e^{K} = e^{-H} = \frac{1}{e^H} = \left(1 - \frac{2GM}{c^2 r}\right)^{-1} \tag{149}$$

On substituting from the last two equations into Eqs. 122 and 123 we get the Schwarzschild metric in its definite form, that is:

$$[g_{\mu\nu}] = \text{diag}\left[\left(1 - \frac{2GM}{c^2 r}\right), -\left(1 - \frac{2GM}{c^2 r}\right)^{-1}, -r^2, -r^2 \sin^2\theta\right] \tag{150}$$

---

[158] We note that $r \neq 0$ because the problem is a vacuum problem in a region exterior to the gravitating body which is positioned at $r = 0$.

[159] We may also use the condition $H = K = 0$ as $M = 0$.

[160] The fact that C_s is equal to the Schwarzschild radius (which has the dimension of length) is consistent with the requirement that C_s/r must be dimensionless since r has the dimension of length. For the definition of Schwarzschild radius, see exercise 14.

4.1.1 Derivation of Schwarzschild Metric

$$[g^{\mu\nu}] \;=\; \mathrm{diag}\left[\left(1-\frac{2GM}{c^2 r}\right)^{-1},\; -\left(1-\frac{2GM}{c^2 r}\right),\; -r^{-2},\; -r^{-2}\sin^{-2}\theta\right] \tag{151}$$

The last two equations represent the Schwarzschild metric tensor in its covariant and contravariant definite form, i.e. it is the required solution of the Schwarzschild problem.

Finally, we should draw the attention to the following important notes:

- The above derivation is based on several simplifications and restrictions; moreover the final solution was in sight (at least in its form) prior to the derivation since we started with a given guess about how the solution should look like. Therefore, it might be suspected that the obtained solution is not general and a different solution might have been obtained if we followed a different course of action. However, despite the aforementioned limitations and preconception it can be shown that the obtained solution is general within the stated assumptions, i.e. all the static spherically symmetric solutions of the vacuum tensorial equation $R_{\mu\nu}=0$ subject to the given boundary condition at infinity (as well as the stated physical setting and assumptions of the Schwarzschild problem) are equivalent to the obtained Schwarzschild solution. The interested reader should refer to the Birkhoff theorem in the literature of general relativity (noting that this theorem contains more generalizations than what we are using and demanding here).[161]
- To obtain a specific expression for C_s (see Eq. 147) and hence determine the definite form of the Schwarzschild metric completely, we use the fact that in the classical limit (see § 5.1) the component g_{00} of the Schwarzschild metric tensor is approximated by:

$$g_{00} \simeq 1 + \frac{2\Phi}{c^2} = 1 - \frac{2GM}{c^2 r} \tag{152}$$

where in the second equality we use $\Phi = -\frac{GM}{r}$ with Φ being the classical gravitational potential (see exercise 2 of § 1.5). Accordingly:

$$C_s = \frac{2GM}{c^2} \tag{153}$$

The details of this derivation are given in § 5.1.

- Regarding the use of the spherical coordinate system to represent the spatial part of the spacetime and the meaning of the Schwarzschild coordinates in general (and indeed the interpretation of the coordinates in general relativity to be more general), we refer the reader to § 6.2, § 6.3.2 and § 6.4.2.
- In the literature of general relativity there are some examples of Schwarzschild geometry generated by gravitating objects with non spherical symmetric matter distribution. If so, the spherical symmetry of the matter distribution is a sufficient but not necessary condition for the generation of Schwarzschild geometry (assuming that other conditions are satisfied) although the derivation of the metric in these examples should follow a different route since our derivation is based on the assumption of spherical symmetry of the source of gravity (in fact we characterized the source of gravity as a static, spherically symmetric and electrically uncharged gravitating body). It is worth noting that the reader should not confuse the spherical symmetry of the matter distribution of the gravitating object with the spherical symmetry of the solution (noting that in our argument in this subsection we based the latter on the former and considering that this should be consistent with being a sufficient but not necessary condition as stated above).

Problems

1. What about the metric of the spacetime inside the gravitating object of the Schwarzschild problem?
   Answer: The metric inside the object is different from the metric outside the object (where the metric outside is the Schwarzschild metric). The reason is that in the outside region the metric belongs to vacuum (i.e. $T_{\mu\nu}=0$ and hence $R_{\mu\nu}=0$) while the metric inside the object is not (i.e. $T_{\mu\nu}\neq 0$ and hence $R_{\mu\nu}\neq 0$ since $R_{\mu\nu}=0$ iff $T_{\mu\nu}=0$ according to exercise 9 of § 3.2). However, the two metrics

---

[161] The Birkhoff theorem may be stated succinctly as follows: any solution of the vacuum Field Equation of a spherically symmetric (non-rotating) gravitating object is given by the Schwarzschild metric. This means that in the vacuum region the solution is static (where static here means time independent metric and time-reversible quadratic form) even if the gravitating object is not static (e.g. pulsating spherical-symmetrically).

should match at the boundary (i.e. the surface of the gravitating object) to ensure the continuity of the metric in the entire spacetime.

We note that in this answer we are assuming that the gravitating object has a finite size and hence the spacetime is divided into inside the object and outside the object. So, if we assume that the gravitating object is point-like then the Schwarzschild metric corresponds to the entire spacetime (excluding the point where the object is located). We should also note that the metric inside the object is not unique since it depends on the equation of state of the object and its physical and geometric properties (e.g. fluid or solid, of uniform density or not, etc.).

2. Show that the Schwarzschild metric is time independent and spherically symmetric.

Answer: The Schwarzschild metric tensor is given by:

$$[g_{\mu\nu}] = \text{diag}\left[1 - \frac{2GM}{c^2 r},\ -\left(1 - \frac{2GM}{c^2 r}\right)^{-1},\ -r^2,\ -r^2 \sin^2\theta\right]$$

As we see, there is no temporal variable in this metric and hence it is time independent. Moreover, all the components depend on r only (noting the constancy of M) and hence it is spherically symmetric.[162] We note that the dependence of the last component on θ is accidental (due to the choice of coordinate system) and hence it does not affect the spherical symmetry of the physical situation (as can be verified by rotating the coordinate system or the gravitating object; also see exercise 6 of § 4.2). In brief, the physical situation dictates spherical symmetry despite this mathematical artifact which suggests otherwise. In fact, this should put a question mark on the inferred time independence and symmetry with respect to ϕ as mathematical consequences rather than physically inferred conclusions that are supported by the mathematical form of the metric.[163] So, the final judgment belongs to physics which should validate or invalidate any feature and artifact suggested by the formalism of mathematics.

Note 1: from a formal perspective the Schwarzschild metric[164] is not ideal since it has several misleading artifacts and features related to the choice of coordinates such as this ostensible spherical non-symmetry and the accidental singularity at the Schwarzschild radius (which will be investigated later). This should highlight the issue (which we advocate continuously) that mathematics is not such a perfect and ideal tool when used in physical sciences and hence we should be careful in reading and interpreting our mathematical formulations and models of physical phenomena (see § 1.10).

Note 2: according to the circulating literature of black holes, the spacetime inside the event horizon of Schwarzschild black hole is neither time independent nor reversible in time (i.e. it is not stationary or static according to the common terminology). This is because nothing can stand still inside the horizon.[165] Accordingly, in the case of black holes the above-claimed time independence should be restricted to outside the event horizon. However, the interpretations and consequences of some of these issues should be treated with caution due to the many fantasies and questionable issues in the circulating physics of black holes. In fact, we may even question the meaning of time dependency and irreversibility in time in these claims although we do not discuss this issue due to its pedantic and hypothetical nature.

Exercises

1. What is the Schwarzschild solution?
2. Summarize the main characteristics of the Schwarzschild solution.
3. Summarize the procedure that we followed to find the Schwarzschild metric.

---

[162] Being spherically symmetric means it is independent of θ and ϕ.

[163] In fact, the mathematical form is obtained based on the above physical considerations and assumptions and hence the mathematical form actually reflects the physics (rather than being suggestive of the physics).

[164] "Schwarzschild metric" here should mean the solution of the Schwarzschild problem that is based on the use of the Schwarzschild coordinate system. In other words, it is "Schwarzschild metric in the Schwarzschild coordinates".

[165] This is also linked to the exchange of the roles of time coordinate and radial coordinate and the change of their nature where t becomes spacelike and r becomes timelike and hence the metric will have time dependency through its dependency on r. The inevitability of moving toward the singularity inside the horizon is attributed to the inevitability of moving in time toward the future since r inside the horizon is timelike.

4. Write down the Schwarzschild metric tensor and the Schwarzschild quadratic form explaining all the symbols used.
5. Write the Schwarzschild metric tensor in terms of the escape speed v_{es} and in terms of the Schwarzschild radius R_S.
6. Verify that the Schwarzschild metric converges to the Lorentz metric of special relativity very far away from the gravitating object.
7. Verify that when the mass vanishes the Schwarzschild metric converges to the Lorentz metric of special relativity.
8. Based on the last two exercises, analyze the relation between special relativity and general relativity and how the latter is seen to converge to the former in the absence of gravity.
9. What are the types of singularity and how you define them?
10. Show that there is a singularity in the Schwarzschild quadratic form. What is the physical interpretation of this? What is the type of this singularity?
11. Show that the non-vanishing Christoffel symbols of the Schwarzschild spacetime (in the Schwarzschild coordinates) are those given by Eq. 126 and verify that these symbols are as given by Eq. 126.
12. Show that the non-vanishing coefficients of the Ricci curvature tensor of the Schwarzschild spacetime (in the Schwarzschild coordinates) are R_{00}, R_{11}, R_{22} and R_{33} and verify that these coefficients are as given by Eqs. 127-130.
13. Show that the differential equation $e^H (1 + rH') = 1$ has the following solution: $e^H = 1 - \frac{C_s}{r}$ where C_s is a constant and the prime represents derivative with respect to r.
14. What is Schwarzschild radius?
15. Verify that the solution found in exercise 13 satisfies the system given by Eqs. 139-141. What this means?
16. Show that the definite form of the Schwarzschild metric that we obtained in the text (Eqs. 150-151) satisfies the vacuum Field Equations $R_{\mu\nu} = 0$.
17. Obtain the non-vanishing Christoffel symbols of the Schwarzschild metric in their definite form (refer to Eq. 126).
18. Use the Schwarzschild metric to obtain a mathematical condition for the weak field approximation (i.e. where Newtonian gravity is a good approximation to general relativity).
19. In § 1.3 we defined weak gravitational field as a field in which the gravitational potential energy of a test particle is negligible in comparison to its mass energy. Try to correlate this definition to the mathematical condition that is obtained in the previous exercise.
20. What is the essence of the Birkhoff theorem?

4.1.2 Geodesic Equation in Schwarzschild Metric

On using the Christoffel symbols that were derived in § 4.1.1 it can be shown (see Problems) that the four components of the tensorial geodesic equation in the Schwarzschild metric are:

$$\frac{d^2 t}{d\lambda^2} + \frac{2GM}{c^2 r^2 F} \frac{dt}{d\lambda} \frac{dr}{d\lambda} = 0 \qquad (154)$$

$$\frac{d^2 r}{d\lambda^2} + \frac{GMF}{r^2} \left(\frac{dt}{d\lambda}\right)^2 - \frac{GM}{c^2 r^2 F} \left(\frac{dr}{d\lambda}\right)^2 - rF \left(\frac{d\theta}{d\lambda}\right)^2 - rF \sin^2\theta \left(\frac{d\phi}{d\lambda}\right)^2 = 0 \qquad (155)$$

$$\frac{d^2 \theta}{d\lambda^2} + \frac{2}{r} \frac{dr}{d\lambda} \frac{d\theta}{d\lambda} - \sin\theta \cos\theta \left(\frac{d\phi}{d\lambda}\right)^2 = 0 \qquad (156)$$

$$\frac{d^2 \phi}{d\lambda^2} + \frac{2}{r} \frac{dr}{d\lambda} \frac{d\phi}{d\lambda} + 2\cot\theta \frac{d\theta}{d\lambda} \frac{d\phi}{d\lambda} = 0 \qquad (157)$$

where $F = 1 - \frac{2GM}{c^2 r}$. It should be obvious that the tensorial geodesic equation is a rank-1 tensor (i.e. vector) equation in 4D spacetime and hence it should have four components.

We will also show in the Problems the following results:

4.1.2 Geodesic Equation in Schwarzschild Metric

(**A**) The above system of geodesic equations (i.e. Eqs. 154-157) can be reduced to the following system of equations:

$$F\frac{dt}{d\lambda} = A \tag{158}$$

$$\frac{d^2 u}{d\phi^2} - \frac{GMA^2}{B^2 F} + \frac{GM}{B^2 c^2 F}\left(\frac{du^{-1}}{d\lambda}\right)^2 + Fu = 0 \tag{159}$$

$$\frac{d\phi}{d\lambda} = Bu^2 \tag{160}$$

where A and B are constants and $u = 1/r$.[166]

(**B**) For massive objects the reduced system of Eqs. 158-160 can be simplified to the following system of equations:

$$F\frac{dt}{d\tau} = A \tag{161}$$

$$\frac{d^2 u}{d\phi^2} + u = \frac{GM}{B^2} + \frac{3GMu^2}{c^2} \tag{162}$$

$$\frac{d\phi}{d\tau} = Bu^2 \tag{163}$$

where we use the proper time parameter τ as affine parameter.

(**C**) For massless objects the reduced system of Eqs. 158-160 can be simplified to the following system of equations:

$$F\frac{dt}{d\lambda} = A \tag{164}$$

$$\frac{d^2 u}{d\phi^2} + u = \frac{3GMu^2}{c^2} \tag{165}$$

$$\frac{d\phi}{d\lambda} = Bu^2 \tag{166}$$

Problems

1. Show that the four components of the tensorial geodesic equation in the Schwarzschild metric are as given by Eqs. 154-157.
   Answer: The tensorial geodesic equation is given by:

$$\frac{d^2 x^\alpha}{d\lambda^2} + \Gamma^\alpha_{\beta\gamma}\frac{dx^\beta}{d\lambda}\frac{dx^\gamma}{d\lambda} = 0$$

Now, if we symbolize $\left(1 - \frac{2GM}{c^2 r}\right)$ with F and use the results of exercise 17 of § 4.1.1 then we have:
- For $\alpha = 0$ the non-vanishing Christoffel symbols are Γ^0_{01} and Γ^0_{10}. Hence:

$$\frac{d^2 x^0}{d\lambda^2} + \Gamma^0_{01}\frac{dx^0}{d\lambda}\frac{dx^1}{d\lambda} + \Gamma^0_{10}\frac{dx^1}{d\lambda}\frac{dx^0}{d\lambda} = 0$$

$$\frac{d^2 x^0}{d\lambda^2} + 2\Gamma^0_{01}\frac{dx^0}{d\lambda}\frac{dx^1}{d\lambda} = 0$$

$$\frac{d^2 x^0}{d\lambda^2} + 2\frac{GM}{c^2 r^2 F}\frac{dx^0}{d\lambda}\frac{dx^1}{d\lambda} = 0$$

$$\frac{d^2 (ct)}{d\lambda^2} + \frac{2GM}{c^2 r^2 F}\frac{d(ct)}{d\lambda}\frac{dr}{d\lambda} = 0$$

---

[166] We should note that A here is not the same as A in the orbital shape equations of § 1.5.1 (see for example Eq. 17) although both are constants.

4.1.2 Geodesic Equation in Schwarzschild Metric

$$\frac{d^2t}{d\lambda^2} + \frac{2GM}{c^2 r^2 F}\frac{dt}{d\lambda}\frac{dr}{d\lambda} = 0$$

which is Eq. 154.

- For $\alpha = 1$ the non-vanishing Christoffel symbols are Γ^1_{00}, Γ^1_{11}, Γ^1_{22} and Γ^1_{33}. Hence:

$$\frac{d^2 x^1}{d\lambda^2} + \Gamma^1_{00}\frac{dx^0}{d\lambda}\frac{dx^0}{d\lambda} + \Gamma^1_{11}\frac{dx^1}{d\lambda}\frac{dx^1}{d\lambda} + \Gamma^1_{22}\frac{dx^2}{d\lambda}\frac{dx^2}{d\lambda} + \Gamma^1_{33}\frac{dx^3}{d\lambda}\frac{dx^3}{d\lambda} = 0$$

$$\frac{d^2 x^1}{d\lambda^2} + \frac{GMF}{c^2 r^2}\left(\frac{dx^0}{d\lambda}\right)^2 - \frac{GM}{c^2 r^2 F}\left(\frac{dx^1}{d\lambda}\right)^2 - rF\left(\frac{dx^2}{d\lambda}\right)^2 - rF\sin^2\theta\left(\frac{dx^3}{d\lambda}\right)^2 = 0$$

$$\frac{d^2 r}{d\lambda^2} + \frac{GMF}{r^2}\left(\frac{dt}{d\lambda}\right)^2 - \frac{GM}{c^2 r^2 F}\left(\frac{dr}{d\lambda}\right)^2 - rF\left(\frac{d\theta}{d\lambda}\right)^2 - rF\sin^2\theta\left(\frac{d\phi}{d\lambda}\right)^2 = 0$$

which is Eq. 155.

- For $\alpha = 2$ the non-vanishing Christoffel symbols are Γ^2_{12}, Γ^2_{21} and Γ^2_{33}. Hence:

$$\frac{d^2 x^2}{d\lambda^2} + \Gamma^2_{12}\frac{dx^1}{d\lambda}\frac{dx^2}{d\lambda} + \Gamma^2_{21}\frac{dx^2}{d\lambda}\frac{dx^1}{d\lambda} + \Gamma^2_{33}\frac{dx^3}{d\lambda}\frac{dx^3}{d\lambda} = 0$$

$$\frac{d^2 x^2}{d\lambda^2} + 2\Gamma^2_{12}\frac{dx^1}{d\lambda}\frac{dx^2}{d\lambda} + \Gamma^2_{33}\frac{dx^3}{d\lambda}\frac{dx^3}{d\lambda} = 0$$

$$\frac{d^2 x^2}{d\lambda^2} + \frac{2}{r}\frac{dx^1}{d\lambda}\frac{dx^2}{d\lambda} - \sin\theta\cos\theta\frac{dx^3}{d\lambda}\frac{dx^3}{d\lambda} = 0$$

$$\frac{d^2 \theta}{d\lambda^2} + \frac{2}{r}\frac{dr}{d\lambda}\frac{d\theta}{d\lambda} - \sin\theta\cos\theta\left(\frac{d\phi}{d\lambda}\right)^2 = 0$$

which is Eq. 156.

- For $\alpha = 3$ the non-vanishing Christoffel symbols are Γ^3_{13}, Γ^3_{31}, Γ^3_{23} and Γ^3_{32}. Hence:

$$\frac{d^2 x^3}{d\lambda^2} + \Gamma^3_{13}\frac{dx^1}{d\lambda}\frac{dx^3}{d\lambda} + \Gamma^3_{31}\frac{dx^3}{d\lambda}\frac{dx^1}{d\lambda} + \Gamma^3_{23}\frac{dx^2}{d\lambda}\frac{dx^3}{d\lambda} + \Gamma^3_{32}\frac{dx^3}{d\lambda}\frac{dx^2}{d\lambda} = 0$$

$$\frac{d^2 x^3}{d\lambda^2} + 2\Gamma^3_{13}\frac{dx^1}{d\lambda}\frac{dx^3}{d\lambda} + 2\Gamma^3_{23}\frac{dx^2}{d\lambda}\frac{dx^3}{d\lambda} = 0$$

$$\frac{d^2 x^3}{d\lambda^2} + \frac{2}{r}\frac{dx^1}{d\lambda}\frac{dx^3}{d\lambda} + 2\cot\theta\frac{dx^2}{d\lambda}\frac{dx^3}{d\lambda} = 0$$

$$\frac{d^2 \phi}{d\lambda^2} + \frac{2}{r}\frac{dr}{d\lambda}\frac{d\phi}{d\lambda} + 2\cot\theta\frac{d\theta}{d\lambda}\frac{d\phi}{d\lambda} = 0$$

which is Eq. 157.

2. Show that the system of Eqs. 154-157 can be reduced to the system of Eqs. 158-160.
   Answer: We start from the system of Eqs. 154-157, that is:

$$\frac{d^2 t}{d\lambda^2} + \frac{2GM}{c^2 r^2 F}\frac{dt}{d\lambda}\frac{dr}{d\lambda} = 0 \tag{167}$$

$$\frac{d^2 r}{d\lambda^2} + \frac{GMF}{r^2}\left(\frac{dt}{d\lambda}\right)^2 - \frac{GM}{c^2 r^2 F}\left(\frac{dr}{d\lambda}\right)^2 - rF\left(\frac{d\theta}{d\lambda}\right)^2 - rF\sin^2\theta\left(\frac{d\phi}{d\lambda}\right)^2 = 0 \tag{168}$$

$$\frac{d^2 \theta}{d\lambda^2} + \frac{2}{r}\frac{dr}{d\lambda}\frac{d\theta}{d\lambda} - \sin\theta\cos\theta\left(\frac{d\phi}{d\lambda}\right)^2 = 0 \tag{169}$$

$$\frac{d^2 \phi}{d\lambda^2} + \frac{2}{r}\frac{dr}{d\lambda}\frac{d\phi}{d\lambda} + 2\cot\theta\frac{d\theta}{d\lambda}\frac{d\phi}{d\lambda} = 0 \tag{170}$$

In the following points we show step by step how the system of Eqs. 167-170 can be reduced to the system of Eqs. 158-160.

4.1.2 Geodesic Equation in Schwarzschild Metric

(a) Eliminating Eq. 169: if we assume that initially the gravitated object is in the plane $\theta = \pi/2$ and the first order variation of its θ dependency vanishes (and hence initially $\cos\theta = 0$ and $\frac{d\theta}{d\lambda} = 0$; refer to the upcoming note) then Eq. 169 becomes:

$$\frac{d^2\theta}{d\lambda^2} = 0$$

On differentiating this equation again and again we conclude that the λ-derivatives of θ of all orders are zero (i.e. $\frac{d\theta}{d\lambda} = \frac{d^2\theta}{d\lambda^2} = \frac{d^3\theta}{d\lambda^3} = \cdots = 0$) and hence θ has no dependency on the affine parameter λ.[167] This means that θ is constant and hence if the gravitated object moves initially in the plane $\theta = \pi/2$ then it will stay in this plane permanently. This result is consistent with the classical mechanics where the trajectory of a gravitated object in a central force field is confined to a plane which is determined by the initial conditions, i.e. the position and velocity of the gravitated object (with λ standing for time). Accordingly, Eq. 169 will reduce to the simple condition $\theta = \pi/2$ and hence the system of Eqs. 167-170 will reduce to the following three equations (noting that $\frac{d\theta}{d\lambda} = 0$, $\sin^2\theta = 1$ and $\cot\theta = 0$ since $\theta = \pi/2$):

$$\frac{d^2t}{d\lambda^2} + \frac{2GM}{c^2r^2F}\frac{dt}{d\lambda}\frac{dr}{d\lambda} = 0 \qquad (171)$$

$$\frac{d^2r}{d\lambda^2} + \frac{GMF}{r^2}\left(\frac{dt}{d\lambda}\right)^2 - \frac{GM}{c^2r^2F}\left(\frac{dr}{d\lambda}\right)^2 - rF\left(\frac{d\phi}{d\lambda}\right)^2 = 0 \qquad (172)$$

$$\frac{d^2\phi}{d\lambda^2} + \frac{2}{r}\frac{dr}{d\lambda}\frac{d\phi}{d\lambda} = 0 \qquad (173)$$

(b) Simplifying Eq. 171: if we note that $\frac{2GM}{c^2r^2} = \frac{dF}{dr}$ (since $F = 1 - \frac{2GM}{c^2r}$) then Eq. 171 can be simplified as follows:

$$\frac{d^2t}{d\lambda^2} + \frac{1}{F}\frac{dF}{dr}\frac{dt}{d\lambda}\frac{dr}{d\lambda} = 0$$

$$\frac{d^2t}{d\lambda^2} + \frac{1}{F}\frac{dF}{d\lambda}\frac{dt}{d\lambda} = 0$$

$$F\frac{d^2t}{d\lambda^2} + \frac{dF}{d\lambda}\frac{dt}{d\lambda} = 0$$

$$\frac{d}{d\lambda}\left(F\frac{dt}{d\lambda}\right) = 0$$

$$F\frac{dt}{d\lambda} = A$$

where in line 2 we use the chain rule, in line 3 we multiply by F, in line 4 we use the product rule, and in line 5 we integrate with respect to λ (with A being a constant). Accordingly, the system of Eqs. 171-173 becomes:

$$F\frac{dt}{d\lambda} = A \qquad (174)$$

$$\frac{d^2r}{d\lambda^2} + \frac{GMF}{r^2}\left(\frac{dt}{d\lambda}\right)^2 - \frac{GM}{c^2r^2F}\left(\frac{dr}{d\lambda}\right)^2 - rF\left(\frac{d\phi}{d\lambda}\right)^2 = 0 \qquad (175)$$

$$\frac{d^2\phi}{d\lambda^2} + \frac{2}{r}\frac{dr}{d\lambda}\frac{d\phi}{d\lambda} = 0 \qquad (176)$$

(c) Simplifying Eq. 176: we have:

$$\frac{d^2\phi}{d\lambda^2} + \frac{2}{r}\frac{dr}{d\lambda}\frac{d\phi}{d\lambda} = 0$$

---

[167] In fact, this may be expressed more technically and elegantly in terms of a Taylor series. In brief, the essential point in the above argument is that since θ and all its derivatives vanish initially then it (with its derivatives) should vanish permanently (which is formally demonstrated by a Taylor expansion about the initial point).

4.1.2 Geodesic Equation in Schwarzschild Metric

$$r^2 \frac{d^2\phi}{d\lambda^2} + 2r \frac{dr}{d\lambda}\frac{d\phi}{d\lambda} = 0$$

$$\frac{d}{d\lambda}\left(r^2 \frac{d\phi}{d\lambda}\right) = 0$$

$$r^2 \frac{d\phi}{d\lambda} = B$$

where in line 2 we multiply by r^2, in line 3 we use the product rule, and in line 4 we integrate with respect to λ (with B being a constant). Accordingly, the system of Eqs. 174-176 becomes:

$$F\frac{dt}{d\lambda} = A \qquad (177)$$

$$\frac{d^2r}{d\lambda^2} + \frac{GMF}{r^2}\left(\frac{dt}{d\lambda}\right)^2 - \frac{GM}{c^2 r^2 F}\left(\frac{dr}{d\lambda}\right)^2 - rF\left(\frac{d\phi}{d\lambda}\right)^2 = 0 \qquad (178)$$

$$r^2 \frac{d\phi}{d\lambda} = B \qquad (179)$$

(d) Simplifying Eq. 178: if we substitute from Eqs. 177 and 179 into Eq. 178 we get:

$$\frac{d^2r}{d\lambda^2} + \frac{GMF}{r^2}\left(\frac{A}{F}\right)^2 - \frac{GM}{c^2 r^2 F}\left(\frac{dr}{d\lambda}\right)^2 - rF\left(\frac{B}{r^2}\right)^2 = 0$$

$$\frac{d^2r}{d\lambda^2} + \frac{GMA^2}{r^2 F} - \frac{GM}{c^2 r^2 F}\left(\frac{dr}{d\lambda}\right)^2 - \frac{FB^2}{r^3} = 0$$

Accordingly, the system of Eqs. 177-179 becomes:

$$F\frac{dt}{d\lambda} = A \qquad (180)$$

$$\frac{d^2r}{d\lambda^2} + \frac{GMA^2}{r^2 F} - \frac{GM}{c^2 r^2 F}\left(\frac{dr}{d\lambda}\right)^2 - \frac{FB^2}{r^3} = 0 \qquad (181)$$

$$r^2 \frac{d\phi}{d\lambda} = B \qquad (182)$$

(e) Developing expression for $\frac{d^2r}{d\lambda^2}$: by the chain and product rules we have:

$$\frac{dr}{d\lambda} = \frac{dr}{d\phi}\frac{d\phi}{d\lambda} \qquad (183)$$

$$\frac{d^2r}{d\lambda^2} = \frac{d^2r}{d\phi^2}\left(\frac{d\phi}{d\lambda}\right)^2 + \frac{dr}{d\phi}\frac{d^2\phi}{d\lambda^2} \qquad (184)$$

Now:

$$\frac{d^2\phi}{d\lambda^2} = \frac{d}{d\lambda}\left(\frac{d\phi}{d\lambda}\right) = \frac{d}{d\lambda}\left(\frac{B}{r^2}\right) = \frac{d}{dr}\left(\frac{B}{r^2}\right)\frac{dr}{d\lambda} = -2\frac{B}{r^3}\frac{dr}{d\lambda}$$

where Eq. 182 is used in step 2 and the chain rule is used in step 3. On substituting from the last equation into Eq. 184 we get:

$$\frac{d^2r}{d\lambda^2} = \frac{d^2r}{d\phi^2}\left(\frac{d\phi}{d\lambda}\right)^2 - 2\frac{B}{r^3}\frac{dr}{d\lambda}\frac{dr}{d\phi}$$

$$= \frac{d^2r}{d\phi^2}\left(\frac{B}{r^2}\right)^2 - 2\frac{B}{r^3}\left(\frac{dr}{d\phi}\frac{d\phi}{d\lambda}\right)\frac{dr}{d\phi}$$

$$= \frac{B^2}{r^4}\frac{d^2r}{d\phi^2} - 2\frac{B}{r^3}\left(\frac{dr}{d\phi}\right)^2 \frac{B}{r^2}$$

4.1.2 Geodesic Equation in Schwarzschild Metric

$$= \frac{B^2}{r^4}\frac{d^2r}{d\phi^2} - \frac{2B^2}{r^5}\left(\frac{dr}{d\phi}\right)^2$$

where in line 2 we substitute from Eqs. 182 and 183, and in line 3 we substitute from Eq. 182. On substituting from the last equation into Eq. 181 we get:

$$\frac{B^2}{r^4}\frac{d^2r}{d\phi^2} - \frac{2B^2}{r^5}\left(\frac{dr}{d\phi}\right)^2 + \frac{GMA^2}{r^2F} - \frac{GM}{c^2r^2F}\left(\frac{dr}{d\lambda}\right)^2 - \frac{FB^2}{r^3} = 0$$

Accordingly, the system of Eqs. 180-182 becomes:

$$F\frac{dt}{d\lambda} = A \qquad (185)$$

$$\frac{B^2}{r^4}\frac{d^2r}{d\phi^2} - \frac{2B^2}{r^5}\left(\frac{dr}{d\phi}\right)^2 + \frac{GMA^2}{r^2F} - \frac{GM}{c^2r^2F}\left(\frac{dr}{d\lambda}\right)^2 - \frac{FB^2}{r^3} = 0 \qquad (186)$$

$$r^2\frac{d\phi}{d\lambda} = B \qquad (187)$$

(**f**) Introducing new dependent variable u: if we introduce a new dependent variable $u = 1/r$ (and hence $r = 1/u$) then we have:

$$\frac{dr}{d\phi} = -\frac{1}{u^2}\frac{du}{d\phi} \qquad (188)$$

$$\frac{d^2r}{d\phi^2} = \frac{2}{u^3}\left(\frac{du}{d\phi}\right)^2 - \frac{1}{u^2}\frac{d^2u}{d\phi^2} \qquad (189)$$

Hence, Eq. 186 can be written as:

$$B^2u^4\left[\frac{2}{u^3}\left(\frac{du}{d\phi}\right)^2 - \frac{1}{u^2}\frac{d^2u}{d\phi^2}\right] - 2B^2u^5\left(-\frac{1}{u^2}\frac{du}{d\phi}\right)^2 + \frac{GMA^2u^2}{F} - \frac{GMu^2}{c^2F}\left(\frac{du^{-1}}{d\lambda}\right)^2 - FB^2u^3 = 0$$

$$2B^2u\left(\frac{du}{d\phi}\right)^2 - B^2u^2\frac{d^2u}{d\phi^2} - 2B^2u\left(\frac{du}{d\phi}\right)^2 + \frac{GMA^2u^2}{F} - \frac{GMu^2}{c^2F}\left(\frac{du^{-1}}{d\lambda}\right)^2 - FB^2u^3 = 0$$

$$-B^2u^2\frac{d^2u}{d\phi^2} + \frac{GMA^2u^2}{F} - \frac{GMu^2}{c^2F}\left(\frac{du^{-1}}{d\lambda}\right)^2 - FB^2u^3 = 0$$

$$\frac{d^2u}{d\phi^2} - \frac{GMA^2}{B^2F} + \frac{GM}{B^2c^2F}\left(\frac{du^{-1}}{d\lambda}\right)^2 + Fu = 0$$

where in equality 1 we substitute from Eqs. 188 and 189 and use $u = 1/r$, in equalities 2 and 3 we simplify, and in equality 4 we divide by $-B^2u^2$. Accordingly, the system of Eqs. 185-187 becomes:

$$F\frac{dt}{d\lambda} = A \qquad (190)$$

$$\frac{d^2u}{d\phi^2} - \frac{GMA^2}{B^2F} + \frac{GM}{B^2c^2F}\left(\frac{du^{-1}}{d\lambda}\right)^2 + Fu = 0 \qquad (191)$$

$$\frac{d\phi}{d\lambda} = Bu^2 \qquad (192)$$

as required.
Note: although the assumption that the gravitated object is initially in the plane $\theta = \pi/2$ is justifiable (since we can choose our coordinate system in such a way that satisfies this condition) the assumption that the first derivative of θ vanishes (i.e. $\frac{d\theta}{d\lambda} = 0$) may not be obvious. Yes, this may be justified by the use of classical mechanics where orbits in central force field are shown to be planar (or equivalently

4.1.2 Geodesic Equation in Schwarzschild Metric

by employing the conservation of angular momentum) although this is much stronger condition (since it will make variations of all orders vanish) and it will make the rest of our argument redundant. However, since the initial point is arbitrary it can be made to satisfy both conditions (i.e. $\theta = \pi/2$ and $\frac{d\theta}{d\lambda} = 0$) although the general validity of this may not be obvious.

We should also note that due to the spherical symmetry of the physical setting the restriction of the motion to the equatorial plane (i.e. $\theta = \pi/2$ condition) does not impose restriction on the general validity of the obtained geodesic equations since the equatorial plane is arbitrary and can be any plane that passes through the center of the gravitating body.

3. Show that for massive objects the system of Eqs. 158-160 can be simplified to the system of Eqs. 161-163.
   Answer: Massive objects follow timelike geodesics and hence the proper time parameter τ can be used as an affine parameter in the geodesic equations, i.e. Eqs. 158-160 (see problem 1 of § 2.9.5). So, we have:

$$F \frac{dt}{d\tau} = A \tag{193}$$

$$\frac{d^2 u}{d\phi^2} - \frac{GMA^2}{B^2 F} + \frac{GM}{B^2 c^2 F} \left(\frac{du^{-1}}{d\tau} \right)^2 + Fu = 0 \tag{194}$$

$$\frac{d\phi}{d\tau} = Bu^2 \tag{195}$$

Now, from the Schwarzschild quadratic form we have:

$$\begin{aligned}
(d\sigma)^2 &= F(cdt)^2 - F^{-1}(dr)^2 - r^2(d\theta)^2 - r^2 \sin^2\theta (d\phi)^2 \\
(d\sigma)^2 &= F(cdt)^2 - F^{-1}(dr)^2 - r^2(d\phi)^2 \\
\frac{(d\sigma)^2}{c^2} &= F(dt)^2 - \frac{F^{-1}}{c^2}(dr)^2 - \frac{r^2}{c^2}(d\phi)^2 \\
(d\tau)^2 &= F(dt)^2 - \frac{F^{-1}}{c^2}(dr)^2 - \frac{r^2}{c^2}(d\phi)^2 \\
1 &= F\left(\frac{dt}{d\tau}\right)^2 - \frac{F^{-1}}{c^2}\left(\frac{dr}{d\tau}\right)^2 - \frac{r^2}{c^2}\left(\frac{d\phi}{d\tau}\right)^2 \\
\left(\frac{dr}{d\tau}\right)^2 &= c^2 F^2 \left(\frac{dt}{d\tau}\right)^2 - F r^2 \left(\frac{d\phi}{d\tau}\right)^2 - c^2 F \\
\left(\frac{du^{-1}}{d\tau}\right)^2 &= c^2 F^2 \left(\frac{dt}{d\tau}\right)^2 - \frac{F}{u^2}\left(\frac{d\phi}{d\tau}\right)^2 - c^2 F \\
\left(\frac{du^{-1}}{d\tau}\right)^2 &= c^2 A^2 - B^2 F u^2 - c^2 F
\end{aligned} \tag{196}$$

where line 1 is the Schwarzschild quadratic form with $F = 1 - \frac{2GM}{c^2 r}$, in line 2 we use $d\theta = 0$ and $\sin^2 \theta = 1$ since $\theta = \pi/2$ (see problem 2), in line 3 we divide by c^2, in line 4 we use the relation between σ and τ, in line 5 we divide by $(d\tau)^2$, in line 6 we multiply by $c^2 F$ with some algebraic manipulation, in line 7 we use $r = 1/u$, and in line 8 we substitute from Eqs. 193 and 195 and simplify. On inserting the expression of $\left(\frac{du^{-1}}{d\tau}\right)^2$ from the last equation into Eq. 194 we get:

$$\frac{d^2 u}{d\phi^2} - \frac{GMA^2}{B^2 F} + \frac{GM}{B^2 c^2 F}(c^2 A^2 - B^2 F u^2 - c^2 F) + Fu = 0$$

$$\frac{d^2 u}{d\phi^2} - \frac{GM u^2}{c^2} - \frac{GM}{B^2} + Fu = 0$$

$$\frac{d^2 u}{d\phi^2} - \frac{GM u^2}{c^2} - \frac{GM}{B^2} + \left(1 - \frac{2GMu}{c^2}\right) u = 0$$

4.1.2 Geodesic Equation in Schwarzschild Metric

$$\frac{d^2u}{d\phi^2} - \frac{GMu^2}{c^2} - \frac{GM}{B^2} + u - \frac{2GMu^2}{c^2} = 0$$

$$\frac{d^2u}{d\phi^2} - \frac{GM}{B^2} + u - \frac{3GMu^2}{c^2} = 0$$

$$\frac{d^2u}{d\phi^2} + u = \frac{GM}{B^2} + \frac{3GMu^2}{c^2}$$

where in line 3 we use $F = 1 - \frac{2GMu}{c^2}$. Hence, the system of Eqs. 193-195 becomes:

$$F\frac{dt}{d\tau} = A$$

$$\frac{d^2u}{d\phi^2} + u = \frac{GM}{B^2} + \frac{3GMu^2}{c^2}$$

$$\frac{d\phi}{d\tau} = Bu^2$$

as required.

4. Show that for massless objects the system of Eqs. 158-160 can be simplified to the system of Eqs. 164-166.

Answer: Massless objects follow lightlike geodesics (or null geodesics) and hence their quadratic form is zero, i.e. $(d\sigma)^2 = 0$. So, from the quadratic form of the Schwarzschild metric we have:

$$0 = F(cdt)^2 - F^{-1}(dr)^2 - r^2(d\theta)^2 - r^2\sin^2\theta(d\phi)^2$$

$$0 = F(cdt)^2 - F^{-1}(dr)^2 - r^2(d\phi)^2$$

$$0 = F\left(\frac{dt}{d\lambda}\right)^2 - \frac{F^{-1}}{c^2}\left(\frac{dr}{d\lambda}\right)^2 - \frac{r^2}{c^2}\left(\frac{d\phi}{d\lambda}\right)^2$$

$$0 = F\left(\frac{dt}{d\lambda}\right)^2 - \frac{F^{-1}}{c^2}\left(\frac{du^{-1}}{d\lambda}\right)^2 - \frac{1}{c^2u^2}\left(\frac{d\phi}{d\lambda}\right)^2$$

$$0 = \frac{A^2}{F} - \frac{F^{-1}}{c^2}\left(\frac{du^{-1}}{d\lambda}\right)^2 - \frac{B^2u^2}{c^2}$$

$$\left(\frac{du^{-1}}{d\lambda}\right)^2 = c^2A^2 - B^2Fu^2$$

where in line 1 we use the Schwarzschild quadratic form with $F = 1 - \frac{2GM}{c^2 r}$, in line 2 we use $d\theta = 0$ and $\sin^2\theta = 1$ since $\theta = \pi/2$ (see problem 2), in line 3 we divide by $c^2(d\lambda)^2$, in line 4 we use $r = 1/u$, in line 5 we substitute from Eqs. 158 and 160 and simplify, and in line 6 we multiply by Fc^2 and manipulate. On inserting the expression of $\left(\frac{du^{-1}}{d\lambda}\right)^2$ from the last equation into Eq. 159 we get:

$$\frac{d^2u}{d\phi^2} - \frac{GMA^2}{B^2 F} + \frac{GM}{B^2 c^2 F}\left(c^2A^2 - B^2Fu^2\right) + Fu = 0$$

$$\frac{d^2u}{d\phi^2} - \frac{GMu^2}{c^2} + Fu = 0$$

$$\frac{d^2u}{d\phi^2} - \frac{GMu^2}{c^2} + \left(1 - \frac{2GMu}{c^2}\right)u = 0$$

$$\frac{d^2u}{d\phi^2} - \frac{GMu^2}{c^2} + u - \frac{2GMu^2}{c^2} = 0$$

$$\frac{d^2u}{d\phi^2} + u - \frac{3GMu^2}{c^2} = 0$$

4.2 Kerr Solution

$$\frac{d^2u}{d\phi^2} + u = \frac{3GMu^2}{c^2}$$

where in line 3 we use $F = 1 - \frac{2GMu}{c^2}$. Hence, the system of Eqs. 158-160 becomes:

$$F \frac{dt}{d\lambda} = A$$

$$\frac{d^2u}{d\phi^2} + u = \frac{3GMu^2}{c^2}$$

$$\frac{d\phi}{d\lambda} = Bu^2$$

as required.

Exercises

1. Summarize the essence of the formalism and strategy of general relativity in determining the trajectories of massive and massless objects in the spacetime under the effect of gravity. Hence, justify the investigation of this subsection and its applications.
2. Compare the system of Eqs. 161-163 to the system of Eqs. 164-166.

4.2 Kerr Solution

This is another exact analytical solution to the Field Equations of general relativity. The Kerr metric describes the geometry of empty spacetime outside a **rotating** "Schwarzschild object" (i.e. static, spherically symmetric and electrically neutral gravitating object)[168] and hence it is an extension to the Schwarzschild solution by adding rotation to the Schwarzschild object. Again, it is a vacuum solution (due to the "empty" attribute) and hence it is a solution to the tensorial equation $R_{\mu\nu} = 0$ (also see § 8.7 and § 8.7.2). The quadratic form $(d\sigma)^2$ of the Kerr solution is given by:

$$(d\sigma)^2 = \left(1 - \frac{R_S r}{\rho^2}\right)(cdt)^2 - \frac{\rho^2}{\Delta}(dr)^2 - \rho^2 (d\theta)^2$$
$$- \sin^2\theta \left(r^2 + a^2 + \frac{R_S r a^2 \sin^2\theta}{\rho^2}\right)(d\phi)^2 + \frac{2R_S r a \sin^2\theta}{\rho^2} cdt\, d\phi \qquad (197)$$

where R_S is the "Schwarzschild radius" (i.e. $R_S \equiv \frac{2GM}{c^2}$), $a = \frac{J}{cM}$ with J being the magnitude of the angular momentum,[169] (ct, r, θ, ϕ) are the Boyer-Lindquist coordinates (see exercise 10), $\rho^2 = a^2 \cos^2\theta + r^2$, and $\Delta = r^2 - rR_S + a^2$. Moreover, the axis of rotation is assumed to be the same as the polar axis of the spatial coordinates.

Problems

1. Show that the Kerr metric converges to the Lorentz metric of special relativity when $r \to \infty$.
   Answer: We have:

$$\lim_{r \to \infty} \frac{R_S r}{\rho^2} = \lim_{r \to \infty} \frac{R_S r}{a^2 \cos^2\theta + r^2} = \lim_{r \to \infty} \frac{R_S}{(a^2/r)\cos^2\theta + r} = 0$$

$$\lim_{r \to \infty} \frac{\rho^2}{\Delta} = \lim_{r \to \infty} \frac{a^2 \cos^2\theta + r^2}{r^2 - rR_S + a^2} = \lim_{r \to \infty} \frac{(a^2/r^2)\cos^2\theta + 1}{1 - (R_S/r) + (a^2/r^2)} = 1$$

$$\lim_{r \to \infty} \rho^2 = \lim_{r \to \infty} (a^2 \cos^2\theta + r^2) = \lim_{r \to \infty} r^2 \left([a^2/r^2]\cos^2\theta + 1\right) = r^2$$

---

[168] We note that what is required for the Kerr problem (and the upcoming Kerr-Newman problem) is axial symmetry of the gravitating object. Also, "static" excludes "rotating".
[169] The parameter a may be called the Kerr parameter or the rotation parameter.

$$\lim_{r \to \infty} \left(r^2 + a^2 + \frac{R_S r a^2 \sin^2 \theta}{\rho^2} \right) = \lim_{r \to \infty} r^2 \left(1 + (a^2/r^2) + \frac{R_S a^2 \sin^2 \theta}{\rho^2 r} \right)$$
$$= \lim_{r \to \infty} r^2 \left(1 + (a^2/r^2) + \frac{R_S a^2 \sin^2 \theta}{(a^2 \cos^2 \theta + r^2) r} \right) = r^2$$

$$\lim_{r \to \infty} \left(\frac{2 R_S r a \sin^2 \theta}{\rho^2} \right) = \lim_{r \to \infty} \left(\frac{2 R_S r a \sin^2 \theta}{a^2 \cos^2 \theta + r^2} \right) = \lim_{r \to \infty} \left(\frac{2 R_S a \sin^2 \theta}{(a^2/r) \cos^2 \theta + r} \right) = 0$$

Hence:

$$\lim_{r \to \infty} (d\sigma)^2 = (1-0)(cdt)^2 - 1(dr)^2 - r^2 (d\theta)^2 - \sin^2 \theta \left(r^2 \right) (d\phi)^2 + 0$$
$$= (cdt)^2 - (dr)^2 - r^2 (d\theta)^2 - r^2 \sin^2 \theta (d\phi)^2$$

which is the Lorentz metric (or rather quadratic form which is based on the metric) of special relativity with spatial spherical coordinates (see exercise 10 of § 2.5). In fact, we should also assume that the Boyer-Lindquist coordinates reduce to the ordinary coordinates of flat space at infinity (which is the case according to the definition of the Boyer-Lindquist coordinates; see exercise 10).

Exercises

1. Summarize the main characteristics of the Kerr solution.
2. Write down the Kerr metric tensor.
3. Show that the Kerr solution converges to the Schwarzschild solution when the gravitating object is not rotating.
4. Show that the Kerr metric converges to the Lorentz metric of special relativity when $M = 0$.
5. Show that the Kerr metric is time independent.
6. Show that the Kerr metric is axially symmetric but not spherically symmetric.
7. Investigate the singularities of the Kerr solution.
8. Is the Kerr line element reversible in time and why?
9. Why is there a $cdt\, d\phi$ mixed term in the Kerr line element?
10. What are the Boyer-Lindquist coordinates?
11. What is the difference between "static" and "stationary" (as part of the terminology used in the discussion of space metric)?

4.3 Reissner-Nordstrom Solution

This is another exact analytical solution to the Field Equations of general relativity. The Reissner-Nordstrom metric describes the geometry of spacetime outside an **electrically charged** "Schwarzschild object" and hence it is an extension to the Schwarzschild solution by adding electric charge to the Schwarzschild object. However, it is not a vacuum solution because the surrounding space is filled with electrostatic field which should generate non-vanishing energy-momentum tensor (see exercise 5 of § 2.14). It is noteworthy that the Reissner-Nordstrom problem (as well as the upcoming Kerr-Newman problem) of an electrically charged gravitating object (which is generally presumed to be a black hole) has mainly theoretical value since the existence of such an electrically charged object is questionable in real astrophysical situations where charge (if it does exist initially) should neutralize (or almost neutralize) in a rather short time due to the interaction with other objects in the vicinity (e.g. gas or cosmic dust) or through emission (if emission is possible).[170] Therefore, we do not investigate this solution (and the upcoming Kerr-Newman solution) in detail.

Exercises

1. Summarize the main characteristics of the Reissner-Nordstrom solution.

---

[170] On a large scale matter is electrically neutral.

4.4 Kerr-Newman Solution

This is another exact analytical solution to the Field Equations of general relativity. The Kerr-Newman metric describes the geometry of spacetime outside a **rotating** and **electrically charged** "Schwarzschild object"[171] and hence it is an extension to the Schwarzschild solution by adding angular momentum and electric charge to the Schwarzschild object. Again, it is not a vacuum solution because the surrounding space is filled with electromagnetic field which should generate non-vanishing energy-momentum tensor.

Problems

1. The gravitating object of the Kerr-Newman solution is characterized physically by mass, angular momentum and electric charge. Comment on this.
   Answer: This should remind us of the theorem (which is commonly called "no hair theorem")[172] that any black hole is totally characterized (from the perspective of an external observer) by only three independent (and intrinsic) physical parameters: mass M, angular momentum J and electric charge Q (see § 8.7). In this sense, the Kerr-Newman metric may be considered as the most general metric for black holes.

Exercises

1. Summarize the main characteristics of the Kerr-Newman solution.
2. Compare the Schwarzschild metric to the following three metrics: Kerr, Reissner-Nordstrom, and Kerr-Newman.
3. Referring to black holes, what is the main physical distinction of the Kerr and Kerr-Newman metrics from the Schwarzschild metric?

---

[171] We note again the requirement of axial (rather than spherical) symmetry and the exclusion of "rotating" from "static".
[172] Hair refers to the detailed physical properties of the object other than those three (i.e. mass, angular momentum and electric charge).

Chapter 5
Classical Limit of General Relativity

According to the correspondence principle, the formalism of general relativity should converge to the Newtonian formalism in the classical limit (i.e. time independent weak gravitational fields sourced by matter only with low-speed gravitated object). This is a necessary (but not sufficient) requirement for the acceptance of general relativity because no correct theory should contradict the predictions of classical physics in its domain of validity. In the following sections we give some examples of this convergence where we will see that the classical formulae approximate the corresponding general relativistic formulae in the domain of validity of classical gravity.

5.1 Convergence to Newtonian Gravity

In this section we show that the Field Equation converges to the Newtonian gravity, as represented by the Poisson equation (i.e. Eq. 14), in the classical limit. The required assumptions for the validity of this classical limit are:

1. The source of gravity includes matter but not energy such as electromagnetic radiation. This is justified by the fact that classical gravity is an attribute of mass in its classical sense and hence energy is not a source of gravity.
2. The gravitational field is weak (see § 1.3 and exercises 18 and 19 of § 4.1.1 for the definition of "weak"). This is because in strong gravitational fields non-classical effects of gravity become tangible.
3. The gravitational field is time independent. This can be justified by the fact that classical gravity is inherently a spatial phenomenon with no inbuilt time dependency (see exercise 2).[173]
4. Any motion of the gravitated object is very slow relative to the speed of light c and hence the temporal component of its velocity 4-vector is much larger than the spatial components, i.e. $\frac{dx^0}{d\tau} \gg \frac{dx^i}{d\tau}$ ($i = 1, 2, 3$).[174] This is because at high speeds (i.e. comparable to c) peculiar special and general relativistic effects take place and this invalidates the classical limit treatment.

Now, since the gravitational field is weak the space is approximately flat Minkowskian and hence the metric tensor $g_{\mu\nu}$ can be approximated by:

$$g_{\mu\nu} \simeq \eta_{\mu\nu} + \varepsilon_{\mu\nu} \tag{198}$$

where $\eta_{\mu\nu}$ is the Minkowski metric tensor (i.e. diag $[1, -1, -1, -1]$) while $\varepsilon_{\mu\nu}$ is a perturbation tensor that represents first order perturbations to the components of the Minkowski metric such that $|\varepsilon_{\mu\nu}| \ll 1$.

The above assumptions for the validity of the classical limit means that the only non-zero component of the energy-momentum tensor $T_{\mu\nu}$ is $T_{00} = \rho c^2$ (see § 2.14) because this component represents scaled mass density (which does exist in this classical setting) while all the other components vanish because they do not represent a classical source of gravity and hence they do not exist according to the given assumptions. In fact, this justification originates from a classical perspective. From a general relativistic perspective the component T_{00} is much larger than any other component of $T_{\mu\nu}$ in the classical limit and hence the other components are negligible in their overall contribution to gravity.[175] Anyway, let ignore

---
[173] Regardless of this justification, this can be seen as a convenient case for demonstrating the convergence.
[174] This condition should become more obvious if we note that the temporal component of the velocity 4-vector is γc while the spatial components are γu^i.
[175] In fact, we may assume an energy-momentum tensor of a homogeneous cloud of dust (possibly in a co-moving frame; see problems 3 and 4 of § 2.14) which seems appropriate in this classical setting (although it may weaken the generality of our argument). Moreover, since we are dealing with classical limit (and possibly in a co-moving frame) then ρ and ρ_0 are essentially the same (and we use ρ for simplicity).

5.1 Convergence to Newtonian Gravity

these trivial and messy details (some of which may be questionable) and try to obtain the classical limit from the 00 component alone since the field formulation of classical gravity (i.e. Poisson equation) has only one component which should correspond to the 00 component of the general relativistic formulation. Accordingly:

$$R_{\mu\nu} = \kappa \left(T_{\mu\nu} - \frac{1}{2} g_{\mu\nu} T \right) \tag{199}$$

$$R_{00} = \kappa \left(T_{00} - \frac{1}{2} g_{00} T \right) \tag{200}$$

$$R_{00} \simeq \kappa \left(\rho c^2 - \frac{1}{2} [\eta_{00} + \varepsilon_{00}] \rho c^2 \right) \tag{201}$$

$$R_{00} = \kappa \left(\rho c^2 - \frac{1}{2} [1 + \varepsilon_{00}] \rho c^2 \right) \tag{202}$$

$$R_{00} = \kappa \left(\rho c^2 - \frac{1}{2} \rho c^2 - \frac{1}{2} \rho c^2 \varepsilon_{00} \right) \tag{203}$$

$$R_{00} = \kappa \left(\frac{1}{2} \rho c^2 - \frac{1}{2} \rho c^2 \varepsilon_{00} \right) \tag{204}$$

$$R_{00} \simeq \frac{1}{2} \kappa \rho c^2 \tag{205}$$

where in line 1 we use the tensorial Field Equation in one of its forms (see exercise 5 of § 3.2), in line 2 we take just the 00 component of the tensorial equation, in line 3 we use $T_{00} = T = \rho c^2$ and $g_{00} \simeq \eta_{00} + \varepsilon_{00}$ (see Eq. 198), in line 4 we use $\eta_{00} = 1$, and in line 7 we use $|\varepsilon_{00}| \ll 1$.

Similarly, from the geometric definition of R_{00} we have (see Eq. 89):

$$R_{00} = \partial_0 \Gamma^\alpha_{0\alpha} - \partial_\alpha \Gamma^\alpha_{00} + \Gamma^\beta_{0\alpha} \Gamma^\alpha_{\beta 0} - \Gamma^\beta_{00} \Gamma^\alpha_{\beta\alpha} \simeq -\partial_\alpha \Gamma^\alpha_{00} \tag{206}$$

This approximation is justified by the fact that $\partial_0 \Gamma^\alpha_{0\alpha} = 0$ due to the presumed time independence while $\Gamma^\beta_{0\alpha} \Gamma^\alpha_{\beta 0}$ and $\Gamma^\beta_{00} \Gamma^\alpha_{\beta\alpha}$ are negligible in our first order approximation since they are of higher order in $\varepsilon_{\mu\nu}$ (refer to § 3.4). Now, Γ^α_{00} is given by (see Eq. 52):

$$\Gamma^\alpha_{00} = \frac{g^{\alpha\beta}}{2} \left(\partial_0 g_{0\beta} + \partial_0 g_{0\beta} - \partial_\beta g_{00} \right)$$

$$= -\frac{g^{\alpha\beta}}{2} \partial_\beta g_{00}$$

$$\simeq -\frac{g^{\alpha\beta}}{2} \partial_\beta \left(\eta_{00} + \varepsilon_{00} \right)$$

$$= -\frac{g^{\alpha\beta}}{2} \partial_\beta \left(1 + \varepsilon_{00} \right)$$

$$= -\frac{g^{\alpha\beta}}{2} \partial_\beta \varepsilon_{00}$$

$$\simeq -\frac{1}{2} \partial^\alpha \varepsilon_{00} \tag{207}$$

where line 2 is because $\partial_0 g_{0\beta} = 0$ due to time independence, in line 3 we substitute for g_{00} using Eq. 198, in line 4 we use $\eta_{00} = 1$, and in line 6 we raise the index.[176] On substituting from Eq. 207 into Eq. 206 we obtain:

$$R_{00} \simeq -\partial_\alpha \left(-\frac{1}{2} \partial^\alpha \varepsilon_{00} \right)$$

---

[176] We note that $g^{\alpha\beta}$ may be replaced by $\eta^{\alpha\beta}$ for index raising (see Eq. 111) as we did in § 3.4. In fact, this could be needed in justifying some of the subsequent steps (see footnote [144]).

5.1 Convergence to Newtonian Gravity

$$= \frac{1}{2}\partial_\alpha \partial^\alpha \varepsilon_{00} \qquad (208)$$

$$= \frac{1}{2}\Box^2 \varepsilon_{00} \qquad (209)$$

$$= -\frac{1}{2}\nabla^2 \varepsilon_{00} \qquad (210)$$

where in line 3 we use the definition of the d'Alembertian operator, and in line 4 we use time independence (noting that $\Box^2 = \frac{1}{c^2}\frac{\partial^2}{\partial t^2} - \nabla^2$; see exercise 15 of § 2.2). On comparing Eq. 210 with Eq. 205 we get:

$$\nabla^2 \varepsilon_{00} = -\kappa \rho c^2 \qquad (211)$$

Now, from the geodesic equation, using the proper time parameter τ as an affine parameter (see problem 1 of § 2.9.5), we have:

$$\frac{d^2 x^\alpha}{d\tau^2} + \Gamma^\alpha_{\beta\gamma} \frac{dx^\beta}{d\tau} \frac{dx^\gamma}{d\tau} = 0 \qquad (212)$$

The term $\Gamma^\alpha_{\beta\gamma} \frac{dx^\beta}{d\tau} \frac{dx^\gamma}{d\tau}$ is a sum of terms all of which are negligible according to our approximation except $\Gamma^\alpha_{00} \frac{dx^0}{d\tau} \frac{dx^0}{d\tau}$. The reason for the dominance of $\Gamma^\alpha_{00} \frac{dx^0}{d\tau} \frac{dx^0}{d\tau}$ over all other terms combined is the aforementioned condition $\frac{dx^0}{d\tau} \gg \frac{dx^i}{d\tau}$ which makes all the terms involving one or two spatial components of the velocity 4-vector negligible in comparison to the term that involves the square of the temporal component, i.e. $\Gamma^\alpha_{00} \frac{dx^0}{d\tau} \frac{dx^0}{d\tau}$. Accordingly, the geodesic equation can be approximated by:

$$\frac{d^2 x^\alpha}{d\tau^2} \simeq -\Gamma^\alpha_{00} \left(\frac{dx^0}{d\tau}\right)^2 \qquad (213)$$

$$\frac{d^2 x^\alpha}{d\tau^2} \simeq -\left(-\frac{1}{2}\partial^\alpha \varepsilon_{00}\right)\left(\frac{dx^0}{d\tau}\right)^2 \qquad (214)$$

$$\frac{d^2 x^\alpha}{d\tau^2} \simeq \frac{1}{2}\left(\frac{dx^0}{d\tau}\right)^2 \partial^\alpha \varepsilon_{00} \qquad (215)$$

$$\frac{d^2 x^\alpha}{d\tau^2} \simeq \frac{c^2}{2}\left(\frac{dt}{d\tau}\right)^2 \partial^\alpha \varepsilon_{00} \qquad (216)$$

$$\frac{d^2 x^\alpha}{dt^2} \simeq \frac{c^2}{2} \partial^\alpha \varepsilon_{00} \qquad (217)$$

where in line 2 we substitute from Eq. 207, in line 4 we use $x^0 = ct$, and in line 5 we use the fact that in the classical limit we have $d\tau \simeq dt$.[177] Now, if we take the spatial components only (which are the only components that exist classically or at least they are the only ones that we need here) we get:

$$\frac{d^2 x^i}{dt^2} \simeq \frac{c^2}{2} \partial^i \varepsilon_{00} \qquad (i = 1, 2, 3)$$

$$= \frac{c^2}{2}[-\nabla \varepsilon_{00}]^i$$

$$= -\frac{c^2}{2}[\nabla \varepsilon_{00}]^i \qquad (218)$$

where in line 2 we use the definition of the contravariant partial differential operator ∂^i (see exercise 15 of 2.2).

---

[177] In the classical limit the gravitational field is weak and hence we have $\left(1 - \frac{2GM}{c^2 r}\right)^{1/2} \simeq 1$ and therefore from Eq. 243 we get $d\tau \simeq dt$ (see § 6.3.1). Also, any special relativistic effects are negligible due to the low-speed limit and hence $\gamma \simeq 1$ and $d\tau \simeq dt$ (see B4). We may also multiply both sides of line 4 with $(d\tau/dt)^2$ and use the chain rule noting that $\frac{d^2 x^\alpha}{d\tau^2}\left(\frac{d\tau}{dt}\right)^2 = \frac{d^2 x^\alpha}{dt^2}$ (see problem 1 of § 2.9.5) although this should require t to be an affine parameter (which may be reasonable in the classical limit). Also see exercise 7.

5.1 Convergence to Newtonian Gravity

Referring to Eq. 12, which is the classical formulation of the gravitational field noting that $[\mathbf{g}]^i = \frac{d^2 x^i}{dt^2}$,[178] we have:

$$\frac{d^2 x^i}{dt^2} = -[\nabla \Phi]^i \tag{219}$$

So, if we equate the right hand side of Eq. 218 (which is the approximate general relativistic formulation as derived from the geodesic equation) to the right hand side of Eq. 219 (which is the classical formulation) we get:

$$\nabla \Phi = \frac{c^2}{2} \nabla \varepsilon_{00} \tag{220}$$

On taking the divergence of the two sides of the last equation we obtain:

$$\nabla \cdot \nabla \Phi = \frac{c^2}{2} \nabla \cdot \nabla \varepsilon_{00} \tag{221}$$

$$\nabla^2 \Phi = \frac{c^2}{2} \nabla^2 \varepsilon_{00} \tag{222}$$

$$\nabla^2 \Phi = \frac{c^2}{2} \left(-\kappa \rho c^2\right) \tag{223}$$

$$\nabla^2 \Phi = -\frac{1}{2} \kappa \rho c^4 \tag{224}$$

where in line 2 we use the definition of the Laplacian operator as the divergence of gradient, and in line 3 we substitute from Eq. 211.

Now, the Poisson equation of classical gravity is given by (refer to Eq. 14):

$$\nabla^2 \Phi = 4\pi G \rho \tag{225}$$

So, if κ in the general relativistic part of Eq. 224 is defined as $\kappa = -\frac{8\pi G}{c^4}$ then Eq. 224 becomes:

$$\nabla^2 \Phi = -\frac{1}{2} \left(-\frac{8\pi G}{c^4}\right) \rho c^4 = 4\pi G \rho \tag{226}$$

which is identical to the Poisson equation of classical gravity. We may also reverse this argument by saying: if the general relativistic formulation should converge to the classical limit then by comparing Eq. 224 to Eq. 225 we get:

$$-\frac{1}{2} \kappa \rho c^4 = 4\pi G \rho \tag{227}$$

$$\kappa = -\frac{8\pi G}{c^4} \tag{228}$$

This should justify what have been stated in § 3.2 about the mathematical expression of κ. In brief, the general relativistic formulation (which is mainly based on the Field Equation as explained above) converges to the classical gravity formulation (which is represented by the Poisson equation as explained in § 1.5) with κ of the general relativistic formulation being defined by Eq. 228.

Finally, it is time to show that $C_s = \frac{2GM}{c^2} \equiv R_S$ which we stated in the final stage of the derivation of the definite form of the Schwarzschild metric in § 4.1.1 without sufficient justification.[179] On integrating the two sides of Eq. 220 we get:

$$\Phi = \frac{c^2}{2} \varepsilon_{00} + A \tag{229}$$

---

[178] In fact, we should need the equivalence principle (in conjunction with Newton's second law) to justify this. However, we should need no more than the classical one and hence there should be no problem.

[179] In fact, this requires the assumption of the applicability of the classical limit approximation to the Schwarzschild solution. Although some of the classical limit assumptions are consistent with the Schwarzschild assumptions (e.g. time independence) some may not (e.g. weak gravitational field) unless we restrict the validity of the Schwarzschild solution which we should not. So, the validity of $C_s = \frac{2GM}{c^2}$ in general may require further assumptions and extensions.

where A is the constant of integration (noting that gravity is a central field). Now, Φ is given classically by $\Phi = -\frac{GM}{r}$ (see exercise 2 of § 1.5) and hence at infinity (i.e. $r \to \infty$) we have $\Phi = 0$. Moreover, at infinity the spacetime is flat and hence $\varepsilon_{00} = 0$. Therefore, we should have $A = 0$, that is:

$$\Phi = \frac{c^2}{2}\varepsilon_{00} \tag{230}$$

$$-\frac{GM}{r} = \frac{c^2}{2}\varepsilon_{00} \tag{231}$$

$$\varepsilon_{00} = -\frac{2GM}{c^2 r} \tag{232}$$

So, from Eq. 198 we get:

$$g_{00} \simeq \eta_{00} + \varepsilon_{00} = 1 - \frac{2GM}{c^2 r} \tag{233}$$

Noting that in § 4.1.1 we obtained (see Eqs. 122 and 147):

$$g_{00} = e^H = 1 - \frac{C_s}{r} \tag{234}$$

we can see (by comparing the last two equations) that $C_s = \frac{2GM}{c^2} \equiv R_S$ as claimed in § 4.1.1.

Problems

1. Summarize the essence of this section.
   Answer: Assuming classical limit conditions (i.e. weak, time independent gravitational field sourced by matter only with low speed):
   (a) From the Field Equation we obtained $R_{00} \simeq \frac{1}{2}\kappa\rho c^2$.
   (b) From the geometric definition of R_{00} we obtained $R_{00} \simeq -\frac{1}{2}\nabla^2 \varepsilon_{00}$.
   (c) On comparing the equations in **a** and **b** we obtained $\nabla^2 \varepsilon_{00} = -\kappa\rho c^2$.
   (d) From the geodesic equation we obtained $\frac{d^2 x^i}{dt^2} \simeq -\frac{c^2}{2}[\nabla\varepsilon_{00}]^i$.
   (e) From the classical formulation we have $\frac{d^2 x^i}{dt^2} = -[\nabla\Phi]^i$.
   (f) On comparing the equations in **d** and **e** we obtained $\nabla\Phi = \frac{c^2}{2}\nabla\varepsilon_{00}$.
   (g) On taking the divergence of the equation in **f** we obtained $\nabla^2\Phi = \frac{c^2}{2}\nabla^2 \varepsilon_{00}$.
   (h) On substituting from the equation in **c** (which is a relativistic equation) in the equation in **g** (which is a classic-relativistic equation) we obtained $\nabla^2\Phi = -\frac{1}{2}\kappa\rho c^4$.
   (i) Hence, if κ in the general relativistic part of the equation in **h** is defined as $\kappa = -\frac{8\pi G}{c^4}$ then the equation in **h** becomes $\nabla^2\Phi = 4\pi G\rho$ which is the classical Poisson equation. In other words, in the classical limit (associated with an appropriate definition of κ) the classical and general relativistic formulations agree.

Exercises

1. What are the main factors that determine the intensity (or strength) of the gravitational field?
2. Considering that classical gravity is basically a spatial phenomenon with no time dependency, can we have time dependency in classical gravity? If so, what is the difference between time dependency in classical gravity and in general relativity?
3. The condition of time independence may be stated by some as slow variation in time. Comment on this.
4. Does the condition of slow speed relative to c contradict the time independence of the gravitational field?
5. Outline, roughly and in general terms, the situations where general relativity can be replaced by classical gravity.
6. Use the approximation $g_{00} \simeq \eta_{00} + \varepsilon_{00} = 1 - \frac{2GM}{c^2 r}$ which is given in the text to propose a formal condition for the validity of classical gravity. Also interpret this condition.
7. Justify the claim that in the classical limit we have $d\tau \simeq dt$ considering both the special and general relativistic effects.

8. Obtain (using a compact method) the classical equation of gravitational field from the geodesic equation of general relativity in the classical limit.

5.2 Planetary Motion

In this section we obtain the Newtonian formulation of planetary motion as a classical limit for the general relativistic formulation. In other words, we will show that in the classical limit the general relativistic formulation converges to the Newtonian formulation as given in § 1.5.1 by Eqs. 15-16 and hence the planetary motion is described by Newton's laws (and indeed by Kepler's laws which are direct consequences of Newton's laws). In fact, the results that we obtained in § 5.1 about the convergence of the Field Equation to the Newtonian gravity in the classical limit should be sufficient to establish the claim that we can obtain the Newtonian formulation of planetary motion as a classical limit for general relativity.[180] However, the content in the present section should establish this fact more vividly and directly. Moreover, it can serve as a double check for the previous results. We should also note that planetary motion here is more general than the motion of planets in the solar system. In fact, it applies to all sorts of motion that meet the given criteria and hence we may call it orbital motion for the sake of generality. However, we prefer "planetary motion" due to its correspondence to the historical investigation of planetary motion in classical physics (see § 1.5.1). Moreover, orbital motion is too general since it extends to orbiting systems that do not meet the conditions of classical limit (e.g. some binary orbiting systems of compact objects with very high orbital speed). So, "planetary motion" here can be seen as a typical example of orbital motion of classical orbiting systems.

It should be obvious that what is required here is to show that in the classical limit the tensorial geodesic equation of general relativity converges to the Newtonian formulation for planetary motion since the geodesic equation describes the trajectory of gravitated free objects in spacetime. Now, the geodesic equation (see Eq. 73) requires the knowledge of the Christoffel symbols which depend on the metric of spacetime (see Eq. 52). So, what metric should we use to describe the geometry of the spacetime of the planetary motion in the classical limit? It should be obvious that the appropriate metric for the spacetime in this case is the Schwarzschild metric (because Schwarzschild object is a good prototype for the gravitating objects in orbiting systems).[181]

In fact, we have already obtained the Christoffel symbols of the Schwarzschild metric in § 4.1.1. Moreover, we have obtained in § 4.1.2 the system of geodesic equations of the Schwarzschild spacetime (i.e. the system of Eqs. 154-157) and simplified this system to the system of Eqs. 161-163 for the case of massive objects which is what we need here since planets are massive objects. So, all we need for demonstrating that the general relativistic formulation converges to the Newtonian formulation in the classical limit is to show that the general relativistic system of Eqs. 161-163 reduces to the Newtonian system of Eqs. 15-16 in this limit and hence all the classical formulations (e.g. Kepler's laws) that we obtained in § 1.5.1 from Eqs. 15-16 are obtainable from the general relativistic formulation in the classical limit. This demonstration is given in the Problems.

Problems

1. Show that in the classical limit the system of Eqs. 161-163 will reduce to the classical system of Eqs. 15-16 (which means that in the classical limit the general relativistic formulation for planetary motion reduces to the corresponding classical formulation).
   Answer: First, we discard Eq. 161 by noting that in the classical limit we have $\frac{dt}{d\tau} \simeq 1$ and hence Eq. 161 is reduced to $F \simeq A$ which is just an approximation for F. Now, since F and A have no presence in the other two equations then Eq. 161 is redundant in the classical limit and hence the system of

---

[180] In fact, the results of § 5.1 are sufficient from a gravitational perspective. So, if we note that classical planetary motion requires other classical results (e.g. Newton's second law) then we may still need (partially at least) this section to establish the above claim.

[181] We should note that some restrictive conditions in the Schwarzschild metric (such as spherical symmetry and non-rotation of the gravitating object) should apply in the considered orbiting systems. These conditions are generally valid (at least approximately) in the common classical orbiting systems.

5.2 Planetary Motion

Eqs. 161-163 reduces to two equations only, that is:

$$\frac{d^2 u}{d\phi^2} + u = \frac{GM}{B^2} + \frac{3GMu^2}{c^2} \tag{235}$$

$$\frac{d\phi}{d\tau} = Bu^2 \tag{236}$$

Next, we reduce Eq. 236 to Eq. 16 using the same argument, i.e. in the classical limit we have $d\tau \simeq dt$ and hence $\frac{d\phi}{d\tau}$ in Eq. 236 becomes $\frac{d\phi}{dt}$. Accordingly, the system of Eqs. 235-236 becomes:

$$\frac{d^2 u}{d\phi^2} + u = \frac{GM}{B^2} + \frac{3GMu^2}{c^2} \tag{237}$$

$$\frac{d\phi}{dt} = Bu^2 \tag{238}$$

Finally, we need to reduce Eq. 237 to Eq. 15. To do this we should discard the term $\frac{3GMu^2}{c^2}$ by showing that in the classical limit this term is negligible in comparison to the term $\frac{GM}{B^2}$. The ratio of $\frac{3GMu^2}{c^2}$ to $\frac{GM}{B^2}$ is:

$$\frac{3GMu^2}{c^2} \times \frac{B^2}{GM} = \frac{3u^2}{c^2} B^2$$

$$= \frac{3u^2}{c^2} \frac{1}{u^4} \left(\frac{d\phi}{dt}\right)^2$$

$$= \frac{3}{c^2 u^2} \left(\frac{d\phi}{dt}\right)^2$$

$$= \frac{3r^2}{c^2} \left(\frac{d\phi}{dt}\right)^2$$

where in line 2 we use Eq. 238 and in line 4 we use $u = 1/r$. Now, if we take the Earth as a typical planet then the average distance from the Sun is $r \simeq 1.4960 \times 10^{11}$ m and the angular speed around the Sun is $\frac{d\phi}{dt} \simeq 1.9924 \times 10^{-7}$ rad/s, and hence the ratio of $\frac{3GMu^2}{c^2}$ to $\frac{GM}{B^2}$ is:

$$\frac{3r^2}{c^2} \left(\frac{d\phi}{dt}\right)^2 \simeq \frac{3 \times \left(1.4960 \times 10^{11}\right)^2}{(3 \times 10^8)^2} \times \left(1.9924 \times 10^{-7}\right)^2 \simeq 2.9614 \times 10^{-8}$$

i.e. in the classical limit the term $\frac{3GMu^2}{c^2}$ is very small compared to the term $\frac{GM}{B^2}$ and hence the term $\frac{3GMu^2}{c^2}$ can be discarded and Eq. 237 becomes:

$$\frac{d^2 u}{d\phi^2} + u = \frac{GM}{B^2}$$

Accordingly, the system of Eqs. 237-238 becomes:

$$\frac{d^2 u}{d\phi^2} + u = \frac{GM}{B^2} \tag{239}$$

$$\frac{d\phi}{dt} = Bu^2 \tag{240}$$

As we see, the system of Eqs. 239-240 (which is obtained from the geodesic equation of general relativity as a classical limit) is identical to the system of Eqs. 15-16 (which is obtained from Newton's laws). This means that in the classical limit the general relativistic formulation for planetary motion reduces to the corresponding classical formulation, as claimed.

Note 1: it may be claimed that the validity of the above argument should require an extra condition that is the B in the classical formulation is equal to the B in the general relativistic formulation. However, we think this is a trivial requirement that can be obtained directly (if needed) from the similarity of the above two formulations in the classical limit plus the correspondence principle. In other words, the two formulations are identical provided that B in the two formulations is the same (similar to what we did in § 5.1 with κ). In fact, Eq. 240 (which is relativistic as well as classic) identifies B as the magnitude of angular momentum per unit mass or twice the areal speed (see § 1.5.1) and hence B should represent the same physical quantity in both formulations.

Note 2: there is an important difference between the classical and general relativistic formulations (apart from the term $\frac{3GMu^2}{c^2}$). This difference is related to the meaning of the coordinates (specifically t and $r = 1/u$) since in general relativity they are just coordinate variables while in classical physics they are physical variables (see § 6.2). However, in the classical limit the gap between the two should be bridged because the spacetime in this limit is essentially flat and hence the coordinate variables reduce to the physical variables (in fact this has already been considered in a sense in the above formulation with regard to t where we used the approximation $d\tau \simeq dt$ since t corresponds to proper time τ according to this approximation).

Note 3: if we note that $r\frac{d\phi}{dt} = v$ (with v representing orbital speed) then the ratio of $\frac{3GMu^2}{c^2}$ to $\frac{GM}{B^2}$ can be expressed as $\frac{3r^2}{c^2}\left(\frac{d\phi}{dt}\right)^2 = \frac{3v^2}{c^2}$ which in the classical limit (where $v \ll c$) should be very small.[182] In fact, this demonstrates the tininess of this ratio in a more general way than the example of the Earth (which is given for pedagogical purposes and to show this fact more realistically and vividly).

Exercises

1. Give some examples of necessary assumptions that should be made about the gravitating object for the validity of employing the classical limit in the planetary motion.
2. Give some examples of necessary assumptions that should be made about the gravitated object for the validity of employing the classical limit in the planetary motion.

---

[182] For example, the orbital speed of the Earth is $v \simeq 29.8$ km/s and hence $\frac{3v^2}{c^2} \simeq 2.96 \times 10^{-8}$ which is similar to the above calculation.

Chapter 6
Frames, Coordinates and Spacetime

In this chapter we investigate frames and coordinates in general relativity and their relation to spacetime. In fact, the materials in this chapter should have been investigated in an early chapter. However, the investigation is delayed to this position in the book because part of the materials requires having specific spacetime metric which was the subject of our investigation in § 4. As we will see, we will use here the Schwarzschild metric (in the Schwarzschild coordinates) that we investigated earlier in § 4.1. The choice of the Schwarzschild metric is partly justified by being the most typical and widely used metric in the applications of general relativity as well as its favorable properties such as simplicity. However, we would like to emphasize that many of the results that we obtain in this chapter depend on the validity of the Schwarzschild metric and hence they may not be valid if the spacetime is metricized by another metric. In other words, the results are "Schwarzschild-based general relativistic results" and not "general relativistic results" unconditionally.[183]

Exercises

1. Discuss the consequences of the dependency of the results that we obtain in this chapter on the Schwarzschild metric.

6.1 Frames in General Relativity

In this section we investigate the different types of frame in general relativity and compare them to the frames in classical mechanics and in special relativity (or what we call Lorentz mechanics). As we will see, several types of frame are needed in general relativity for analyzing various physical situations from different perspectives and drawing the required physical consequences.

It is obvious that the frames in classical mechanics and in special relativity are global as they cover the entire spacetime because the spacetime is flat and uniform in its metric properties and hence there is no necessity or advantage in using local frames (since this requires using numerous local frames to cover the entire spacetime) when a single global frame that covers the entire spacetime is sufficient to fulfill the intended objectives and achieve the purpose of reference frame. However, the situation in general relativity is very different because the spacetime is globally curved and locally flat and we are supposed to apply the rules of special relativity at local level. So, it is more sensible and advantageous to use local frames to coordinate local patches of spacetime and develop the required physics as dictated by the plan of general relativity.

Accordingly, in the development of the concepts, arguments and formalism of general relativity (as well as its application) we may consider (motivated mainly by pragmatic purposes as well as physical requirements) three main types of local frame:

(**a**) Frame at infinity (i.e. very far away from any source of gravitation). Since this frame does exist in a flat part of the spacetime then the rules of special relativity should apply in the frame due to the actual absence of gravity. Whether this frame is moving uniformly or at rest (considering the relativity of these attributes although in a sense they are absolute relative to the source of gravity) is irrelevant in this context because the frame is subject to the rules of special relativity regardless of its kinematical state. However, the frame should not be accelerating (as indicated by "uniformly") because otherwise it will be like a gravitational frame due to the presumed equivalence between gravity and acceleration (or

---

[183] We should note that some of the general relativistic results related to the Schwarzschild metric may even depend on the Schwarzschild coordinates specifically. In fact, this is a general feature in general relativity and is not specific to the Schwarzschild metric and coordinates (see § 10.1.7). This should add more limitation to general relativistic results in general.

6.1 Frames in General Relativity

alternatively because special relativity is restricted to inertial frames). Accordingly, when we talk in this book about a frame (or observer) at infinity (or far from the source of gravity) we generally assume it to be inertial unless we indicate otherwise.[184]

(**b**) Freely falling frame in a gravitational field. Although this frame does exist in a curved part of the spacetime the rules of special relativity should also apply in the frame locally. The reason is that the gravity is effectively absent due to its neutralization by the acceleration of free fall according to the equivalence principle. In fact, we may say: the weak equivalence principle establishes the status of the frame as inertial by canceling the effect of gravity while the strong equivalence principle ensures the local application of special relativity.

(**c**) Frame in a gravitational field but it is not in a state of free fall, i.e. the frame has a propelling mechanism that prevents free fall under the influence of gravity and hence it is moving non-uniformly relative to a freely falling frame in its location. This frame should either be at rest in the spacetime relative to the source of gravitation or be moving in a non free fall fashion. In the first case (i.e. at rest) the frame is equivalent to an accelerating frame while in the second case the frame is either equivalent to an accelerating frame or it is actually an accelerating frame.[185] In both cases, the rules of special relativity should not apply because the frame essentially is neither inertial nor equivalent to inertial. More details about this type of frame is given in the Problems.

It is noteworthy that in the above classification of frames we are assuming that the frames are non-rotating; otherwise the frame of any one of the above three types will be accelerating and hence it will not be subject to the rules of special relativity (also see § 6.5). As we will see, this should lead to the conclusion of the necessity of the existence of an absolute frame in the background of the spacetime of general relativity (similar to what we found in B4 about this necessity in the spacetime of special relativity) and this conclusion has many physical and epistemological implications and consequences that will be investigated in the future.

Problems

1. Provide more details about the local frames of type c.
   Answer: There are different views in the literature about this messy issue. In the following points we provide some details:
   • Some authors seem to suggest that the frames of this type are also subject to the rules of special relativity (apparently with no condition). But this is obviously wrong (if the application is unconditional like the application in the flat spacetime of special relativity) because this effectively nullifies the effect of gravity.
   • Some authors suggest that special relativity should apply if we assume the existence of fictitious forces that influence the objects in its surrounding in such a manner that makes the frame effectively inertial. However, for this to apply we may need to assume that such an assumption is physically viable and there is no absolute frame and hence everything is relative.
   • It is also suggested that since a stationary gravitational frame[186] is equivalent to an accelerating frame (according to the equivalence principle) then the rules of special relativity should apply to the "equivalent instantaneous rest frames" which are inertial frames by definition (see B4). So, by considering the instantaneous rest frames (which are inertial), special relativity should apply in this series of frames "event by event". However, the validity and legitimacy of this interpretation requires direct experimental evidence (as well as proper explanation and application). In fact, the equivalence of the observations in an accelerating frame and in its instantaneous inertial rest frame can be challenged and refuted at least in some cases (e.g. the accelerometers in the two frames should have different readings) and hence the validity of this interpretation is debatable (at least in its generality).
   • It is also proposed[187] that in a stationary gravitational frame we should first apply special relativity

---

[184] In fact, in most cases we even assume it to be at rest to exclude special relativistic effects caused by motion.

[185] In both cases we are considering the acceleration relative to the frame of the source of gravitation and not relative to an absolute frame. Accordingly, the frame is equivalent to an accelerating frame if it is moving uniformly.

[186] "Stationary gravitational frame" here and in the next point should be a typical instance of the frames of type c and not the only instance of this type.

[187] This proposal (which is inline with the *Principle of General Covariance*; see problem 3 of § 1.8.2) seems to be the most

in the freely falling local frame (which is effectively inertial coordinating a flat patch of spacetime) and then we transform the special relativistic laws to the stationary frame by a general coordinate transformation (and hence what actually applies in the stationary frame is a "modified version" of the special relativistic laws in which the effect of gravity is incorporated). However, the validity of this depends on the validity of the strong equivalence principle (whose general validity is questionable; in fact observational evidence seems to be against it; see § 10.1.1 for example). It should also depend on the existence and availability of such a transformation (which is not obvious at least in general especially when we consider non-stationary frames of this type).

Finally, we should note that there are many details about the status of the frames of type c and the potential challenges to the above views and interpretations. However, most of these details are not very useful and some may be indeterminate within the framework of the theory and hence the value of any investigation or analysis will be largely speculative. Therefore, we prefer not to go through these consuming details (although some of these issues will be investigated further rather briefly in the future; see for example § 10.1.1 and § 10.1.11). Also, some of the above views may differ from others in presentation and form rather than in essence and content (due for instance to the use of misleading or lax phrasing or different methods of presentation and expression).

Exercises

1. The frames in classical mechanics and in special relativity are usually global while the frames in general relativity are generally local. Why?
2. Is it necessary that the frames in general relativity are local? Why?
3. If there are frames in general relativity that are not subject to the rules of special relativity (i.e. gravitational frames that are not in a state of gravitational free fall) then how are these frames treated in general relativity? Can this be a source of defect or limitation in the theory of general relativity?
4. What are the cases in which special relativity applies in the spacetime of general relativity?
5. For the application of special relativity, what conditions we need to assume about the spacetime and the reference frame? Try to link this to the application of special relativity locally in freely falling frames in the spacetime of general relativity.
6. Discuss the status of accelerating frame at infinity (i.e. very far away from any source of gravity) in the spacetime of general relativity.
7. In the text we considered three main types of local frame in the spacetime of general relativity. Try to elaborate on this.

6.2 Coordinates in General Relativity

In this section we investigate the significance and interpretation of coordinates in general relativity. We will use in our investigation (partly in this section and mostly in the subsequent sections which continue this investigation) the Schwarzschild coordinates as an example and case study and hence we investigate their interpretation and highlight their significance. This should help us to appreciate the meaning of coordinates in general relativity in general despite the restricted nature of the investigation.

The main issue regarding the meaning of coordinates in general relativity and their interpretation is related to the metrical significance of the coordinates of spacetime. In other words, when we use (ct, r, θ, ϕ) in Schwarzschild spacetime for example do t stand for the physical time as measured by a clock and r stand for the physical distance as measured by a stick? Or we should rather consider t and r as mere coordinate variables (or useful symbolic tools or labels or markers) for obtaining the physical time and distance through the fundamental relations in the spacetime which are embedded in the metric of the spacetime? The answer (according to general relativity) is that the coordinates are just symbolic tools to label and distinguish points (or events) in spacetime and hence they do not have direct metrical significance or physical value. Therefore, the physical variables such as time and length should be obtained from the metric of the spacetime through the established relations where through these relations (and only through these relations) the coordinates provide physical meaning and significance. For example, in the relation

---

rational, acceptable and consistent with the framework of the relativity theories.

6.2 Coordinates in General Relativity

$d\tau = \left(1 - \frac{2GM}{c^2 r}\right)^{1/2} dt$ (which will be investigated thoroughly in § 6.3.1) we have coordinate variables r and dt and a physical variable $d\tau$ and hence the function of this metrical relation is to convert the coordinate variables r and dt to a physical variable $d\tau$, i.e. we use r and dt (which are labels to specific point and interval in spacetime) in this metrical relation to obtain the physical variable $d\tau$.

An obvious consequence of the interpretation of coordinates in general relativity is that we will have coordinate time versus physical time and have coordinate length versus physical length. Although this might seem sensible from a formal or mathematical perspective, it requires justification and physical rationalization as well as addressing some practical problems. This issue will be assessed further later on (in this section as well as elsewhere such as § 10.1.6). However, the simple justification from a general relativistic viewpoint is that the qualification of coordinates as representative of physical variables comes from our experiences and presumptions in flat spaces where coordinates have direct physical significance, but this is not the case in general relativity where the spacetime is curved and hence the coordinates do not necessarily represent physical quantities although they are related to the physical quantities through the metric.

Regarding the issue of the Schwarzschild metric, we used the spatial variables (r, θ, ϕ) to represent the spatial coordinates in the Schwarzschild spacetime. It should be noted that in principle the (r, θ, ϕ) coordinates in the Schwarzschild spacetime are not the same as the (r, θ, ϕ) coordinates in the familiar spherical coordinate system of a Euclidean 3D space despite the strong similarity between the two. The reason is that the (r, θ, ϕ) of the spherical system represent coordinate variables of a flat space (and hence they represent physical variables with metrical significance) while the (r, θ, ϕ) of the Schwarzschild spacetime represent coordinate variables of a curved space (and hence they are useful labels that should be used in conjunction with the metrical relations to obtain the physical variables). Therefore, in principle the (r, θ, ϕ) of the spherical system are not the same as the (r, θ, ϕ) of the Schwarzschild spacetime. Yes, since the Schwarzschild spacetime is asymptotically flat at infinity then the two become identical (asymptotically) at infinity and this should provide the missing link between the two that justifies the claim that the spherical coordinates underlie (in a sense) the spatial part of the Schwarzschild spacetime. We will also see that the difference between the two sets of coordinates (i.e. Schwarzschild and spherical) actually occurs only in the radial variable (i.e. r) and that is why we used "in principle" in the above statements.

We should remark that although the distinction between non-metrical coordinates and metrical variables may be understandable from a pure theoretical perspective, in some practical situations (when solving real world problems; see for example § 8.3) this distinction is difficult to comply with and implement. The reason is that in some of the formulae that contain non-metrical coordinates we do not have values for these non-metrical coordinates and hence we find ourselves obliged to use metrical values instead. For example, when we find t or r (which are coordinate variables) in a formula that is supposed to be used to solve a real world problem we may not know the value of these coordinate variables and hence we will have no choice but to use the actual physical time and radial distance (i.e. the metrical variables) because in real life we have access only to physical quantities (e.g. the time of a physical event that we observe is a physical time and hence when we want to use the time of this event in a metrical relation to obtain another physical quantity that depends on coordinate time we have no choice but to use the physical time instead of the coordinate time that is required as an input to the metrical relation). This should cast a shadow over the practicality of this distinction and the feasibility of the general relativistic methodology.

We believe this could be a demonstration of the unsuitability of using a geometric approach to describe physical phenomena. In other words, in a purely mathematical (or geometrical) theory we have no such a problem because we are dealing with entirely abstract spaces and objects. Similarly, in a purely physical theory we have no such a problem because we are dealing with entirely physical entities and hence no such inconsistency can arise. But when we try to correlate the abstract entities of a geometric theory to physical entities in the real world we face such a departure and conflict between the abstract and the physical. In fact, the root of this problem is in the use of curved spacetime in modeling physical gravity. As we will discuss later, we think gravity should be treated like any other physical phenomenon, i.e. it is a physical phenomenon that takes place in a flat spacetime rather than an attribute of a curved spacetime. In our view, this general relativistic paradigm of modeling gravity as a curvature of spacetime (according to the principle of metric gravity; see § 1.8.4) rather than a physical phenomenon in an ordinary flat spacetime is the origin of several problems (apart from being non-intuitive and unnecessary). Of course, general relativists do not agree on the above criticism and they have certain approaches and strategies to overcome the aforementioned difficulties. However, we will meet in the future some examples related to this criticism and the defence of general relativists and we will leave the final judgment to the readers (also see § 10.1.6).

Exercises

1. What distinguishes general relativity from classical mechanics and special relativity and hence makes the coordinates in general relativity of symbolic rather than metrical or physical value?
2. Can we fix the problems that potentially arise from the premise that the coordinates of spacetime in general relativity have no metrical or physical significance?
3. Assess the coordinates of rectilinear and curvilinear systems of 3D flat Euclidean space, e.g. Cartesian, cylindrical and spherical coordinate systems.
4. Use a specific metrical relation as an example to elucidate the issue that in real life we have access only to physical (rather than coordinate) variables.
5. Address the problem that modeling gravity as a physical phenomenon in a flat spacetime (rather than a geometric attribute of a curved spacetime) may be impossible or non-physical if the actual physical spacetime is not flat.

6.3 Time in Schwarzschild Spacetime

To analyze the meaning and properties of time in general relativity we use the Schwarzschild spacetime as a typical general relativistic spacetime and use stationary[188] frames of type c (refer to § 6.1) located at different positions in the Schwarzschild spacetime. In the following subsections we investigate a number of situations and settings to highlight several important issues about time in general relativity.

6.3.1 Relation between Coordinate Time and Proper Time

Let have a stationary frame at a certain position in the Schwarzschild spacetime and consider events that vary in time only. Now, since the variation is in time only then the variation in the spatial coordinates is null, i.e. $dr = d\theta = d\phi = 0$. Therefore, the Schwarzschild quadratic form (see § 4.1.1) will reduce to:[189]

$$(d\sigma)^2 = \left(1 - \frac{2GM}{c^2 r}\right)(cdt)^2 \qquad (241)$$

$$(d\tau)^2 = \left(1 - \frac{2GM}{c^2 r}\right)(dt)^2 \qquad (242)$$

$$d\tau = \left(1 - \frac{2GM}{c^2 r}\right)^{1/2} dt \qquad (243)$$

where in line 2 we divide by c^2 and use the relation $(d\tau)^2 = (d\sigma/c)^2$. The significance of Eq. 243 is that:
(a) At finite r and for $\frac{2GM}{c^2 r} < 1$ the proper time interval $d\tau$ is shorter than the coordinate time interval dt by the factor $\left(1 - \frac{2GM}{c^2 r}\right)^{1/2}$.
(b) At infinite r (i.e. $r \to \infty$) the proper time interval $d\tau$ is equal to the coordinate time interval dt.
In brief, in the Schwarzschild spacetime the proper time interval is shorter than the coordinate time interval but it converges to the coordinate time interval at infinity (see § 6.3.3).

Problems

---

[188] Stationary here means at rest spatially relative to the source of gravity. The term "stationary" may also be used in the literature to mean time independent (i.e. there is no variation with respect to time).
[189] In the following we assume $\frac{2GM}{c^2 r} < 1$ which is a valid assumption in general. Apart from being a mathematical necessity (to avoid having complex value when $\frac{2GM}{c^2 r} > 1$ and singularity when $\frac{2GM}{c^2 r} = 1$), this condition excludes what is on and inside the event horizon of a black hole noting that M is the mass contained inside r.

1. Justify point (a) and point (b) about the significance of Eq. 243.
   Answer: For finite r and $\frac{2GM}{c^2 r} < 1$ we should have $\left(1 - \frac{2GM}{c^2 r}\right)^{1/2} < 1$. This justifies point (a).
   If $r \to \infty$ then $\frac{2GM}{c^2 r} \to 0$ and hence $\left(1 - \frac{2GM}{c^2 r}\right)^{1/2} \to 1$. This justifies point (b).

Exercises
1. What the condition $\frac{2GM}{c^2 r} < 1$ means?
2. What you note about the formula $d\tau = \left(1 - \frac{2GM}{c^2 r}\right)^{1/2} dt$?

6.3.2 Interpretation of Coordinate Time

If $r \to \infty$ in Eq. 243 then we have:
$$d\tau_\infty = dt \qquad (244)$$
where we use the subscript ∞ to indicate the proper time at infinity. The significance of Eq. 244 is that:
(a) The coordinate time is equal to the proper time of a (stationary)[190] observer at infinity. In fact, this can be used as a definition for the Schwarzschild temporal coordinate t and hence we say: the coordinate time t is the proper time of a (stationary) observer at infinity.
(b) Since the proper time of a (stationary) observer at infinity is the same for any frame in the Schwarzschild spacetime then the coordinate time (i.e. the Schwarzschild temporal coordinate t) is global and it is common to all frames. We note that this point may sound trivial but we made it explicit because it will be needed later. We should also note that the priority of being global may be reversed and hence we may state that the Schwarzschild temporal coordinate is global and hence the proper time of a (stationary) observer at infinity is global according to Eq. 244 (see § 6.3.3). We may also consider these as two independent facts.

Exercises
1. Is it useful to consider events at infinity which cannot be reached?

6.3.3 Gravitational Time Dilation

Let have two stationary frames: O_1 at r_1 and O_2 at r_2 where $r_1 < r_2$. From Eq. 243 we have:
$$d\tau_1 = \left(1 - \frac{2GM}{c^2 r_1}\right)^{1/2} dt \qquad \text{and} \qquad d\tau_2 = \left(1 - \frac{2GM}{c^2 r_2}\right)^{1/2} dt \qquad (245)$$
and hence $d\tau_1 < d\tau_2$ because $r_1 < r_2$ (see Problems). The significance of this is that the proper time interval of a frame that is deeper in a gravitational well is shorter than the corresponding proper time interval of a frame that is higher in the gravitational well (where this comparison is calibrated by the coordinate time interval dt which is global). In other words, time runs slower closer to the source of gravity where the gravitational field is stronger. So, we have gravitational time dilation where a frame farther from the source of gravity (i.e. at larger r) observes the time of a frame nearer to the source of gravity (i.e. at smaller r) to run slower (or alternatively a frame nearer to the source of gravity observes the time of a frame farther from the source of gravity to run faster). We should note that dt in Eq. 245 is not subscripted with 1 and 2 because coordinate time is common to all frames since it is global according to point (b) in § 6.3.2.

Problems
1. Determine if the factor $\left(1 - \frac{2GM}{c^2 r}\right)$ changes (i.e. increases or decreases) in the same sense as r or in the opposite sense. What about the factor $\left(1 - \frac{2GM}{c^2 r}\right)^{1/2}$?
   Answer: It is obvious that if r increases/decreases then $\frac{2GM}{c^2 r}$ decreases/increases and hence $\left(1 - \frac{2GM}{c^2 r}\right)$ increases/decreases. This means that the factor $\left(1 - \frac{2GM}{c^2 r}\right)$ changes in the same sense as r, i.e. $\left(1 - \frac{2GM}{c^2 r}\right)$ is an increasing function of r. This should also apply to $\left(1 - \frac{2GM}{c^2 r}\right)^{1/2}$ because taking

---
[190] The condition "stationary" is added here to exclude special relativistic effects caused by motion.

the square root does not affect the sense of inequality, i.e. if $a > b$ then $\sqrt{a} > \sqrt{b}$ (with a and b being non-negative).

This should justify the statement in the text that $d\tau_1 < d\tau_2$ because $r_1 < r_2$.

Exercises

1. The statement: "the proper time interval of a frame that is deeper in a gravitational well is shorter than the corresponding proper time interval of a frame that is higher in the gravitational well" may be criticized by saying: you are comparing the time in two different local frames which have no common standard of calibration and hence the comparison is baseless. Discuss this issue.

6.3.4 Gravitational Frequency Shift

An effect that is directly related to the gravitational time dilation is the gravitational frequency shift. It is obvious from the definition of frequency ν as the reciprocal of the periodic time T (i.e. $\nu = 1/T$) that any physical phenomenon that is periodic (and hence it has periodic time) like electromagnetic waves should have a frequency where the two are linked by this reciprocal relation. So, if the periodic time was influenced by gravity in a certain manner (i.e. dilation or contraction) then the frequency should necessarily be influenced by gravity in the opposite manner (i.e. contraction or dilation).[191] In other words, if the periodic time is gravitationally dilated then the frequency should be gravitationally contracted and vice versa. Accordingly, a wave that is ascending in a gravitational well should be red shifted while a wave that is descending in a gravitational well should be blue shifted (see exercise 1).

In quantitative terms, if we substitute from Eq. 244 into Eq. 243 then we have:

$$d\tau = \left(1 - \frac{2GM}{c^2 r}\right)^{1/2} d\tau_\infty \tag{246}$$

$$T = \left(1 - \frac{2GM}{c^2 r}\right)^{1/2} T_\infty \tag{247}$$

$$\frac{1}{\nu} = \left(1 - \frac{2GM}{c^2 r}\right)^{1/2} \frac{1}{\nu_\infty} \tag{248}$$

$$\nu_\infty = \left(1 - \frac{2GM}{c^2 r}\right)^{1/2} \nu \tag{249}$$

where in line 2 we use the periodic time T at a given local stationary frame to represent the proper time interval $d\tau$ in that frame (and similarly for T_∞ with $d\tau_\infty$), and in line 3 we use the relation $T = 1/\nu$. Noting that ν is the frequency of the wave in the frame at r, Eq. 249 is the quantitative representation of the aforementioned fact that a wave ascending in a gravitational well is red shifted while a wave descending in a gravitational well is blue shifted. This fact may be more clarified and generalized by considering the relation between the frequency in two stationary frames: O_1 at r_1 with frequency ν_1 and O_2 at r_2 with frequency ν_2 where $r_1 < r_2$. From Eq. 249 we have:

$$\nu_\infty = \left(1 - \frac{2GM}{c^2 r_1}\right)^{1/2} \nu_1 = \left(1 - \frac{2GM}{c^2 r_2}\right)^{1/2} \nu_2 \tag{250}$$

and hence:

$$\frac{\nu_1}{\nu_2} = \left(1 - \frac{2GM}{c^2 r_2}\right)^{1/2} \left(1 - \frac{2GM}{c^2 r_1}\right)^{-1/2} \tag{251}$$

Now, since $r_1 < r_2$ then $\left(1 - \frac{2GM}{c^2 r_1}\right)^{1/2} < \left(1 - \frac{2GM}{c^2 r_2}\right)^{1/2}$ (see § 6.3.3) and hence the last equation implies that $\nu_2 < \nu_1$ (i.e. the frequency ν_1 at r_1 is red shifted to the frequency ν_2 at r_2 by ascending in the

---

[191] The use of contraction and dilation with respect to frequency is for the purpose of imitation; otherwise it may be more appropriate to use decrease and increase instead.

gravitational well, or alternatively the frequency ν_2 at r_2 is blue shifted to the frequency ν_1 at r_1 by descending in the gravitational well).

Exercises

1. Justify the following statement: "A wave that is ascending in a gravitational well should be red shifted while a wave that is descending in a gravitational well should be blue shifted".

6.3.5 Comparison with Classical Mechanics and Special Relativity

In classical mechanics, the time is global, absolute and independent of any local metric property[192] and hence none of the general relativistic implications that we investigated in the previous subsections (e.g. gravitational time dilation) do exist[193] because all the general relativistic implications are based on the locality and relativity of time (i.e. physical or metrical time) and its dependence on the local metric properties of the spacetime.

Regarding special relativity, the spacetime of special relativity is flat coordinated by global frames (i.e. one global frame for each observer) where each frame is distinguished from all other frames by its relative uniform motion with respect to the other frames or with respect to absolute frame if we believe in absolute frame (noting that the frames in special relativity are inertial). As a consequence, we have the following:
(a) In special relativity the time is global and absolute in each frame (although it is relative across frames) unlike the time in general relativity which is local (i.e. it varies from one location of spacetime to another) and relative (i.e. it depends on the local frame). Moreover, the time in special relativity is independent of the local metric properties of the spacetime because these metric properties are uniform and they equally apply to all parts of the spacetime. Yes, when we compare the time in two special relativistic frames then the time in special relativity is also local in this sense (since each frame has its own time) and relative. Moreover, it depends on the "local metric properties" of the spacetime in this sense.
(b) In special relativity time dilation (as observed across two global frames in relative motion with respect to each other) is caused by motion and hence it is a function of speed (i.e. it is a kinematical effect) while in general relativity time dilation is caused by gravity and hence it is a function of location (or local metric properties of spacetime), i.e. it is a gravitational effect. Yes, there should also be non-gravitational time dilation in the spacetime of general relativity (in addition to the gravitational time dilation) if the frame is moving and not stationary. Supposedly, the determination of this non-gravitational time dilation should in principle follow the rules of special relativity (and hence this non-gravitational time dilation can be labeled as special relativistic time dilation). However, this depends on several factors and considerations (see for example § 6.1; also refer to B4) which are not a priority for our investigation here due to the general relativistic nature of this investigation.

Exercises

1. Compare the time of occurrence and the time of observation in special relativity and in general relativity.
2. Compare between time dilation in special relativity and time dilation in general relativity.

6.4 Length in Schwarzschild Spacetime

To analyze the meaning and properties of length (or spatial interval or distance) in general relativity let use the Schwarzschild spacetime again as a typical general relativistic spacetime and use stationary[194] frames of type c (refer to § 6.1) located at different positions in the Schwarzschild spacetime. In the following subsections we investigate several situations and settings to highlight a number of important issues about length in general relativity.

---

[192] In fact, the paradigm of metric space does not exist in the theoretical framework of classical mechanics or at least it is irrelevant.
[193] In fact, they do not exist as effects of the locality and relativity of time although some may exist as results of other classically established principles like the conservation of energy which can classically explain gravitational frequency shift (at least in some cases and circumstances).
[194] Again, stationary here means at rest spatially relative to the source of gravity.

6.4.1 Relation between Spatial Coordinates and Proper Length

Let have a stationary frame at a certain position in the Schwarzschild spacetime and consider events that vary in space only. Now, since the variation is in space only then the variation in the temporal coordinate is null, i.e. $dt = 0$. Therefore, the Schwarzschild quadratic form (see § 4.1.1) will reduce to:

$$(d\sigma)^2 = -\left(1 - \frac{2GM}{c^2 r}\right)^{-1} (dr)^2 - r^2 (d\theta)^2 - r^2 \sin^2\theta (d\phi)^2 = -(ds)^2 \qquad (252)$$

where ds is the length (i.e. 3D spatial interval in its infinitesimal form or 3D line element) which can be correctly described as the proper length (since it is the length measured at the same time and hence it is "proper" like the "proper" time interval which is the time interval measured at the same spatial location).[195] Now, since we are investigating length then let use $(ds)^2$ instead of $(d\sigma)^2$, that is:

$$(ds)^2 = \left(1 - \frac{2GM}{c^2 r}\right)^{-1} (dr)^2 + r^2 (d\theta)^2 + r^2 \sin^2\theta (d\phi)^2 \qquad (253)$$

As we see, Eq. 253 is the same as the quadratic form of spherical coordinate system of a Euclidean space apart from the factor $\left(1 - \frac{2GM}{c^2 r}\right)^{-1}$. In brief, Eq. 253 implies that the quadratic form $(ds)^2$ of the proper length ds is modified from its spherical form by the factor $\left(1 - \frac{2GM}{c^2 r}\right)^{-1}$ in its radial component where this component varies by this factor as r varies (see exercise 1).

However, this is a gross analysis because length as formulated by Eq. 253 is rather complex to analyze in one go due to the involvement of several variables. So, let analyze length in the Schwarzschild spacetime by using Eq. 253 but with some simplification by exploiting the fact that length in spherical coordinates (which in a sense underlie the Schwarzschild coordinates) can be resolved into a circumferential[196] component (where r is held constant and hence $dr = 0$) and a radial component (where θ and ϕ are held constant and hence $d\theta = d\phi = 0$). In other words, the length is resolved into a component embedded in the r coordinate surface (which is the circumferential component) and a component along the r coordinate curve (which is the radial component). So, our analysis of length in the Schwarzschild spacetime will be simplified in the following discussion by analyzing these two components separately.

Considering the circumferential component, we have $dr = 0$ and hence Eq. 253 becomes:

$$(ds)^2 = r^2 (d\theta)^2 + r^2 \sin^2\theta (d\phi)^2 \qquad (254)$$

which is the same as the quadratic form in spherical coordinates on an r coordinate surface (i.e. on a sphere of constant r). This means that in the Schwarzschild spacetime there is no direct effect of gravity on the circumferential component of length and hence the r coordinate surfaces in the Schwarzschild spacetime are similar to their counterparts in a spherical coordinate system (see exercise 2).

Considering the radial component, we have $d\theta = d\phi = 0$ and hence Eq. 253 becomes:

$$(ds)^2 = \left(1 - \frac{2GM}{c^2 r}\right)^{-1} (dr)^2 \qquad (255)$$

$$ds = \left(1 - \frac{2GM}{c^2 r}\right)^{-1/2} dr \qquad (256)$$

which is the same as the line element in spherical coordinates along an r coordinate curve (i.e. the intersection of θ and ϕ coordinate surfaces) but with modification by the factor $\left(1 - \frac{2GM}{c^2 r}\right)^{-1/2}$. The significance of Eq. 256 is that:

(a) At finite r and for $\frac{2GM}{c^2 r} < 1$ the proper length ds (in its pure radial form) is longer than the coordinate radial length dr by the factor $\left(1 - \frac{2GM}{c^2 r}\right)^{-1/2}$.

---

[195] We note that in this general relativistic context "proper" may be used rather differently from "proper" in its special relativistic sense.

[196] The circumferential component may be labeled as azimuthal or transversal (although these might be less general).

6.4.2 Interpretation of Spatial Coordinates

(b) At infinite r (i.e. $r \to \infty$) the proper length ds (in its pure radial form) is equal to the coordinate radial length dr.

Problems

1. Justify point (a) and point (b) about the significance of Eq. 256.
   Answer: For finite r and $\frac{2GM}{c^2 r} < 1$ we should have $\left(1 - \frac{2GM}{c^2 r}\right)^{-1/2} > 1$. This is because $\left(1 - \frac{2GM}{c^2 r}\right) < 1$ and hence $\left(1 - \frac{2GM}{c^2 r}\right)^{1/2} < 1$ and therefore $\left(1 - \frac{2GM}{c^2 r}\right)^{-1/2} > 1$ according to the rules of inequalities. This justifies point (a).
   If $r \to \infty$ then $\frac{2GM}{c^2 r} \to 0$ and hence $\left(1 - \frac{2GM}{c^2 r}\right)^{-1/2} \to 1$. This justifies point (b).

Exercises

1. Determine if the factor $\left(1 - \frac{2GM}{c^2 r}\right)^{-1}$ changes (i.e. increases or decreases) in the same sense as r or in the opposite sense (assuming $\frac{2GM}{c^2 r} < 1$). What about the factor $\left(1 - \frac{2GM}{c^2 r}\right)^{-1/2}$?
2. Comment on our statement: the r coordinate surfaces in the Schwarzschild spacetime are similar to their counterparts in a spherical coordinate system.

6.4.2 Interpretation of Spatial Coordinates

For very large r (i.e. $r \to \infty$) we have $\left(1 - \frac{2GM}{c^2 r}\right)^{-1} = 1$ and hence Eq. 253 becomes:

$$(ds)^2 = (dr)^2 + r^2 (d\theta)^2 + r^2 \sin^2\theta (d\phi)^2 \tag{257}$$

which is no more than the quadratic form of spherical coordinate system of 3D Euclidean space. This should provide an interpretation for the spatial coordinates of Schwarzschild spacetime. In fact, this shows that the spatial Schwarzschild coordinates converge to the spherical coordinates at infinity.

To clarify the situation further, we consider the radial component alone (as represented by Eq. 256) in our interpretation of the spatial coordinates since the circumferential component is not affected by gravity as we found earlier in § 6.4.1 (and also because of what we observed here about the significance of θ and ϕ). Accordingly, if $r \to \infty$ in Eq. 256 then we have:

$$ds_\infty = dr \tag{258}$$

where we use the subscript ∞ to indicate the proper length at infinity. The significance of Eq. 258 is that:
(a) The coordinate radial length dr is equal to the proper length ds (in its pure radial form) of a (stationary) observer at infinity. In fact, this can be used as a definition for the Schwarzschild radial coordinate r and hence we say: the radial coordinate r is the measure of proper length (in its pure radial form) of a (stationary) observer at infinity.
(b) Since the proper length ds of a (stationary) observer at infinity is the same for any frame in the Schwarzschild spacetime then the radial coordinate r in the Schwarzschild spacetime is global and it is common to all frames. This may be reversed and hence we infer from the presumed global nature of the radial coordinate r the global nature of the proper length at infinity. We may also consider these as two independent facts.

Exercises

1. Use the spatial quadratic form of the Schwarzschild metric to interpret the spatial Schwarzschild coordinates.
2. The Schwarzschild metric converges to the flat Lorentz metric at infinity. What is the significance of this on the meaning and interpretation of the Schwarzschild coordinates?

6.4.3 Gravitational Length Contraction

Let have two stationary frames: O_1 at r_1 and O_2 at r_2 where $r_1 < r_2$, and let consider the radial component alone in our analysis since the circumferential component is not affected (at least directly) by gravity as we found earlier. From Eq. 256 we have:

$$ds_1 = \left(1 - \frac{2GM}{c^2 r_1}\right)^{-1/2} dr \qquad \text{and} \qquad ds_2 = \left(1 - \frac{2GM}{c^2 r_2}\right)^{-1/2} dr \qquad (259)$$

and hence $ds_1 > ds_2$ because $r_1 < r_2$ (see Problems). We note that dr is not subscripted with 1 and 2 because the radial coordinate r (which the differential dr is based on) is common to all frames since it is global according to point (b) in § 6.4.2. The significance of Eq. 259 is that the proper length in a frame that is deeper in a gravitational well is longer than the corresponding proper length in a frame that is higher in the gravitational well. In other words, length is contracted farther from the source of gravity where the gravitational field is weaker. So, we have gravitational length contraction in the radial direction where a frame nearer to the source of gravity (i.e. at smaller r) observes the length in a frame farther from the source of gravity (i.e. at larger r) to be contracted (or alternatively a frame farther from the source of gravity observes the length in a frame nearer to the source of gravity to be dilated).

We should finally remark that gravitational length contraction may not be seen in the literature of general relativity to be as natural and authentic as gravitational time dilation (in fact some authors describe it as apparent). This may be justified by some by the non-metrical (or non-physical) significance of coordinates in general relativity.[197] However, this justification can be refuted by its applicability even to gravitational time dilation. In fact, the supposed uniformity of spacetime in all its dimensions (whether temporal or spatial) dictates that any temporal effect (like time dilation) should have a corresponding spatial effect (like length contraction) and vice versa and hence any consistent physical theory about spacetime (as general relativity is supposed to be) should guarantee the equivalence between temporal and spatial effects and consequences. So, gravitational length contraction should be as natural and genuine as gravitational time dilation (at least according to the Schwarzschild metric whose implications are supposedly supported experimentally in certain physical situations) and hence if gravitational time dilation is natural and genuine then gravitational length contraction should also be natural and genuine while if gravitational length contraction is not natural and genuine then gravitational time dilation should also be not natural and genuine. In fact, if a spacetime theory does not meet this condition (i.e. equivalence and correspondence between temporal and spatial effects) then its consistency and soundness should be questioned because it discriminates between the dimensions of the presumably-uniform spacetime.[198]

Problems

1. Justify the following: if $r_1 < r_2$ then $ds_1 > ds_2$.
   Answer: The factor $\left(1 - \frac{2GM}{c^2 r}\right)^{-1/2}$ changes in the opposite sense to r, i.e. $\left(1 - \frac{2GM}{c^2 r}\right)^{-1/2}$ is a decreasing function of r (see exercise 1 of § 6.4.1). This means that if $r_1 < r_2$ then we should have $\left(1 - \frac{2GM}{c^2 r_1}\right)^{-1/2} > \left(1 - \frac{2GM}{c^2 r_2}\right)^{-1/2}$ and hence from Eq. 259 we conclude that $ds_1 > ds_2$, as required.

Exercises

1. Discuss how the issue of gravitational length contraction is dealt with in the literature of general relativity.

6.4.4 Comparison with Classical Mechanics and Special Relativity

In classical mechanics, the space is absolute and flat (in fact it is Euclidean) and hence length is global and independent of any local metric property. Therefore, none of the general relativistic implications that we investigated in the previous subsections (e.g. gravitational length contraction) do exist because all the

---

[197] In fact, this seems to be based on comparing the coordinate and metrical variables (e.g. dr and ds_1 in Eq. 259). But this is not necessary because as we stated earlier the comparison can be between the metrical variables (e.g. ds_1 and ds_2 in Eq. 259) at two locations (one of which could be at infinity, where the coordinate and metrical variables agree, although this is not necessary).

[198] In fact, there are many issues related to this discussion some of which have been touched on (from a special relativistic perspective) in B4 such as the difference between space and time in having freedom of movement in space but not in time. However, we see no urgency in discussing these issues here. We should also note that "uniform" here means in its dimensions (i.e. all dimensions have the same contribution and significance to the manifold) as indicated earlier.

general relativistic implications are based on the locality and relativity of length and its dependence on the local metric properties of the spacetime.

Regarding special relativity, the spacetime of special relativity is flat coordinated by global inertial frames (i.e. one global frame for each observer) where each frame is distinguished from all other frames by its relative uniform motion with respect to the other frames (or with respect to absolute frame if we believe in absolute frame). As a consequence, we have the following:

(**a**) In special relativity the length is global and absolute in each frame unlike the length in general relativity which is local (i.e. it varies from one location of spacetime to another) and relative (i.e. it depends on the local frame). Moreover, it is independent of the local metric properties of the spacetime because these metric properties are uniform and apply equally to all parts of the spacetime. Yes, when we compare the length in two special relativistic frames then the length in special relativity is also local in this sense (since each frame has its own length) and relative.

(**b**) In special relativity length contraction (as observed across two global frames in relative motion with respect to each other) is caused by motion and hence it is a function of speed (i.e. it is kinematical) while in general relativity length contraction is caused by gravity and hence it is a function of location (or local metric properties of spacetime), i.e. it is gravitational. Yes, there should also be non-gravitational length contraction in the spacetime of general relativity (in addition to the gravitational length contraction) if the frame is moving and not stationary. Supposedly, the determination of this non-gravitational length contraction should in principle follow the rules of special relativity (and hence this non-gravitational length contraction can be labeled as special relativistic length contraction).

Exercises
1. Is there a similarity between length contraction in special relativity and in general relativity? If so, what?
2. Comment on the fact that physical effects like time dilation and length contraction occur in special relativity only across global frames (each of which covers the entire spacetime) while in general relativity they occur across local frames (each of which covers only part of the spacetime).

6.5 General Relativity and Absolute Frame

The issue of absolute frame is central to the theory of special relativity (or what we call Lorentz mechanics) but it does not seem to be central to the theory of general relativity. We discuss in the following points some potential reasons for this difference:

1. In the spacetime of special relativity the demand for an absolute frame is natural (as we demonstrated in B4) because we have a single global frame that covers the entire spacetime. But in the spacetime of general relativity what we have in reality is a collection of local frames pieced together and hence the demand for absolute frame seems less obvious and urgent because we are always focusing on a certain part of the spacetime. However, this difference should not lead to the conclusion that the issue of absolute frame is irrelevant to general relativity. The reason is that the spacetime of general relativity is a modified version of the spacetime of special relativity and hence this demand in special relativity should be inherited in general relativity.
2. In special relativity, the existence (or non-existence) of an absolute frame is a fundamental issue because special relativity is primarily a theory about space and time and their transformations. This issue may seem less important (or even irrelevant) in general relativity which primarily is a theory of gravity because gravity is essentially an interaction between objects and this interaction can be described satisfactorily by using any local frame that embraces the interacting objects without worrying about a global absolute frame.[199] This should apply even to the classical theory of gravity in which gravity is a force and hence the issue of absolute frame is a secondary factor that should not require particular attention. However, this may be refuted by the fact that general relativity (especially in its extended

---

[199] Interaction may suggest classical meaning since gravity in general relativity is an action of matter on the geometry of spacetime with the gravitated object being affected by this geometry. However, this should still be an interaction between the gravitating and gravitated objects through the geometry of spacetime. Anyway, this is just a matter of language and expression and hence it should not affect the essence of the point.

6.5 General Relativity and Absolute Frame

form as a "General Theory" that includes special relativity) is essentially a theory of space and time and hence it is like special relativity in this regard.[200] This should be endorsed by the fact that gravity in general relativity (according to the principle of metric gravity; see § 1.8.4) is an effect of the distortion of spacetime rather than a force. We should also mention the need of the equivalence principle (which is fundamental to the theory of general relativity even as a gravity theory) to a sensible and realistic definition of acceleration which requires absolute frame, as will be discussed later.

Hence, we can conclude that the issue of absolute frame is central even to the theory of general relativity and therefore it is important to pose the following question: does general relativity require the existence of an absolute frame of reference to make sense of its formalism and provide a logical interpretation? In other words, is there an epistemological necessity for the formalism of general relativity to have (or not have) a unique absolute frame, and what is the significance of the existence of absolute frame on general relativity? So, let inspect this issue in more details.

According to general relativity, a frame in a state of gravitational free fall (i.e. relative to the source of gravity) is locally equivalent to an inertial frame. It should be obvious that the validity of this claim is based on the denial of the existence of absolute frame because if we have an absolute frame then the entire system (i.e. the source of gravity plus the freely falling frame) could be accelerating relative to the absolute frame and this accelerated motion is equivalent to the existence of another source of gravity relative to which the frame is not in a state of free fall and hence it is not inertial (even locally). In brief, the general relativistic claim that "freely falling frame is locally inertial" requires the denial of absolute frame (noting that free fall is defined relative to the source of gravity).

This is about the issue of freely falling frame. Regarding the issue of non freely falling frame (regardless of being moving or at rest relative to the source of gravity), we just repeat what we said in the case of freely falling frame about the possibility of the existence of an absolute frame that contains the entire gravitational system. Accordingly, a frame that is not in a state of gravitational free fall (i.e. relative to the source of gravity) could be actually in a state of "free fall" and hence it is inertial. This is because it can be accelerating relative to the absolute frame in such a way that its acceleration annuls the gravitational effect of the source of gravity. So, in both cases (i.e. freely falling frame and non freely falling frame) the validity of the equivalence of free fall to inertiality (which is based on the equivalence of acceleration to gravity which is the essence of the equivalence principle) requires the denial of absolute frame.

Now, this denial could be sensible and acceptable as long as we restrict our attention to gravity. However, if we look to the wider picture where we are supposed to believe in the validity of special relativity (and especially if we incorporate special relativity in the framework of general relativity) then our position cannot be maintained unless we extend this denial to special relativity (since we cannot deny it in general relativity and accept it in special relativity). As we saw in B4, the sensibility of special relativity decisively depends on the existence of absolute frame because even the domain of special relativity (which is inertial frames) cannot be defined in a physically realistic and sensible way without the existence of absolute frame. In fact, even the formalism of special relativity necessitates the existence of absolute frame if it should have a sensible and logical interpretation and avoid obvious paradoxes and contradictions. So, if we accept the existence of absolute frame (where this existence in our view is necessitated by special relativity which is supposed to be a valid theory) then neither the freely falling frame is necessarily inertial nor the non freely falling frame is necessarily non-inertial and this should mean invalidation of the equivalence principle which is one of the pillars of general relativity.

Anyway, the existence of absolute frame is denied by general relativity (or at least by general relativists or the majority of them), and this denial is inconsistent with the fact that in special relativity accelerating frames were excluded from the theory because the theory is restricted to inertial frames. This exclusion means that there is at least an implicit confession in special relativity (which is still valid in general relativity) of the existence of accelerating frames in spacetimes in which no gravity does exist. The significance of this is that acceleration is an attribute that is determined and defined by the spacetime alone regardless of the existence or non-existence of any source of gravity, i.e. the existence of acceleration

---

[200] As discussed earlier, general relativity in its extended form has a gravitational component and a special relativistic component, and hence even if the gravitational component does not need absolute frame (which we reject) the special relativistic component should need absolute frame.

6.5 General Relativity and Absolute Frame

is independent of the existence of gravitation. Accordingly, acceleration (which is equivalent to gravitation) can exist even in spacetimes in which gravity does exist (where this acceleration is defined relative to a frame other than the frame of the gravitating object). As shown in B4 and indicated above, the existence of accelerating frames requires the existence of absolute frame and we see nothing in general relativity that can remove this requirement. In other words, the mere inclusion of gravity in general relativity and the claim of the equivalence between gravity and acceleration does not lead to the abolishment of the logic of the absolute frame that we established in our discussion of special relativity.

The situation may be clarified further by considering the regions of spacetime of general relativity far away from matter where the spacetime is essentially the same as the spacetime of special relativity and hence it should be possible to have accelerating frames without any reference to gravitating matter (whose existence is irrelevant in those regions). In fact, considerations like this may be behind the claim that the spacetime of special relativity is a fictitious idealization, but this should lead to questioning even the validity of general relativity itself since special relativity is supposed to be at the heart of general relativity due to the local application of special relativity in the spacetime of general relativity. In fact, we will see later that special relativity (or what we call Lorentz mechanics) is not an idealization at least from a practical perspective if not from a theoretical perspective.

We note that the literature of general relativity is full of examples that indicate the necessity of the existence of an absolute frame if we have to logicalize this theory and make sense of it. For example, the famous thought experiment of a rocket accelerating in free space indicates the necessity of the existence of absolute frame because in the absence of absolute frame the acceleration of the rocket and the observed "quasi-gravitational" field cannot be explained.[201] The necessity of absolute frame even in general relativity can also be inferred from the fact that the frame in this thought experiment and its alike is assumed to be irrotational which cannot be explained unless we assume the existence of an absolute frame relative to which the rotation can have significance and real physical effect. In other words, the absolute nature of rotation implies the absolute nature of spacetime and hence the existence of an absolute frame.

These examples may be challenged by the claim that acceleration and rotation are in reference to another frame and hence we do not need a global and unique absolute frame. But this can be refuted by the fact that the other frame is not an arbitrary frame. In other words, while the acceleration and rotation relative to a certain category of frames have accelerating and rotating effects the acceleration and rotation relative to another category of frames have no such effect and hence we should have a physically realistic sense of acceleration and rotation by having an absolute frame relative to which the acceleration and rotation have real physical effects.

In brief, the effect of "accelerating/rotating" and "non-accelerating/non-rotating" is a real physical effect that can be observed and measured regardless of any particular reference frame except a unique reference frame which is the "absolute non-accelerating/non-rotating frame". The uniqueness of this absolute frame should be obvious since only those frames that are non-accelerating/non-rotating relative to this frame do not observe accelerating/rotating effects while all the other frames experience such effects where these effects are quantified by the magnitude and direction or sense of the acceleration/rotation relative to this unique frame. Since such a frame can be determined by the sole entity of spacetime then the existence of any particular source of gravity in the spacetime should not affect its accelerating and rotating effects.[202]

Problems

---

[201] For example, we read in one of the textbooks (see page 114 of Lambourne et al. in the References): Another thought experiment involves a rocket in a region in which there is no gravitational field. If the rocket is accelerated with a uniform acceleration of magnitude g, no sufficiently localized experiment within the rocket can distinguish between the consequences of the acceleration and the gravitational field on the surface of the Earth. An object released from rest within the rocket would accelerate downwards, just as an object on Earth would do. (End of quote)
So we may ask: why the "object released from rest" should not stay at rest in this accelerating frame (as it should do if the frame is inertial) or move uniformly if acceleration is defined relative to an arbitrary frame (noting that no real physical effect can depend on acceleration relative to an arbitrary frame)? Similar challenges can be directed to other examples and thought experiments like the bending of light in a frame that is accelerating with respect to another frame in which the light follows a straight trajectory (see for example page 116 of Lambourne et al.).

[202] The phrase "sole entity of spacetime" should not imply taking a definite view about Mach principle and that is why we added "particular" (see § 6.5.1).

1. Explain briefly the significance of the existence of absolute frame on the equivalence principle and general relativity.
   Answer: The existence of absolute frame means that in the spacetime of general relativity we have accelerated motion (which is equivalent to a gravity that is not annulled by free fall relative to a source of gravity) regardless of the existence or non-existence of any matter or energy in the spacetime. Accordingly, even a freely falling frame (i.e. relative to the source of gravity) can be an accelerating frame (i.e. relative to the absolute frame) and hence it is equivalent to a gravitational frame according to the principle of equivalence (see § 1.8.2). Similarly, even a non freely falling frame (e.g. stationary frame relative to the source of gravity) can be an inertial frame (if the effect of gravity is annulled by the acceleration relative to the absolute frame). Therefore, the existence of absolute frame should impose restrictions on the validity of the equivalence principle, i.e. the validity of this principle requires the inclusion of the effect of any potential acceleration relative to the absolute frame. This should also raise a question about the wisdom (or even correctness) of modeling gravity as a curvature in spacetime and if this curvature approach is needed (or even sensible) at all because if the spacetime of general relativity is defined by this background absolute frame (which should sensibly and naturally coordinate a flat spacetime) then gravity can be a physical phenomenon in a flat spacetime (rather than a geometric attribute of a curved spacetime) and hence gravity (like any other force) is a physical phenomenon contained in the spacetime rather than being an attribute of the spacetime itself. In other words, the classical paradigm of force and gravitational field is sufficient (or even necessary) for modeling gravity while the general relativistic paradigm of curvature and metric gravity is redundant (or even wrong). If so, then all we need to generalize special relativity (assuming this generalization is viable)[203] is to look for a theory that generalizes the known "inertial" physical laws and extend them from the domain of inertial frames (where special relativity is supposed to be valid) to the domain of non-inertial frames and hence we have only two types of frame (i.e. inertial and accelerating).

2. By analyzing the concept of acceleration in general relativity show that the theory is inconsistent in some of its aspects.
   Answer: We note that acceleration in some aspects of the theory is referred to free space (as seen for instance in the rationale of the accelerating rocket in free space which is supposed to be equivalent to gravitational frame) while in some other aspects it is referred to the source of gravity (as seen for instance in the rationale of the free fall state). In other words, while the presence of an absolute frame is required in the former to make sense of the inferred physical consequences (otherwise real physical effects in the accelerating rocket for instance should depend on its reference to an arbitrary frame; see the upcoming note), the absence of an absolute frame is required in the latter to make sense of the inferred physical consequences (otherwise a stationary frame for instance can be inertial). This inconsistency casts a shadow in particular on the equivalence principle and its logical and physical consequences which permeate the entire theory of general relativity.
   Note: let have two frames, A and B, which are accelerating relative to each other. Now, if a massive object in space is seen stationary in one frame then it should be accelerating in the other frame. This means that an object released from rest within an accelerating rocket (which we can label as frame A) will not necessarily accelerate downwards unless the rocket is accelerating relative to a frame (say frame B) relative to which the object is not accelerating (say at rest). This means that for the object to accelerate downwards, frame B should be a specific frame (i.e. a frame in which the object is at rest) and not arbitrary frame. So, frame B should be an absolute frame (or have a certain relation to an absolute frame).

Exercises

1. Discuss the proposition that general relativity should require (at least in some of its details) an absolute frame of reference because acceleration (which is supposed to be equivalent to gravity according to the equivalence principle) does not have sensible interpretation without such a frame of reference.

---

[203] If the generalization is not viable then we need to look for an entirely different theory that represents the physical laws in accelerating frames (assuming that all types of accelerating frames have the same laws; otherwise we need to look for a theory for each type of accelerating frames).

2. Discuss the claim that absolute frame is irrelevant to general relativity because the motion in spacetime is explained by the concept of geodesics in spacetime.
3. What is the significance of bending of light signals in accelerating frames with regard to the invariance of the speed of light and the necessity of absolute frame?
4. Discuss some of the implications of the existence of an absolute frame in the spacetime of general relativity.
5. Discuss the issue of absolute frame as an epistemological or philosophical issue and its relation to science from these perspectives.
6. What is the significance of the conservation of momentum and angular momentum on the existence of absolute frame?
7. Is there physical evidence for the existence of absolute frame?
8. Challenge general relativity on the basis of the existence and non-existence of absolute frame.
9. Follow the logic of the equivalence principle to analyze a gravitational system in which the gravitating object is at rest or moving uniformly relative to absolute frame (i.e. the frame of the gravitating object is inertial) while the gravitated object falls freely.
10. What is the significance of the existence of orbiting systems (e.g. solar system) on the absoluteness of rotation and the existence of absolute frame?

6.5.1 Absolute Frame and Mach Principle

Apart from its relation to the issue of absolute frame (which in our view is a central issue to general relativity), Mach principle has a link to the historical development of general relativity and this is a good reason on its own for investigating this principle in this book. According to this principle, which is attributed to Mach although it seems to have origin in the writings of other philosophers prior to Mach, forces and accelerations experienced by objects in non-uniform motion are determined by the distribution of matter in the Universe rather than by the existence of an abstract absolute frame of reference. In fact, this principle was originally proposed as an alternative to the hypothesis of the existence of an absolute frame (or space) in the Newton's bucket argument where absolute rest and motion (such as the rotation of bucket) can be defined with respect to this absolute frame according to the Newtonian view while according to the Mach principle the effect seen in the bucket as a result of rotation is not because of the existence of an absolute frame of reference but because of the existence of matter in particular quantity and distribution in the Universe that surrounds the bucket (where the existence and distribution of matter in the surrounding Universe may be defined primarily by reference to the "distant stars").

Apart from the purely philosophical contemplations and futile disputes that are of little interest to us as physicists, we do not see a fundamental contradiction between the hypothesis of the existence of an absolute frame to which absolute spacetime is referred and the hypothesis of the effect of the distribution of matter in the surrounding Universe in defining the state of rest and motion and determining the agents of force and acceleration. What we should be interested in as physicists is a practical operational definition and determination of a frame of reference that is absolute in a sense to define the state of rest and motion sensibly, unambiguously and absolutely. Whether this frame belongs to an abstract absolute spacetime or it originates from the existence and distribution of matter in the surrounding Universe is irrelevant from a physical perspective. After all, we cannot remove this surrounding matter to have physical experiences in a "clean" spacetime that is void of matter. In brief, what is physically important is the existence of a unique frame relative to which rest, motion, acceleration and force can be uniquely and sensibly defined and determined. Whether this frame is the frame of absolute spacetime (due to its intrinsic properties inline with the Newtonian view) or the frame that is established by the particular distribution of matter in the Universe (and hence it can be seen as an extrinsic property inline with the Machian view) is unimportant and physically irrelevant.

It should be remarked that in the literature of the relativity theories we can find numerous proposals of direct and indirect physical tests and consequences that could in principle distinguish between the Newtonian paradigm of absolute space and Mach principle. However, many of these proposed tests and consequences are arguable in their validity and significance as they normally depend on particular

6.5.1 Absolute Frame and Mach Principle

theoretical frameworks whose validity are not universally approved. In fact, many of these proposals are based on personal stereotypes and convictions rather than on rigorous logic and solid science. Anyway, this should not affect our assertion about the irrelevance of the above controversy between the Newtonian and Machian paradigms because this irrelevance is related specifically to the Newtonian and Machian paradigms on the large scale of the Universe and are not concerned with any particular (and normally local) effects and consequences of these paradigms. In our view, there is no viable physical test that can distinguish between these two paradigms on this large scale and hence the difference from this perspective is irrelevant. Moreover, even if such a test does exist it has no practical consequence on the paradigm of "absolute frame" in its generic sense that in principle can include both the Newtonian absolute space and the Machian world (i.e. a world subject to the Mach principle).

Exercises

1. Describe the essence of Mach principle in a few words.
2. Consider two spheres S_1 and S_2 rotating relative to one another and assume that S_2 bulges at its equator, how do we explain this difference? According to Einstein "No answer can be admitted as epistemologically satisfactory, unless the reason given is an observable fact of experience ... Newtonian mechanics does not give a satisfactory answer to this question. It pronounces as follows: The laws of mechanics apply to the space R_1 in respect to which the body S_1 is at rest, but not to the space R_2 in respect to which the body S_2 is at rest. But the privileged space R_1 ... is a merely factitious cause, and not a thing that can be observed". Discuss this issue.
3. Assess (in general terms) the claimed evidence in support and against Mach principle and its link to general relativity.

Chapter 7
Physics of General Relativity

The purpose of this chapter is to give a general view about the physics of general relativity and assess its nature with some comparisons with the physics in special relativity and occasionally with classical physics. The chapter also includes some of the commonly used formulations and equations in general relativity about the physics of motion in spacetime. However, before we go through these details it is important to make a general remark about the nature of the physics of general relativity and its relation to the physics of special relativity which is the precursor to general relativity (noting that this remark has mainly theoretical value with no practical impact on the forthcoming investigation). As we know, the distinctive feature of the physics of general relativity is the presence of gravity as a metric agent that shapes the spacetime and hence influences the physics that takes place in it. So, it is important to formulate a general recipe for incorporating the effect of this agent in the physics and hence create a link between the physics in the presence of gravity (i.e. general relativity) and the physics in the absence of gravity (i.e. special relativity). This recipe is essentially based on the strong equivalence principle and is formulated in its most elegant form in the *Principle of General Covariance* (see § 1.8.2) which dictates that the effect of gravity on any physical system involving physical quantities and relations can be determined by formulating the laws of the system in its special relativistic tensorial form followed by applying a general coordinate transformation with the replacement of $\eta_{\mu\nu}$ with $g_{\mu\nu}$ and the replacement of ordinary derivatives with their equivalent tensor derivatives. Accordingly, we can say that the physics of general relativity is a modified version of the physics of special relativity. In other words, the physics of general relativity is a curved (or distorted) version of the flat (or straight) version of the physics of special relativity. This should be appreciated in the light of the principle of metric gravity (see § 1.8.4) where the role of gravity according to this principle is to curve the spacetime and hence "distort" the "straight" laws of the flat spacetime of special relativity.

7.1 Coordinates of Spacetime

Coordinates of spacetime in general relativity (assuming mostly a Schwarzschild metric in Schwarzschild coordinates) have been investigated thoroughly earlier (see § 6) and hence we do not repeat. The main fact that should be kept in mind from our previous investigation is that in the spacetime of general relativity we have coordinate variables (like coordinate time t) and physical variables (like proper time τ) where the two are linked through metrical relations, i.e. relations based on the Schwarzschild metric in our case. This fact should be considered in all other physical quantities and relations that depend on the coordinates of spacetime. Accordingly, we note the following:
• The physical significance of spacetime coordinates in general relativity is different from that in special relativity and in classical physics since in special relativity and classical physics the coordinates have direct physical significance and metrical value (due to the flat nature of the underlying spacetime), while in general relativity (where the spacetime is curved) the coordinates are mere labels and markers to identify points and events in spacetime and hence we need metrical relations to convert these coordinates to physically significant quantities.
• The physical observations in the spacetime are presumably independent of the employed coordinate systems. Accordingly, the choice of coordinate systems is usually a matter of preference and convenience, and hence in principle any coordinate system is legitimate to use as long as it fulfills its basic function. However, in most cases there are advantages and disadvantages in using certain coordinate systems in certain cases, e.g. using systems that demonstrate symmetry or do not lead to singularity (see exercise 11 of § 2.1). In fact, the issue of the choice of coordinate systems in general relativity is more sensitive and grave and can be more than a matter of having advantages and disadvantages. The reason is that

there are many examples in the physics of general relativity where certain coordinate systems can lead to misleading results and suggest wrong conclusions. Accordingly, in general relativity the choice of coordinate systems and the interpretation of their results require extra care and attention, unlike special relativity and classical mechanics where the situation is usually clear and straightforward.

7.2 Time Interval and Length

Time interval and length are physical quantities that basically represent relations between coordinates in spacetime (primarily in flat spacetime and through metrical relations in curved spacetime). Therefore, time interval and length in the spacetime of general relativity should be extracted from the metrical relations that are based on the coordinates of spacetime, as explained in § 7.1. In fact, we have investigated the core of these issues in § 6.3 and § 6.4 and hence we do not repeat. However, we would like to highlight the following important points:

• Considering the presumed validity of special relativity in the spacetime of general relativity (regardless of the details), time interval and length (as well as the physical quantities that depend on them) can be affected non-gravitationally (i.e. kinematically) as well as gravitationally and hence in some cases we should consider the combined result of these special and general relativistic effects. In fact, in some disciplines, like relativistic cosmology, we may even need to consider effects that are cosmological (i.e. neither gravitational nor kinematical) as well.

• Time interval and length are coordinate dependent and hence they are functions of the 4-position in spacetime. Accordingly, they are relative and have no absolute sense on a global scale[204] unlike their status in classical mechanics (and even in special relativity where they are global and absolute in each inertial frame). The significance of this is that physical quantities that depend on them (such as energy) are also relative and have no absolute significance on a global scale, and this has a direct impact on essential physical principles, like the conservation of energy, which lose their significance on a global level in the spacetime of general relativity (see for example § 7.9).

Problems

1. In § 6 we accepted calibration of basic physical quantities (mainly time and length) using local standards (mainly at infinity) while here we claim that there is no global or absolute physical significance of physical quantities that depend on coordinates due to the dependence of these quantities on the local metrical properties. How to reconcile these seemingly contradictory views?
   Answer: Our view is that although the calibration by local standards at infinity gives a legitimate basis for comparison, it is not sufficient to give a universal physical significance to coordinate-based physical quantities in curved spacetimes, and hence the lack of global physical significance cannot be avoided in curved spacetimes despite our ability to make sensible comparisons by the calibration at infinity. In brief, being able to make comparisons thanks to calibration by local standards at infinity does not imply global physical significance. Moreover, the calibration at infinity may apply to certain metric spaces (e.g. Schwarzschild) but not to all metric spaces and hence in the metric spaces that have no sensible calibration at infinity there should be no global physical significance.[205] In fact, even if this applies to some metric spaces it should be enough for denying global physical significance in general due to the violation of the general validity of global physical significance.
   Anyway, this (i.e. seeming contradiction) is an issue for general relativity to address and tackle and hence it is not our problem although in the forthcoming parts of the book we will keep advocating and

---

[204] Although they have no absolute sense on a global scale they can still be calibrated by using local standards (e.g. at infinity) to make useful comparisons, as explained earlier. We should also note that concepts and quantities related to time and length (such as simultaneity) should also lose their global significance and unique meaning and hence they become somewhat ambiguous (although we can still make useful ordering relations based on using local standards and references).

[205] In fact, this is based on considering (implicitly) that the physical quantities at infinity have a sort of global significance (due to the flat nature of spacetime there as in special relativity) unlike other locations in spacetime. However, we should note that "at infinity" is also local and hence the calibration at infinity is like the calibration at any other locality (which obviously has no global significance).

building on our view by considering this as an issue of lack of global physical significance even though useful comparisons (by calibrating by local standards) can still be made.

Exercises
1. Why coordinate dependent quantities lose their global significance despite the ability to calibrate by local standards (e.g. at infinity)?
2. Discuss, briefly, the physical effects that influence time interval and length in general relativity and compare them to those in special relativity.

7.3 Frequency

Frequency in general relativity has its normal and common meaning although it is subject to gravitational effects. In § 6.3.4 (and later in § 8.4) we investigated (rather thoroughly) frequency and how it is affected by gravity in the Schwarzschild spacetime and hence we do not repeat. However, it is important to note the following points about the frequency in general relativity:
• The gravitational effects that influence the frequency in the spacetime of general relativity should also influence the energy through the Planck's relation and this should affect the principle of energy conservation (at least as applied to electromagnetic frequency and energy).
• Frequency may also be affected kinematically (as a consequence of the application of special relativity) and even cosmologically.
• Frequency has no global physical significance thanks to its dependency on physical time (which in general relativity is local and metric dependent).

Exercises
1. Compare frequency in general relativity with frequency in special relativity and in classical physics.

7.4 Mass

Mass in general relativity is a common and unique concept that can be described loosely as an invariant scalar physical quantity that contributes (among other physical quantities) to the formation of the geometry (and hence the metric) of spacetime. In fact, mass in general relativity inherits all its basic properties (e.g. its equivalence to energy according to the relation $E_0 = mc^2$) from special relativity (and to some extent even from classical physics) with the addition of its new qualification as an agent of gravity. However, as discussed in several places in the book the classical concept of gravitated mass (as such) is made redundant by the paradigms of geodesic and free fall gravitation where (free) gravitated objects follow geodesic pathways in the spacetime regardless of their "gravitated" mass (and indeed regardless of even being massive or massless). In fact, even the concept of gravitating mass (as such) could be redundant since the redundancy of the gravitated mass could lead to the redundancy of gravitating mass due to their binary relation. Accordingly, the fundamental role of mass (and hence its qualification) in general relativity as a gravity theory is essentially to create the geometry of spacetime and determine its metric properties and structure.

It should be remarked that the apparently global and absolute nature of mass in general relativity may be inconsistent with the local nature of energy (see § 7.8) as can be seen vividly in the mass-energy equivalence relation (i.e. $E_0 = mc^2$ where E_0 is supposedly local while m is supposedly global)[206] unless we assume that certain types of energy (e.g. energy of rest mass) is global while other types of energy (e.g. energy of electromagnetic radiation) is local. However, this seems nonsensical and accordingly this issue could indicate an inconsistency in the theoretical structure of general relativity. In this context, we should note that the Poincare mass-energy relation of special relativity still holds in general relativity although the contribution of some types of energy to mass may not hold. For example, position-dependent gravitational potential energy does not contribute to mass although this may not be undisputed issue (in fact such type of energy does not exist in a strict sense within the theoretical framework of general

---
[206] It may be claimed that the local nature of E_0 and the global nature of m are not inconsistent if we note that c in general relativity is local.

relativity). This could mean that the mass-energy relation must hold in general relativity exactly as in special relativity with no additional contribution from gravity (which is what distinguishes the spacetime of general relativity from the spacetime of special relativity) although other factors should be taken into consideration.[207] In fact, a close inspection to these issues and their alike could lead to serious challenges to general relativity and the consistency of its theoretical framework.

Exercises
1. Why mass in general relativity should be global while quantities like length and time interval should be local?
2. Compare mass in general relativity with mass in special relativity.
3. Discuss the issue of the nature of mass in general relativity (as being local or global) from the perspective of proper and improper mass (assuming the validity of special relativity in the spacetime of general relativity).

7.5 Velocity, Speed and Acceleration

Although the basic definitions of the concepts of velocity and speed in general relativity are essentially the same as in classical mechanics and Lorentz mechanics, these concepts should have no global significance thanks to their dependence on length, time interval and direction which are not well defined globally in general relativity. The obvious example of this is the speed of light which (according to general relativity) is constant locally but not globally (due to the lack of global significance of distance and time). So, in essence the constancy of the speed of light is not valid in general relativity although this is justified by the distinction between local and global speed (see § 10.1.13).[208] Like velocity and speed, acceleration in general relativity is not well defined globally due to its dependence on distance, time and direction. We should also refer the reader to § 7.13 about geodesic deviation which is related to acceleration.

Exercises
1. It may be claimed that mathematical constructions and techniques like parallel transport can give global significance to physical quantities that depend on coordinates and orientations. Discuss this claim briefly.

7.6 Force

General relativity does not include the paradigm of gravitational force in its theoretical framework and formalism since it is replaced by the paradigms of geodesic motion and free fall gravitation which are based on the principle of metric gravity. Yes, there is a place for the paradigm of non-gravitational force in its theoretical framework and formalism (since non-gravitational forces, like electromagnetic, do exist in general relativity). However, force (like acceleration) should have no global significance in general relativity due to its dependency on local quantities and entities like time and direction (unlike its status in special relativity).

7.7 Momentum

Both linear and angular momentum should have no global physical significance in general relativity because direction and orientation are essential in the definition of linear and angular momentum, whereas in the curved spacetime of general relativity direction and orientation are not well defined globally (despite the availability of certain mathematical techniques such as parallel transport to make useful comparisons). Momentum also depends on other locally defined entities (such as speed) which degrade its global status and significance.

---

[207] The absence of additional contribution from gravity should be inline with the fact that in general relativity gravity (which what distinguishes general relativity from special relativity) is a spacetime attribute and not a force.

[208] To be more consistent we may need to say: the constancy (and non-constancy) of the speed of light in general relativity is meaningless at global level (due to the absence of global physical standard of calibration).

7.8 Energy and Work

Due to their dependency on time and length, energy and work have no global significance in general relativity and hence they are well defined only locally. We should also note that in general relativity the paradigm of gravitational potential energy does not exist in a strict sense.

Exercises
1. Give an example of the relativity of work in the spacetime of general relativity (e.g. Schwarzschild spacetime).

7.9 Conservation Principles

As pointed out earlier, physical quantities and entities like time, length and direction have no absolute and global significance in the framework of general relativity and hence any physical quantity that depends on these quantities and entities (such as energy and momentum) should also lose their absolute and global significance. Accordingly, all the conservation principles that depend on these quantities and entities should become invalid (or rather meaningless) at global level.[209] This is inline with the fact that the conservation principles of energy and momentum are consequences of the homogeneity of time and space (respectively) and such homogeneity does not exist in curved spacetime. This should similarly apply to the conservation of angular momentum which is a consequence of the isotropy of space.

We should also note that the conservation principles (of energy and momentum in particular) are closely related (at least within a classical context) to the Newton's laws of motion and hence any (special or general) relativistic impact on one of these should affect the other. Also, the general relativistic paradigm of geodesic motion of free objects (with the incorporation of gravity in the structure of spacetime instead of being a force) strips gravity from its classical dynamical significance and this may facilitate the connection between the law of gravity and the laws of motion (as represented classically by Newton's gravity law and Newton's motion laws). Some of these issues will be touched on mildly in the questions.

Exercises
1. Discuss the principles of conservation of linear and angular momentum within the framework of general relativity.
2. According to the relativity theories the 4-velocity (and hence the 4-momentum which is the 4-velocity times the supposedly-constant mass) of a free massive test object is a constant tangent vector to its world line which is a geodesic. Discuss this issue and speculate about its significance.
3. Give a technical meaning to the conservation of energy and momentum of free massive objects in the spacetime of general relativity.

7.10 Orbital Motion in Terms of Constants of Motion

Motion in general is commonly described and formulated classically in terms of the constants of the motion, namely energy and momentum. A similar approach is followed in general relativity where motion is described and formulated in terms of similar "constants of the motion". The main advantage of this method is its simplicity relative to other more complicated methods like solving the geodesic equations. Due to the exceptional importance of orbital motion this method is commonly followed in the general relativistic formulation of orbital motion and hence it is common in the literature of general relativity to develop and derive formulae for orbital motion in terms of energy and momentum.

In the following we derive the "energy" equation for the variation of the radial coordinate r assuming Schwarzschild metric. This equation (which may be seen to represent the conservation of energy and angular momentum of freely falling massive objects in gravitational fields although this conservation is not strictly valid globally) describes the orbital motion of massive test object in the gravitational field of

---

[209] As we will see later (refer for example to § 10.1.3), we can identify two main types of violation of energy conservation: one is related to global insignificance of energy and the other is related to creation and annihilation.

7.10 Orbital Motion in Terms of Constants of Motion

a Schwarzschild gravitating object outside Schwarzschild radius (i.e. $r > \frac{2GM}{c^2}$). For timelike trajectory of a massive object we have $(d\sigma)^2 \equiv c^2 (d\tau)^2 \neq 0$ and hence:

$$(d\sigma)^2 = c^2 (d\tau)^2 \tag{260}$$

$$F(cdt)^2 - F^{-1}(dr)^2 - r^2(d\theta)^2 - r^2 \sin^2\theta (d\phi)^2 = c^2 (d\tau)^2 \tag{261}$$

$$Fc^2 \left(\frac{dt}{d\tau}\right)^2 - F^{-1}\left(\frac{dr}{d\tau}\right)^2 - r^2\left(\frac{d\theta}{d\tau}\right)^2 - r^2 \sin^2\theta \left(\frac{d\phi}{d\tau}\right)^2 = c^2 \tag{262}$$

where in line 2 we use Schwarzschild metric (with $F = 1 - \frac{2GM}{c^2 r}$) while in line 3 we divide by $(d\tau)^2 \neq 0$. Now, since the massive object is free then the trajectory is a geodesic described by the system of Eqs. 161-163. So, if we combine the system of Eqs. 161-163 with Eq. 262 (noting that in the system of Eqs. 161-163 we have $\theta = \pi/2$, $d\theta/d\tau = 0$ and $u = 1/r$) then on substituting from Eqs. 161 and 163 into Eq. 262 we get:

$$F^{-1}c^2 A^2 - F^{-1}\left(\frac{dr}{d\tau}\right)^2 - \frac{B^2}{r^2} = c^2 \tag{263}$$

$$c^2 A^2 - \left(\frac{dr}{d\tau}\right)^2 - \frac{B^2}{r^2} F = c^2 F \tag{264}$$

$$c^2 A^2 - \left(\frac{dr}{d\tau}\right)^2 - \frac{B^2}{r^2}\left(1 - \frac{2GM}{c^2 r}\right) = c^2\left(1 - \frac{2GM}{c^2 r}\right) \tag{265}$$

$$\left(\frac{dr}{d\tau}\right)^2 + \frac{B^2}{r^2}\left(1 - \frac{2GM}{c^2 r}\right) - \frac{2GM}{r} = c^2 (A^2 - 1) \tag{266}$$

where in line 2 we multiply with F, in line 3 we substitute $\left(1 - \frac{2GM}{c^2 r}\right)$ for F, and in line 4 we manipulate the equation algebraically.

Now, physically B is the magnitude of the angular momentum per unit mass of the falling object (see § 1.5.1), that is $B = J/m$. Regarding the physical significance of A we can use the Lorentzian formulation where we have $dt/d\tau = \gamma$ to have a clue. Accordingly, from Eq. 161 (noting that γ is the energy scale factor as can be seen from the equation of total energy $E = mc^2 \gamma$) we see that A is proportional to the energy of the object, that is:

$$E = \alpha A \tag{267}$$

where α is the proportionality factor. Now, for an object at rest at infinity (i.e. $dr/d\tau = 0$ and $r \to \infty$) we have from Eq. 266:

$$0 = c^2 (A^2 - 1) \tag{268}$$

and hence $A^2 = 1$ (resulting in $A = 1$ since A is positive as can be deduced from Eq. 161 noting that $dt/d\tau > 0$ and $F > 0$ for $r > \frac{2GM}{c^2}$). So, from Eq. 267 (applied to such object, i.e. at rest at infinity) we get:

$$E_0 \equiv mc^2 = \alpha \tag{269}$$

where m is the invariant mass of the object and E_0 is its rest energy. On substituting from Eq. 269 into Eq. 267 we get:

$$E = mc^2 A \quad \to \quad A = \frac{E}{mc^2} \tag{270}$$

where E is the total energy of the object. The last equation means that physically A represents the ratio of the total energy of the object to its rest energy. On substituting from $B = J/m$ and $A = E/(mc^2)$ into Eq. 266 we obtain:

$$\left(\frac{dr}{d\tau}\right)^2 + \frac{J^2}{m^2 r^2}\left(1 - \frac{2GM}{c^2 r}\right) - \frac{2GM}{r} = c^2 \left(\frac{E^2}{m^2 c^4} - 1\right) \tag{271}$$

Problems

1. Derive a similar equation to Eq. 266 for massless objects.
   Answer: We follow a similar method to the method of derivation of Eq. 266. Massless objects follow lightlike geodesics (or null geodesics) and hence their quadratic form $(d\sigma)^2$ is zero, that is:

$$(d\sigma)^2 = F(cdt)^2 - F^{-1}(dr)^2 - r^2(d\theta)^2 - r^2\sin^2\theta(d\phi)^2 = 0$$

With no loss of generality we can consider the motion to be in the equatorial plane $\theta = \pi/2$ and hence the last equation becomes:

$$F(cdt)^2 - F^{-1}(dr)^2 - r^2(d\phi)^2 = 0$$

$$c^2 F\left(\frac{dt}{d\lambda}\right)^2 - F^{-1}\left(\frac{dr}{d\lambda}\right)^2 - r^2\left(\frac{d\phi}{d\lambda}\right)^2 = 0$$

where in line 1 we use $d\theta = 0$ and $\sin\theta = 1$ while in line 2 we divide by $(d\lambda)^2$. Now, if we substitute from Eqs. 164 and 166 into the last equation we get:

$$c^2 F\left(\frac{A}{F}\right)^2 - F^{-1}\left(\frac{dr}{d\lambda}\right)^2 - r^2(Bu^2)^2 = 0$$

$$c^2 \frac{A^2}{F} - F^{-1}\left(\frac{dr}{d\lambda}\right)^2 - \frac{B^2}{r^2} = 0$$

$$c^2 A^2 - \left(\frac{dr}{d\lambda}\right)^2 - \frac{B^2}{r^2} F = 0$$

$$\left(\frac{dr}{d\lambda}\right)^2 + \frac{B^2}{r^2}\left(1 - \frac{2GM}{c^2 r}\right) = c^2 A^2 \qquad (272)$$

where in line 2 we use $u = 1/r$, in line 3 we multiply with F, and in line 4 we use $F = 1 - \frac{2GM}{c^2 r}$. Eq. 272 for massless objects is the equivalent of Eq. 266 for massive objects.

Exercises
1. Obtain Eq. 162 from Eq. 266.[210]
2. Obtain Eq. 165 from Eq. 272.
3. Derive an equation for the radial motion (i.e. motion with constant θ and constant ϕ) of free massive object using Eq. 266.

7.11 Effective Potential in Orbital Motion

Following the footsteps of classical mechanics, it is customary to develop a formulation for the "effective gravitational potential" of a massive gravitated object in the gravitational field of a gravitating object according to the framework of general relativity although the concept of gravitational potential does not exist in general relativity in a strict sense. To do so, we start from the "energy" equation of orbital motion, i.e. Eq. 271. We should remark first that although the term on the right hand side of Eq. 271 does not represent energy it represents a quantity that is indicative of the energy of the orbiting object and hence Eq. 271 can be regarded as "energy" equation for the orbital motion. Now, if we divide Eq. 271 by 2 we obtain:

$$\frac{1}{2}\left(\frac{dr}{d\tau}\right)^2 + \frac{J^2}{2m^2 r^2}\left(1 - \frac{2GM}{c^2 r}\right) - \frac{GM}{r} = \frac{c^2}{2}\left(\frac{E^2}{m^2 c^4} - 1\right) \qquad (273)$$

$$\frac{1}{2}\left(\frac{dr}{d\tau}\right)^2 + \frac{J^2}{2m^2 r^2} - \frac{GM}{r} - \frac{GMJ^2}{c^2 m^2 r^3} = \frac{c^2}{2}\left(\frac{E^2}{m^2 c^4} - 1\right) \qquad (274)$$

---

[210] In fact, this should serve as verification for the derived energy equation.

7.12 Radial Trajectories in Spacetime

To have a better insight into the significance of Eq. 274, let compare it with its Newtonian counterpart. In Newtonian mechanics the energy equation for orbital motion is given by:

$$\frac{1}{2}\left(\frac{dr}{dt}\right)^2 + V_N = E_m \tag{275}$$

$$\frac{1}{2}\left(\frac{dr}{dt}\right)^2 + \frac{J^2}{2m^2r^2} - \frac{GM}{r} = E_m \tag{276}$$

where V_N is the Newtonian effective potential (which is characterized by being a function of r only for a fixed J) while E_m is the total energy of the orbiting object per unit mass (i.e. E/m). So, if we tolerate the difference between E_m in Eq. 276 and $\frac{c^2}{2}\left(\frac{E^2}{m^2c^4} - 1\right)$ in Eq. 274 (as well as the difference between t and τ and the difference between the radial distance r and the radial coordinate r) then Eqs. 274 and 276 are identical apart from the term $-\frac{GMJ^2}{c^2m^2r^3}$ (which is also a function of r only for a fixed J). Accordingly, if we extend the paradigm of "effective potential" (which is a classical paradigm) to general relativity then the general relativistic effective potential V_{GR} in orbital motion is given by:

$$V_{GR} = \frac{J^2}{2m^2r^2} - \frac{GM}{r} - \frac{GMJ^2}{c^2m^2r^3} \tag{277}$$

$$V_{GR} = V_N - \frac{GMJ^2}{c^2m^2r^3} \tag{278}$$

Exercises
1. Discuss briefly gravitational potential energy in general relativity.
2. Use the equivalence principle to argue that gravitational potential energy has no real meaning in general relativity.

7.12 Radial Trajectories in Spacetime

In this section we briefly investigate the general relativistic formulations for the radial trajectory of massless and massive free objects where we are assuming (as usual) Schwarzschild geometry.[211] These formulations are usually given as implicit correlations between time (t or τ) and radial coordinate r, i.e. $t(r)$ or $\tau(r)$. For massless object the Schwarzschild quadratic form becomes (noting that for massless object $d\sigma = 0$ and for radial trajectory $d\theta = d\phi = 0$):

$$\left(1 - \frac{R_S}{r}\right)(cdt)^2 - \left(1 - \frac{R_S}{r}\right)^{-1}(dr)^2 = 0 \tag{279}$$

$$c^2\left(\frac{dt}{dr}\right)^2 = \left(1 - \frac{R_S}{r}\right)^{-2} \tag{280}$$

$$c\frac{dt}{dr} = \pm\left(1 - \frac{R_S}{r}\right)^{-1} \tag{281}$$

where the plus sign is for motion away from the gravitating object (since r increases as t increases) while the minus sign is for motion toward the gravitating object (since r decreases as t increases). On integrating the last relation we obtain:

$$ct = \pm\left[r + R_S \ln(r - R_S)\right] \tag{282}$$

where the (arbitrary) constant of integration is set to zero and $r > R_S$. The last equation implicitly correlates the radial coordinate r to the temporal coordinate t. It is noteworthy that Eq. 282 indicates that infinite coordinate time is required for a massless object to fall to the event horizon of a black hole (noting that coordinate time represents the proper time of an observer at infinity).

---
[211] In fact, we also restrict our attention in the following formulations to the region $r > R_S$ (to avoid the mess of black holes at and inside the Schwarzschild radius).

7.13 Geodesic Deviation

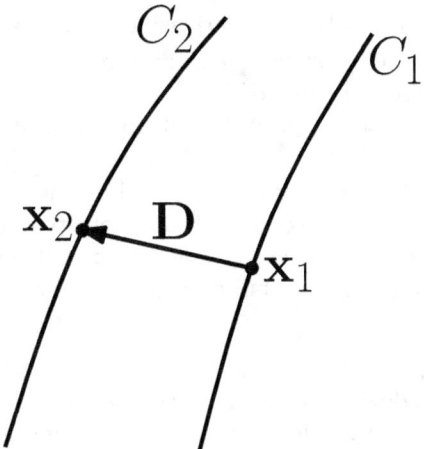

Figure 3: A simple sketch representing two points (or events) in spacetime ($\mathbf{x}_1$ and $\mathbf{x}_2$) on two adjacent geodesics (C_1 and C_2) where these points are connected by a displacement 4-vector $\mathbf{D}$ (i.e. $D^\mu = x_2^\mu - x_1^\mu$). The geodesics C_1 and C_2 are parameterized by a single natural (or affine) parameter λ and the points $\mathbf{x}_1$ and $\mathbf{x}_2$ correspond to the same value of λ.

By a similar argument, it can be shown (see Exercises) that the radial trajectory of massive object falling freely from rest in the Schwarzschild spacetime is given (implicitly) by:

$$t = -2 \left[\frac{r^{3/2} + 3R_S \sqrt{r}}{3c\sqrt{R_S}} + \frac{R_S}{2c} \ln\left(\frac{\sqrt{r} - \sqrt{R_S}}{\sqrt{r} + \sqrt{R_S}} \right) \right] + C \tag{283}$$

where C is a constant (which can be set to zero).

Exercises
1. Show that Eq. 282 is obtained by integrating Eq. 281.
2. Find an analytical expression for the proper time (as a function of r) for a massive object dropped from rest at an initial radial coordinate r_i far away from the source of gravity (and hence it is freely falling radially toward the gravitating object).
3. What is the 4-velocity of the object in the previous exercise?
4. Find an analytical expression for the coordinate time (as a function of r) for a massive object dropped from rest at an initial radial coordinate r_i far away from the source of gravity (and hence it is freely falling radially toward the gravitating object).

7.13 Geodesic Deviation

In general relativity, geodesic deviation essentially describes and quantifies the relative acceleration between two freely falling particles along two adjacent geodesics. In a sense, it is a replacement of the classical paradigms of field gradient and tidal force and acceleration. Let have two neighboring geodesics, C_1 and C_2, and they are parameterized by a single natural (or affine) parameter λ (see Figure 3). Let $D^\mu(\lambda)$ be a displacement 4-vector that connects two corresponding points (i.e. having the same λ value) $x_1^\mu(\lambda)$ on C_1 and $x_2^\mu(\lambda)$ on C_2.[212] The geodesic deviation for these points as λ varies is then given by the condition (which quantifies the change in the displacement 4-vector as a function of λ):

$$\frac{\delta^2 D^\alpha}{\delta \lambda^2} + R^\alpha{}_{\beta\mu\gamma} D^\mu \frac{dx^\beta}{d\lambda} \frac{dx^\gamma}{d\lambda} = 0 \tag{284}$$

---
[212] To make an analogy and avoid confusion we call D^μ *displacement* vector although it may be more appropriate to call it *deviation* or *separation* vector.

7.13 Geodesic Deviation

where $\frac{\delta^2}{\delta\lambda^2}$ symbolizes the second order absolute derivative, $R^\alpha{}_{\beta\mu\gamma}$ is the Riemann-Christoffel curvature tensor of the spacetime, and x^β, x^γ are the spacetime coordinates of the points on the geodesics and where all these quantities correspond to a given value of λ. It is obvious that the equation of geodesic deviation solely depends on the metric tensor of spacetime and hence it is intrinsic. We note that the equation may be written in a number of different forms (e.g. $\frac{\delta^2 D^\alpha}{\delta\lambda^2} = R^\alpha{}_{\beta\gamma\mu}\frac{dx^\beta}{d\lambda}\frac{dx^\gamma}{d\lambda}D^\mu$) using the anti-symmetry properties of the Riemann-Christoffel curvature tensor (and this is the main reason for appearing in the literature in confusingly different forms). We should also note that the equation is a tensorial relation and hence the "tidal gravitational field" is a tensor (even though the gravitational field itself is not since it can be transformed away and vanish by free fall according to the equivalence principle).[213]

Exercises

1. Analyze the equation of geodesic deviation (i.e. Eq. 284) in flat spacetime.
2. Derive the equation of geodesic deviation (i.e. Eq. 284).
3. The general relativistic paradigm of geodesic deviation is commonly considered in the literature to be a substitute for the classical paradigms of gravitational gradient and tidal force. Comment on this.

---

[213] In fact, some of these issues can be questioned.

Chapter 8
Consequences and Predictions

In this chapter we investigate the theoretical consequences and predictions of the theory of general relativity. We will also investigate in some cases how these consequences and predictions are derived from the theory and how they compare to the corresponding theoretical predictions of classical physics. The chapter will also include solved problems and exercises about the applications of these theoretical results.

8.1 Perihelion Precession of Mercury

The precession (or advance) of the perihelion of Mercury is an orbital phenomenon caused by the slow change in the orientation of the long axis of the Mercury orbit. Accordingly, the perihelion does not have a fixed position in space relative to the Sun but it gradually shifts by rotating around the Sun and hence the orbit does not close (see Figure 4). The perihelion position of Mercury advances by about 5600 arcseconds per century; most of which arise from precession of equinoxes and from perturbations caused by gravitational interaction with other planets in the solar system. Classical mechanics can explain almost all this 5600 arcseconds advance except 43 arcseconds per century.[214] Calculations based on general relativity can accurately predict this extra precession and hence almost all the remaining 43 arcseconds can be accounted for by general relativity.

Although the extra precession of perihelion in planetary motion is commonly regarded as a purely general relativistic effect, there are Newtonian formulations of this precession although some of these formulations may not agree quantitatively with the prediction of general relativity. In this context, it is worth noting that the general relativistic formula that correctly predicts the extra precession of Mercury perihelion has been derived earlier in the late 19th century by a German school master called Gerber using a classical approach (known as retarded gravitational potential) and hence general relativity is not the first or the only theory that could explain this extra 43 arcseconds advance. The reader is referred to the paper of Ian McCausland "Anomalies in the History of Relativity" for more interesting details about this issue and the debate surrounding it. It should be noted that this formula has also been obtained later from non-classical gravitational theories other than general relativity (e.g. the theories of Whitehead and Birkhoff).

We should also note that we title this section as "Perihelion Precession of *Mercury*" specifically because of its historical and observational significance. Otherwise, it is just an instance of the periastron[215] precession that is predicted by general relativity to occur in any orbital motion whether of Mercury or of another planet or indeed of any orbiting object. Accordingly, the following investigation and analysis should have general applicability (within the stated conditions) although it is largely phrased and presented in the context of Mercury and planetary motion.

Using the geodesic equations in the Schwarzschild metric (i.e. Eqs. 193-195) it can be shown (see Problems) that the extra precession per revolution in the orbit of a gravitated object in the gravitational field of a static and spherically symmetric gravitating object is given by:

$$\delta\phi = \frac{3G^2 M^2 T^2}{2\pi c^2 a^4 (1-e^2)} \quad (285)$$

where $\delta\phi$ is the extra precession per revolution in radians, G is the gravitational constant, M is the mass of the gravitating object (i.e. the Sun in the case of planetary motion), T is the orbital period of the

---

[214] We are considering the commonly recognized classical theory and hence we do not include marginal theories that claim to explain this extra 43 arcseconds classically.

[215] Periastron is similar to perihelion but perihelion is specific to the Sun (i.e. Helios in Greek and hence helion) while periastron is for stars in general (and hence astron).

8.1 Perihelion Precession of Mercury

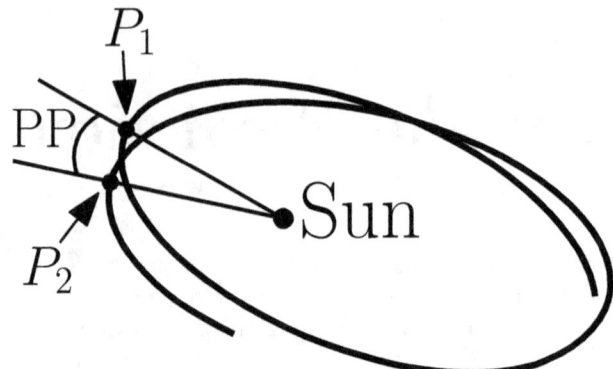

Figure 4: A simple sketch demonstrating the precession of perihelion PP of planetary orbit around the Sun (where P_1 and P_2 are two consecutive perihelia).

gravitated object, c is the characteristic speed of light, a is the mean distance between the gravitated object and the gravitating object,[216] and e is the eccentricity of the orbit.

The above formula gives the extra precession per revolution. So, to calculate the extra precession per century we multiply the extra precession per revolution by the number of revolutions per century where this number can be easily obtained from dividing the time of a century (expressed in a suitable time unit like second) by the time of orbital period T (expressed in the same time unit). Accordingly, the extra precession per century is given by:

$$\delta\phi_c = \frac{3G^2M^2T^2}{2\pi c^2 a^4 (1-e^2)} \times N_c = \frac{3G^2M^2T^2}{2\pi c^2 a^4 (1-e^2)} \times \frac{T_c}{T} = \frac{3G^2M^2TT_c}{2\pi c^2 a^4 (1-e^2)} \qquad (286)$$

where $\delta\phi_c$ is the extra precession per century in radians, N_c is the number of revolutions per century and T_c is the time of a century. For example, if T and T_c are in days then $T_c = 36525$ days.

As we see, Eq. 285 is made of a product of a common factor (i.e. $\frac{3G^2M^2}{2\pi c^2}$) and a planetary factor that distinguishes each planet from the others (i.e. $\frac{T^2}{a^4(1-e^2)}$). Similarly, Eq. 286 is made of a product of a common factor (i.e. $\frac{3G^2M^2}{2\pi c^2}$) and a planetary factor (i.e. $\frac{T^2 N_c}{a^4(1-e^2)}$) where this planetary factor is a product of the previous planetary factor (i.e. $\frac{T^2}{a^4(1-e^2)}$) and the number of revolutions per century (i.e. N_c). Accordingly, the relative size of the extra precession per revolution is determined by the size of the planetary factor (i.e. $\frac{T^2}{a^4(1-e^2)}$) while the relative size of the extra precession per century is determined by the size of the aforementioned planetary factor and the number of revolutions per century.

In Table 1 we present the orbital parameters (specifically T, a and e) of the planets in the solar system and calculate the extra precession per revolution $\delta\phi$ and the extra precession per century $\delta\phi_c$ (using Eqs. 285 and 286). As we see, both these extra precessions are decreasing according to the order of the planets in their proximity to the Sun.

Problems

1. Derive the general relativistic formula for the extra precession of planetary perihelion in the classical limit, and show that the predicted precession by this formula is almost identical to the observed precession of perihelion of Mercury.
   Answer: The precession of perihelion is a planetary motion problem and hence it should be investigated as a geodesic problem in the Schwarzschild metric. So, it is closely related to our investigation in § 4.1.2 and hence we can use the results that we developed in § 4.1.2. More specifically, we can start from Eqs. 193-195 in problem 3 of § 4.1.2. So, if we use the Schwarzschild quadratic form that

---

[216] The mean distance is represented by the semi-major axis of the ellipse that approximates the shape of the orbit.

8.1 Perihelion Precession of Mercury

Table 1: The planets in the solar system with their orbital parameters (T, a, e) and extra precessions $(\delta\phi, \delta\phi_c)$ according to Eqs. 285 and 286 where T is the orbital period (in days), a is the mean distance (in 10^9 m), e is the eccentricity of orbit, $\delta\phi$ is the extra precession per revolution (in arcseconds) and $\delta\phi_c$ is the extra precession per century (in arcseconds). For the purpose of comparison, we also include the number of revolutions per century N_c. We note that the numbers in this table are not highly accurate.

	T	a	e	$\delta\phi$	$\delta\phi_c$	N_c
Mercury	88	57.9	0.206	0.1035	42.9515	415.06
Venus	225	108.2	0.007	0.0530	8.61620	162.55
Earth	365	149.6	0.017	0.0383	3.83293	100.01
Mars	687	227.9	0.094	0.0254	1.35031	53.17
Jupiter	4331	778.6	0.049	0.0074	0.06208	8.43
Saturn	10747	1433.5	0.057	0.0039	0.01342	3.40
Uranus	30589	2872.5	0.046	0.0020	0.00237	1.19
Neptune	59800	4495.1	0.011	0.0013	0.00077	0.61
Pluto	90560	5906.4	0.244	0.0010	0.00042	0.40

we developed in that problem (i.e. Eq. 196 to be specific) then we have:

$$F\left(\frac{dt}{d\tau}\right)^2 - \frac{F^{-1}}{c^2}\left(\frac{dr}{d\tau}\right)^2 - \frac{r^2}{c^2}\left(\frac{d\phi}{d\tau}\right)^2 = 1$$

$$F\left(\frac{A}{F}\right)^2 - \frac{F^{-1}}{c^2}\left(\frac{dr}{d\tau}\right)^2 - \frac{r^2}{c^2}\left(\frac{B}{r^2}\right)^2 = 1$$

$$\frac{A^2}{F} - \frac{F^{-1}}{c^2}\left(\frac{dr}{d\tau}\right)^2 - \frac{B^2}{c^2 r^2} = 1$$

$$\frac{A^2}{F} - \frac{F^{-1}}{c^2}\left(\frac{dr}{d\phi}\frac{d\phi}{d\tau}\right)^2 - \frac{B^2}{c^2 r^2} = 1$$

$$\frac{A^2}{F} - \frac{F^{-1}}{c^2}\left(\frac{dr}{d\phi}\frac{B}{r^2}\right)^2 - \frac{B^2}{c^2 r^2} = 1$$

$$\frac{A^2}{F} - \frac{F^{-1} B^2}{c^2 r^4}\left(\frac{dr}{d\phi}\right)^2 - \frac{B^2}{c^2 r^2} = 1$$

$$\frac{c^2 A^2 r^4}{B^2} - \left(\frac{dr}{d\phi}\right)^2 - Fr^2 = \frac{c^2 F r^4}{B^2}$$

$$\left(\frac{dr}{d\phi}\right)^2 = -\frac{c^2 F r^4}{B^2} + \frac{c^2 A^2 r^4}{B^2} - Fr^2 \qquad (287)$$

where in line 2 we use Eqs. 193 and 195 (noting that $u = 1/r$), in line 3 we simplify, in line 4 we use the chain rule, in line 5 we use Eq. 195, and in line 7 we multiply by $\frac{c^2 F r^4}{B^2}$. Now, since $u = 1/r$ (and hence $r = 1/u$) then we have:

$$\frac{dr}{d\phi} = -\frac{1}{u^2}\frac{du}{d\phi} \qquad (288)$$

and thus Eq. 287 can be written as:

$$\left(-\frac{1}{u^2}\frac{du}{d\phi}\right)^2 = -\frac{c^2 F r^4}{B^2} + \frac{c^2 A^2 r^4}{B^2} - Fr^2$$

$$\frac{1}{u^4}\left(\frac{du}{d\phi}\right)^2 = -\frac{c^2 F}{B^2 u^4} + \frac{c^2 A^2}{B^2 u^4} - \frac{F}{u^2}$$

8.1 Perihelion Precession of Mercury

$$\begin{aligned}
\left(\frac{du}{d\phi}\right)^2 &= -\frac{c^2 F}{B^2} + \frac{c^2 A^2}{B^2} - Fu^2 \\
\left(\frac{du}{d\phi}\right)^2 &= -\frac{c^2}{B^2}\left(1 - \frac{2GMu}{c^2}\right) + \frac{c^2 A^2}{B^2} - u^2\left(1 - \frac{2GMu}{c^2}\right) \\
\left(\frac{du}{d\phi}\right)^2 &= -\frac{c^2}{B^2} + \frac{2GM}{B^2}u + \frac{c^2 A^2}{B^2} - u^2 + \frac{2GM}{c^2}u^3 \\
\left(\frac{du}{d\phi}\right)^2 &= \frac{2GM}{c^2}u^3 - u^2 + \frac{2GM}{B^2}u - \frac{c^2}{B^2} + \frac{c^2 A^2}{B^2} \\
\left(\frac{du}{d\phi}\right)^2 &= \frac{2GM}{c^2}\left(u^3 - \frac{c^2}{2GM}u^2 + \frac{c^2}{B^2}u - \frac{c^4}{2GMB^2} + \frac{c^4 A^2}{2GMB^2}\right)
\end{aligned} \qquad (289)$$

where in line 2 we use $r = 1/u$ and in line 4 we use $F = 1 - \frac{2GMu}{c^2}$ while the other lines are based on simple algebraic manipulation. As we see, the expression inside the parentheses on the right hand side of Eq. 289 is a cubic expression in u and hence according to the rules of polynomials it can be factorized into three factors where each factor corresponds to one root of the cubic.[217] Now, the roots of the above cubic should include u_0 and u_1 that correspond to the perihelion distance r_0 and the aphelion distance r_1 respectively. The reason is that at perihelion and aphelion r_0 and r_1 are minimum and maximum respectively and hence $\frac{dr}{d\phi} = 0$ which (according to Eq. 288) makes $\frac{du}{d\phi} = 0$ and this should cause the cubic in Eq. 289 to vanish (since $\frac{2GM}{c^2} \ne 0$) which means that u_0 and u_1 are roots of this cubic. Now, according to the rules of polynomials that govern the relation between the coefficients and the roots the sum of the roots of the cubic in Eq. 289 is $\frac{c^2}{2GM}$.[218] Therefore, the third root should be $\frac{c^2}{2GM} - u_0 - u_1$. Hence, Eq. 289 can be written as:

$$\begin{aligned}
\left(\frac{du}{d\phi}\right)^2 &= \frac{2GM}{c^2}(u-u_0)(u-u_1)\left(u - \frac{c^2}{2GM} + u_0 + u_1\right) \\
\left(\frac{du}{d\phi}\right)^2 &= (u-u_0)(u-u_1)\left(\frac{2GM}{c^2}u - 1 + \frac{2GM}{c^2}u_0 + \frac{2GM}{c^2}u_1\right) \\
\left(\frac{du}{d\phi}\right)^2 &= (u-u_0)(u-u_1)\left(-1 + \frac{2GM}{c^2}[u + u_0 + u_1]\right) \\
\left(\frac{du}{d\phi}\right)^2 &= (u_0-u)(u-u_1)\left(1 - \frac{2GM}{c^2}[u + u_0 + u_1]\right) \\
\left|\frac{du}{d\phi}\right| &= \sqrt{(u_0-u)(u-u_1)}\left(1 - \frac{2GM}{c^2}[u + u_0 + u_1]\right)^{1/2} \\
\left|\frac{d\phi}{du}\right| &= \frac{\left(1 - \frac{2GM}{c^2}[u + u_0 + u_1]\right)^{-1/2}}{\sqrt{(u_0-u)(u-u_1)}} \\
\left|\frac{d\phi}{du}\right| &\simeq \frac{1 + \frac{GM}{c^2}[u + u_0 + u_1]}{\sqrt{(u_0-u)(u-u_1)}}
\end{aligned} \qquad (290)$$

where in line 5 we have $(u_0 - u) \ge 0$ and $(u - u_1) \ge 0$ because u_0 is maximum (since $r_0 = 1/u_0$ is minimum) and u_1 is minimum (since $r_1 = 1/u_1$ is maximum) and hence $\left(1 - \frac{2GM}{c^2}[u + u_0 + u_1]\right) \ge 0$ according to line 4, while in line 7 we use a truncated power series (or binomial series) approximation.[219]

---

[217] If the roots of a cubic polynomial in x (with a leading x^3 term) are x_1, x_2, x_3 then the polynomial can be factorized as $(x - x_1)(x - x_2)(x - x_3)$.
[218] For a cubic of the form $x^3 + bx^2 + cx + d$ with roots x_1, x_2, x_3 we have $x_1 + x_2 + x_3 = -b$.
[219] We have:
$$(1-x)^{-1/2} = 1 + \frac{1}{2}x + \frac{3}{8}x^2 + \frac{5}{16}x^3 + \cdots \qquad (-1 < x < 1)$$

8.1 Perihelion Precession of Mercury

Now, our objective from developing a general relativistic formula for the extra precession of planetary perihelion is to find the extra change in ϕ caused by a change of u during a complete revolution of the planet around the Sun. In practical terms, we need first to find $\Delta\phi$ which represents the total change in ϕ during a full revolution (i.e. when the orbit changes from u_0 to u_1 then back to u_0) and then we find $\delta\phi$ which is the extra change in ϕ (i.e. the excess over 2π) and this $\delta\phi$ is the required extra precession. So, we should first find $\Delta\phi$ when the orbit changes from u_0 to u_1 then back to u_0 (i.e. $u_0 \to u_1 \to u_0$) by integrating $|d\phi|$, that is:

$$\Delta\phi = \int |d\phi| \simeq 2 \int_{u=u_0}^{u=u_1} \frac{1 + \frac{GM}{c^2}[u + u_0 + u_1]}{\sqrt{(u_0 - u)(u - u_1)}} |du| \qquad (291)$$

where in the second step we used Eq. 290 plus the presumed symmetry of the orbit.[220] To evaluate the integral in the last equation we need to find an expression for u which means that we need to have a solution for the planetary motion problem where u is given as a function of ϕ.[221] Now, in § 5.2 (see problem 1) we demonstrated that in the classical limit the general relativistic formulation for the planetary motion reduces to the classical system of Eqs. 15-16, while in § 1.5.1 (see problem 2 and Eq. 17) we found that the solution of the classical system of Eqs. 15-16 is:

$$r = \frac{1}{u} = \frac{e/A}{e\cos\phi + 1} = \frac{B^2}{GM(e\cos\phi + 1)}$$

where A is a positive constant and $e = \frac{AB^2}{GM}$. Now, since in the present problem we want to derive the general relativistic formula for the extra precession of planetary perihelion in the classical limit for the purpose of comparison to observation (since our observations of planetary motion fall well within the classical limit) then we can use this solution as a good approximation. Also, in § 1.5.1 (see problem 3) we found that:

$$r_0 = \frac{1}{u_0} = \frac{B^2}{GM(1+e)} \qquad \text{and} \qquad r_1 = \frac{1}{u_1} = \frac{B^2}{GM(1-e)}$$

Accordingly:

$$1 + \frac{GM}{c^2}[u + u_0 + u_1] = 1 + \frac{GM}{c^2}\left[\frac{GM(e\cos\phi + 1)}{B^2} + \frac{GM(1+e)}{B^2} + \frac{GM(1-e)}{B^2}\right]$$

$$= 1 + \frac{G^2M^2}{c^2 B^2}(e\cos\phi + 3)$$

$$|du| = \left|d\left(\frac{GM(e\cos\phi + 1)}{B^2}\right)\right| = \frac{GMe}{B^2}|\sin\phi||d\phi|$$

$$u_0 - u = \frac{GM(1+e)}{B^2} - \frac{GM(e\cos\phi + 1)}{B^2} = \frac{GMe(1-\cos\phi)}{B^2}$$

$$u - u_1 = \frac{GM(e\cos\phi + 1)}{B^2} - \frac{GM(1-e)}{B^2} = \frac{GMe(1+\cos\phi)}{B^2}$$

$$\sqrt{(u_0 - u)(u - u_1)} = \frac{GMe}{B^2}\sqrt{1 - \cos^2\phi} = \frac{GMe}{B^2}|\sin\phi|$$

---

[220] Although the orbit is not elliptic (because it is not closed) it can still be symmetric (considering a full revolution) or at least it is almost symmetric (if it is not exactly symmetric).
[221] In fact, we also need to find expressions for u_0 and u_1 (but they are based on the expression of u).

8.1 Perihelion Precession of Mercury

On substituting from the last equations into Eq. 291 we get:[222]

$$
\begin{aligned}
\Delta\phi &\simeq 2\int_{\phi=0}^{\phi=\pi} \frac{1 + \frac{G^2M^2}{c^2B^2}(e\cos\phi + 3)}{\frac{GMe}{B^2}|\sin\phi|}\frac{GMe}{B^2}|\sin\phi|\,d\phi \qquad (292)\\
&= 2\int_{\phi=0}^{\phi=\pi}\left[1 + \frac{G^2M^2}{c^2B^2}(e\cos\phi + 3)\right]d\phi\\
&= 2\pi + \frac{2G^2M^2e}{c^2B^2}[\sin\phi]_0^\pi + \frac{G^2M^2}{c^2B^2}6\pi\\
&= 2\pi + 0 + \frac{6\pi G^2M^2}{c^2B^2}\\
&= 2\pi + \frac{6\pi G^2M^2}{c^2B^2}\\
&= 2\pi + \delta\phi
\end{aligned}
$$

where in line 1 the use of $d\phi$ (instead of $|d\phi|$) is justified by the limits. As we see, during a complete revolution of the planet around the Sun (i.e. when the orbiting planet undergoes a change from a perihelion r_0 to an aphelion r_1 then back to the next perihelion r_0 and hence it completes a full revolution) there is an extra precession $\delta\phi$ in the planetary orbit of $\frac{6\pi G^2M^2}{c^2B^2}$. Now, we have:

$$B = 2\frac{dA}{dt} = 2\times\frac{\pi ab}{T} = \frac{2\pi a^2\sqrt{1-e^2}}{T}$$

where we used formulae that we developed earlier (see § 1.5.1). Therefore, the extra precession in the planetary orbit in each revolution is:

$$\delta\phi = \frac{6\pi G^2M^2}{c^2B^2} = \frac{6\pi G^2M^2}{c^2\left(\frac{2\pi a^2\sqrt{1-e^2}}{T}\right)^2} = \frac{6\pi G^2M^2 T^2}{c^2 4\pi^2 a^4(1-e^2)} = \frac{3G^2M^2T^2}{2\pi c^2 a^4(1-e^2)}$$

which is the same as Eq. 285.

Regarding the observed extra precession of perihelion of Mercury, the time period T of Mercury is about 88 days (i.e. about 7.6032×10^6 s), its semi-major axis a is about 5.790905×10^{10} m, and its eccentricity e is about 0.2056. Hence, the extra precession of its perihelion in one revolution $\delta\phi_M$ is:

$$\delta\phi_M \simeq \frac{3\times(6.674\times 10^{-11})^2\times(1.989\times 10^{30})^2\times(7.6032\times 10^6)^2}{2\pi\times(3\times 10^8)^2\times(5.790905\times 10^{10})^4\times(1-0.2056^2)} \simeq 5.018\times 10^{-7}\text{ rad}$$

Now, in one century (i.e. 100 years) there are 36525 days and hence Mercury should revolve around the Sun about 415 times. Therefore, the extra precession of its perihelion in one century is about:

$$415\times 5.018\times 10^{-7}\text{ rad} \simeq 0.0002083\text{ rad} \simeq 42.95\text{ arcsec}$$

The observed extra precession of perihelion of Mercury is about 43 arcseconds per century (with a reported uncertainty of about 1%). Hence, the agreement between the observed value and the prediction of general relativity is excellent (but see the following note and § 9.1).

Note: in the above derivation and calculation we used the formula:

$$B = 2\frac{dA}{dt} = 2\times\frac{\pi ab}{T} = \frac{2\pi a^2\sqrt{1-e^2}}{T}$$

In fact, this formula is valid for flat classical spacetime[223] (which is not the case in general relativity) and this will put a big question mark on the validity of the above derivation and calculation. Accordingly, the "excellent" agreement between the observation and general relativistic prediction is highly

---

[222] Regarding the change of the limits of integration, refer to the note in the end of this answer.

[223] This may be seen more prominently in the final formula where we use a (which is a physical distance in flat space) rather than a quantity like r (which is a coordinate variable that usually appears in general relativistic formulae in curved spacetime). In fact, all the quantities A, t, a, b, T as well as the relation $A = \pi ab$ are classical and belong to a flat space-time (as can be noticed from importing them from the classical formulation of § 1.5.1).

suspicious.[224] In fact, if we take the effect of the curvature of spacetime (by observing the difference between coordinate variables and physical variables) then there should be no agreement (let alone be excellent). This issue will be investigated further in § 9.1 where we will clarify the issue that being in the classical limit (as demanded by the question and followed in the derivation) does not justify the neglect of the effect of spacetime curvature due to the violation of locality and the accumulation of tiny relativistic effects over extended tempo-spatial interval (see exercise 8).

We should also note that in changing the limits of integration between Eq. 291 and Eq. 292 we are also using an approximation as if there is no extra precession.[225] This could be an acceptable approximation for one revolution but it may not be acceptable for many revolutions when we calculate the extra precession in a century. This should be a cause for more uncertainty about the "excellence" of the agreement between the observation and general relativistic prediction.

Exercises
1. What "perihelion" and "aphelion" mean?
2. What "precession" means? Is it appropriate term for describing the slow shift in the position of perihelion?
3. What we mean by "extra precession"?
4. What are the main causes of the precession of perihelion of Mercury that can be accounted for by classical physics?[226]
5. What distinguishes Mercury from other planets in the solar system and hence makes it prominent with regard to its extra precession of perihelion?
6. What you note about $\delta\phi$, $\delta\phi_c$ and N_c in Table 1?
7. Illustrate the precession of perihelion by a simple sketch.
8. As we noted earlier (see § 1.1), the extra precession of perihelion of Mercury may be seen as an exception to the common rule that the characteristic general relativistic effects that distinguish general relativity from classical gravity are generally observable in strong gravitational fields. Discuss this issue.

8.2 Light Bending by Gravity

Light bending (or light deflection) by the gravitational field of a gravitating body, like the Sun, is one of the major predictions of general relativity because according to this theory the bending of spacetime by matter and energy should result in bending the geodesic paths of the spacetime and hence the physical objects, whether massive like material particles or massless like photons, which follow these geodesic paths in their free fall (i.e. in the absence of forces other than gravity) should move along curved trajectories. The bending of light in gravitational fields may also be justified qualitatively by the equivalence principle since light bends in accelerating frames and hence it should also bend in gravitational frames. In fact, light bending by gravity can be derived even from the Newtonian theory (see Exercises) and hence it is not an effect that is specific to general relativity. However, the Newtonian prediction is quantitatively different from the prediction of general relativity.[227]

Using the geodesic equations of null geodesic in the Schwarzschild metric, it can be shown (see Problems) that the deflection of light by the gravity of a static and spherically symmetric gravitating body is given by:

$$\delta\phi = \frac{4GM}{c^2 d} \tag{293}$$

where $\delta\phi$ is the angle of deflection in radians, G is the gravitational constant, M is the mass of the gravitating body, c is the characteristic speed of light and d is the distance of closest approach of light to

---
[224] The suspicion is actually not about the agreement but about the general relativistic nature of the prediction.
[225] Noting that the total extra precession in a revolution is an accumulative contribution of tiny precessions over the entire orbit, this approximation should be acceptable considering that $\delta\phi$ is very tiny compared to full revolution (i.e. 2π).
[226] As indicated earlier, we are excluding here classical theories that claim to account for the extra precession.
[227] We are referring here to the commonly recognized Newtonian prediction which is based on derivations like the one that we provided in the Exercises. However, some theories based on the Newtonian gravity seem to have different predictions; some of which may agree with the prediction of general relativity.

8.2 Light Bending by Gravity

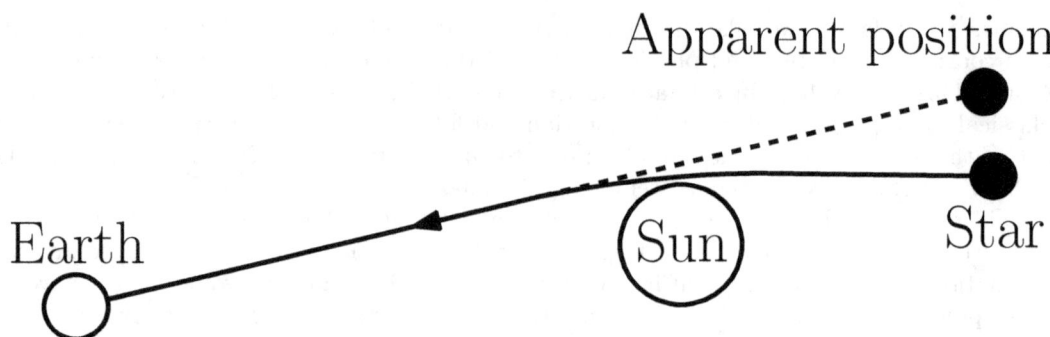

Figure 5: A simple sketch depicting the bending of a light ray from a star as it passes by the Sun toward the Earth and hence it is deflected by the gravity of the Sun.

the center of the gravitating body.

Light bending by the gravitational field of a gravitating body may be demonstrated by a number of astronomical phenomena. For example, it is demonstrated in the bending of light rays originating from stars by the gravitational field of the Sun which results in the apparent displacement of the stars from their real position (see Figure 5). Because the faint light of stars cannot be observed in the close proximity of the bright light of the Sun, these light bending observations are usually conducted during solar eclipse where the bright light of the Sun is masked by the Moon. A historical example for this sort of light bending observations is the famous 1919 expedition by Eddington and his team which led to the general acceptance and dominance of general relativity and the rise of Einstein to worldwide fame.

Light bending by the gravitational field of a gravitating body may also be demonstrated by a number of astronomical phenomena that are generically labeled as gravitational lensing where a gravitating object (such as a black hole or a galaxy or a cluster of galaxies) acts like a lens by gathering light rays from an observed object along multiple paths around the gravitating (or lensing) object and directing them toward the observer and hence the observed object appears with multiple images surrounding the gravitating object. For instance, it may be demonstrated by the so-called Einstein cross where 4 images of an astronomical object, such as a quasar, are formed by the deflection of the light rays that originate from the object and follow 4 different paths in their way around the gravitating object. Gravitational lensing may also be demonstrated by the so-called Einstein ring which is similar to the Einstein cross but instead of forming 4 separate images a ring is formed by a continuous bundle of rays that are symmetrically distributed around the lensing object. Einstein ring is supposed to be formed when the observer, observed object and the lensing object are aligned along a straight line and hence the coming rays form a circular symmetry around the lens, while Einstein cross is supposed to be formed when they are not aligned. However, the formation of a cross (rather than other shapes) requires further justification (see § 9.2).

Problems

1. Derive the general relativistic formula for light bending by stars (e.g. Sun).

   Answer: The spacetime surrounding a star (as a gravitating body) is reasonably described by the Schwarzschild metric and hence we should use this metric here. Moreover, the problem of light bending is a problem of motion along geodesic trajectories in spacetime because the light is in free fall under the influence of gravity alone. Therefore, this problem should be investigated as a geodesic problem in the Schwarzschild metric and hence we can use the results that we obtained in § 4.1.2. Now, since light is massless the system of geodesic equations that is suitable for investigating this problem is the system of Eqs. 164-166. In fact, all we need from that system is Eq. 165 because the essence of light bending is represented by a correlation between r ($= 1/u$) and ϕ (i.e. how ϕ changes as r changes) and that is what Eq. 165 is about. So, our investigation here starts from Eq. 165, that is:

$$\frac{d^2u}{d\phi^2} + u = \frac{3GMu^2}{c^2} \tag{294}$$

8.2 Light Bending by Gravity

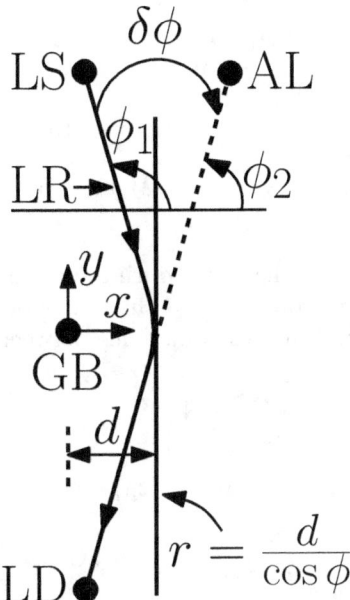

Figure 6: A simple sketch depicting the deflection of a light ray LR emitted from a light source LS as the ray passes by a gravitating body GB (which is at the origin of coordinates) toward a light detector LD and hence the light ray is deflected by an angle $\delta\phi$ causing the image of LS to be visually displaced and thus it is seen at an apparent location AL. The other symbols in the sketch are explained in the text.

In brief, what we need for finding a formula for light bending by gravity is to solve Eq. 294 and hence infer the angular deflection in the path of light that is caused by the gravitational field of the star.
Now, if we consider a light ray grazing a typical star that represents a gravitating object then we have $u \gg \frac{3GMu^2}{c^2}$. This can be verified by taking the ratio of $\frac{3GMu^2}{c^2}$ to u and using the Sun to represent the typical star, that is:

$$\frac{3GMu^2}{c^2 u} = \frac{3GMu}{c^2} = \frac{3GM}{c^2 r} \simeq \frac{3 \times 6.674 \times 10^{-11} \times 1.989 \times 10^{30}}{(3 \times 10^8)^2 \times 6.9551 \times 10^8} \simeq 6.362 \times 10^{-6}$$

where r represents the radius of the Sun since the ray is grazing (in fact r should be radial coordinate). Therefore, as a first approximation the term $\frac{3GMu^2}{c^2}$ can be ignored in solving Eq. 294 and hence we solve the equation $\frac{d^2 u}{d\phi^2} + u = 0$ instead. The solution of the latter equation is (see Exercises):

$$u = \frac{1}{r} = \frac{\cos\phi}{d} \qquad \left(-\frac{\pi}{2} < \phi < \frac{\pi}{2}\right) \qquad (295)$$

where d is a constant. As we will see, d represents the distance of closest approach of the light ray to the center of gravitating object and hence in the case of grazing ray it represents the radius (or radial coordinate) of the gravitating star. We should note that the solution can also be given as $u = \frac{\sin\phi}{d}$ with $0 < \phi < \pi$. In fact, for a polar coordinate system in a standard form (i.e. where ϕ is the angle that $r\mathbf{e}_r$ makes with the positive x axis of the corresponding Cartesian system) $u = \frac{\cos\phi}{d}$ represents a straight line parallel to the y-axis and intercepting the x axis at d (see Figure 6) while $u = \frac{\sin\phi}{d}$ represents a straight line parallel to the x-axis and intercepting the y axis at d.
Now, Eq. 295 is an equation of a straight line in polar coordinates (see Exercises) with d being the perpendicular distance of the straight line from the origin of coordinates where the center of the gravitating object is located (see Figure 6) and hence our first approximation produces no deflection. However, we can seek a better solution by substituting this approximate solution into the u^2 term of

8.2 Light Bending by Gravity

Eq. 294, that is:
$$\frac{d^2 u}{d\phi^2} + u = \frac{3GM \cos^2 \phi}{c^2 d^2}$$

It can be easily verified (see Exercises) that a solution of the last equation is:
$$u = \frac{\cos \phi}{d} + \frac{GM \left(2 - \cos^2 \phi\right)}{c^2 d^2}$$

Now, to find the total deflection of the light ray during its journey from infinity (i.e. $r \to \infty$) on one end to infinity on the other end (i.e. from $u = 0$ at the source of light to $u = 0$ at the detector of light), we set $u = 0$ in the last equation and find the difference between ϕ at one end and ϕ at the other end, that is:

$$\frac{\cos \phi}{d} + \frac{GM \left(2 - \cos^2 \phi\right)}{c^2 d^2} = 0$$
$$\frac{\cos \phi}{d} + \frac{2GM}{c^2 d^2} - \frac{GM \cos^2 \phi}{c^2 d^2} = 0$$
$$\frac{GM}{c^2 d} \cos^2 \phi - \cos \phi - \frac{2GM}{c^2 d} = 0 \qquad (296)$$

Now, if we note that $\frac{GM}{c^2 d} \cos^2 \phi$ is very small for a light ray grazing a typical star (see Exercises) then we can ignore this term and hence we get an approximate equation: $-\cos \phi - \frac{2GM}{c^2 d} \simeq 0$ whose solution is:
$$\cos \phi \simeq -\frac{2GM}{c^2 d} \qquad (297)$$

and hence:
$$\phi_1 \simeq \frac{\pi}{2} + \frac{2GM}{c^2 d} \qquad \text{and} \qquad \phi_2 \simeq \frac{\pi}{2} - \frac{2GM}{c^2 d} \qquad (298)$$

These two possible values of ϕ correspond to the two ends of the light ray (i.e. source of light at infinity on one end and detector of light at infinity on the other end). In fact, the value $\phi_2 \simeq \frac{\pi}{2} - \frac{2GM}{c^2 d}$ corresponds to $\phi \simeq -\frac{\pi}{2} - \frac{2GM}{c^2 d}$ or similarly to $\phi \simeq \frac{3\pi}{2} - \frac{2GM}{c^2 d}$ (see Figure 6) and that is why it does not give the correct sign. More justification is given in the upcoming note 2 in the end of this answer. Now, if we take the difference between ϕ at infinity on one end where the light source is located and ϕ at infinity on the other end where the light detector is located (see Figure 6) then we get the total deflection, that is:

$$\delta \phi = \phi_1 - \phi_2 \simeq \frac{\pi}{2} + \frac{2GM}{c^2 d} - \left(\frac{\pi}{2} - \frac{2GM}{c^2 d}\right) = \frac{4GM}{c^2 d}$$

which is the general relativistic formula that is given in the text (see Eq. 293).

Note 1: the reader may feel uncomfortable about the last approximation where we discarded the term $\frac{GM}{c^2 d} \cos^2 \phi$. So, we show in this note that this approximation is very good. Eq. 296 (which can be written as $GM \cos^2 \phi - c^2 d \cos \phi - 2GM = 0$) is a quadratic equation in $\cos \phi$ and hence from the quadratic formula we get:

$$\cos \phi = \frac{c^2 d \pm \sqrt{c^4 d^2 + 8 G^2 M^2}}{2GM} \qquad (299)$$

It is obvious that the root with plus sign is unacceptable because it is too large and hence it should be rejected since $-1 \leq \cos \phi \leq +1$. Now, if we consider a grazing light ray and use the Sun as a typical star then from the approximate root (i.e. Eq. 297) we get:

$$\cos \phi \simeq -\frac{2GM}{c^2 d} \simeq -\frac{2 \times 6.674 \times 10^{-11} \times 1.989 \times 10^{30}}{(3 \times 10^8)^2 \times 6.9551 \times 10^8} \simeq -4.2413596 \times 10^{-6}$$

8.2 Light Bending by Gravity

while from the exact root (i.e. Eq. 299 with the minus sign) we get:

$$\cos\phi = \frac{c^2 d - \sqrt{c^4 d^2 + 8G^2 M^2}}{2GM} \simeq -4.2413599 \times 10^{-6}$$

As we see, the approximate root is virtually identical to the exact root.

Note 2: the transition from Eq. 297 to Eq. 298 may not be obvious. However, we should note first that we have two different solutions because the original equation is quadratic and although we already rejected one root on mathematical basis we should admit the existence of two physical roots. In fact, we can demonstrate formally that ϕ_1 and ϕ_2 of Eq. 298 are the principal solutions, but instead we can follow a simpler approach by using the trigonometric identity $\cos(\alpha \pm \beta) = \cos\alpha\cos\beta \mp \sin\alpha\sin\beta$ with the sine approximation of small angle (i.e. $\sin\beta \simeq \beta$) to conclude that ϕ_1 and ϕ_2 in Eq. 298 produce Eq. 297 (and hence ϕ_1 and ϕ_2 are solutions to Eq. 297), that is:

$$\cos\phi_1 = \cos\left(\frac{\pi}{2} + \frac{2GM}{c^2 d}\right) = \cos\frac{\pi}{2}\cos\frac{2GM}{c^2 d} - \sin\frac{\pi}{2}\sin\frac{2GM}{c^2 d} \simeq -\frac{2GM}{c^2 d}$$

$$\cos\phi_2 = \cos\left(\frac{3\pi}{2} - \frac{2GM}{c^2 d}\right) = \cos\frac{3\pi}{2}\cos\frac{2GM}{c^2 d} + \sin\frac{3\pi}{2}\sin\frac{2GM}{c^2 d} \simeq -\frac{2GM}{c^2 d}$$

where we used the aforementioned alternative expression for ϕ_2 (to obtain the correct sign) and where the smallness of $\frac{GM}{c^2 d}$ (and hence $\frac{2GM}{c^2 d}$) is justified in exercise 7.

Note 3: the provision $u \gg \frac{3GMu^2}{c^2}$ can be justified more generally by the following:

$$\frac{3GMu^2}{c^2 u} = \frac{3GMu}{c^2} = \frac{3GM}{c^2 r} = \frac{3}{2}\frac{2GM}{c^2 r} = \frac{3}{2}\frac{R_S}{r}$$

where R_S is the Schwarzschild radius and r represents the star radius (or radial coordinate) or larger. Now, for ordinary stars $R_S \ll r$ and hence the provision is justified. The reader should note that the "grazing" condition that associates this provision (and its alike) in the above derivation represents the worst case and hence the provision becomes stronger in the non-grazing case.

Note 4: the validity of the above derivation should be restricted to the weak gravitational field case where the spacetime is approximately flat. This may be suggested by a number of indications such as defining d as "distance", using polar coordinates (also see exercise 5) and accepting straight line solution as a first approximation (which is based on $u \gg \frac{3GMu^2}{c^2}$ and the smallness of $\frac{GM}{c^2 d}$). In fact, this restriction should be suggested by the question itself since "stars" usually generate weak gravitational fields (if we exclude neutron stars). This is also suggested by the use of terms like "typical star" and "ordinary star" (as well as other indications in the solutions of the upcoming Exercises).

Note 5: the above derivation is based on a number of approximations and compromising assumptions (e.g. classical gravitational systems of weak gravity). Accordingly, the derived formula should have limited validity from these perspectives (at least as far as this derivation is concerned).

Note 6: we defined d as the distance of closest approach. But considering the initial straight-line solution, it should be more appropriate to define it as the distance of impact (or impact parameter) where the supposed straight line is defined by the initial direction of the ray. However, we adopted the "distance of closest approach" definition for convenience in the derivation and application (as well as possible ease of comparison to the upcoming classical derivation; see exercise 8). Anyway, for the considered weak gravitational field case (where the deflection of light is very tiny) the difference between the two definitions should be practically negligible.

Exercises
1. Is light bending by gravity restricted to light (i.e. photons)?
2. Discuss the following statement: "The bending of light in gravitational fields may also be justified qualitatively by the equivalence principle since light bends in accelerating frames and hence it should also bend in gravitational frames".
3. What is the significance of light bending by gravity?

4. Show that $u = \frac{\cos\phi}{d}$ is a solution to the equation $\frac{d^2 u}{d\phi^2} + u = 0$.
5. Verify that (**a**) the polar equation $\frac{1}{r} = \frac{\cos\phi}{d}$ (with $-\frac{\pi}{2} < \phi < \frac{\pi}{2}$) is an equation of a straight line parallel to the y axis, and (**b**) the polar equation $\frac{1}{r} = \frac{\sin\phi}{d}$ (with $0 < \phi < \pi$) is an equation of a straight line parallel to the x axis.
6. Verify that:
$$u = \frac{\cos\phi}{d} + \frac{GM\left(2 - \cos^2\phi\right)}{c^2 d^2} \tag{300}$$
is a solution to the following equation:
$$\frac{d^2 u}{d\phi^2} + u = \frac{3GM \cos^2\phi}{c^2 d^2} \tag{301}$$
7. Show that for a light ray grazing a typical star, $\frac{GM}{c^2 d} \cos^2\phi$ is very small (i.e. close to zero).
8. Derive a classical formula for light bending by stars (e.g. Sun) and hence show that the classical deflection is half the general relativistic deflection.
9. Since light is massless and classical gravity is a property of mass (according to Newton's gravity law), how can light be deflected by gravity in classical physics?
10. Calculate the deflection of a light ray grazing the Sun.
11. What is the significance of the apparent displacement of the image of an observed object when the light originating from the object bends?
12. Do you think "gravitational lensing" is a suitable term for the described phenomenon?
13. Make a simple sketch depicting gravitational lensing.
14. What can gravitational lensing reveal about the lensing object and the lensed (or observed) object?
15. Show formally that the deflection of a gravitational lens decreases with the increase of the distance of closest approach of the deflected ray.
16. Show that the deflection of light by gravity can be considerable only if the gravitating object is compact (e.g. black hole).
17. Show that in gravitational lensing the lensed object cannot be observed when the distance of the observer from the lens is less than a certain amount, and give an estimate of this amount.

8.3 Gravitational Time Dilation

The essence of gravitational time dilation in simple terms is that clocks slow down as they descend in a gravitational well and speed up as they ascend in a gravitational well. In fact, gravitational time dilation (assuming a Schwarzschild spacetime) was investigated rather thoroughly in § 6.3.3 and hence we do not repeat the investigation here.

Problems

1. Derive a formula for the relation between proper time intervals in a Schwarzschild spacetime as recorded by two local stationary frames that are at different radial distances from the source of gravity. How this formula becomes when one frame is at infinity, i.e. very far away from the source of gravity?
   Answer: If $d\tau_1$ and $d\tau_2$ are the proper time intervals recorded by frame O_1 at r_1 and frame O_2 at r_2 (where r_1 and r_2 are the radial distances from a source of gravity of mass M and r_1 and r_2 are greater than $\frac{2GM}{c^2}$) then from Eq. 245 we have:
$$\frac{d\tau_2}{d\tau_1} = \left(1 - \frac{2GM}{c^2 r_2}\right)^{1/2} \left(1 - \frac{2GM}{c^2 r_1}\right)^{-1/2} \tag{302}$$
Now, if O_2 is at infinity (i.e. $r_2 \to \infty$) then this formula becomes:
$$d\tau_\infty \equiv d\tau_2 = d\tau_1 \left(1 - \frac{2GM}{c^2 r_1}\right)^{-1/2} \tag{303}$$

8.3 Gravitational Time Dilation

i.e. the proper time interval $d\tau_\infty$ that is recorded in the absence of gravity (i.e. at infinity) is longer than the proper time interval $d\tau_1$ that is recorded in the presence of gravity, i.e. at finite radial distance r_1.[228] In other words, the clock at infinity runs faster than the other clock which means that the effect of gravity is to slow down clocks.

Note: r_1 and r_2 in the formula represent radial coordinates but in the answer we assumed that they represent radial distances (as required by the question). So, if we have to be consistent then we should obtain the radial coordinates that correspond to the radial distances (which are supposedly available from the physical setting according to the question) and use these radial coordinates in the formula (assuming that obtaining the radial coordinates from the available radial distances is possible). This should highlight the practical difficulties that are usually faced in solving general relativistic problems that involve coordinate variables (whereas in reality we have physical variables). In fact, we can easily avoid this difficulty by changing the question to: "at different radial coordinates" instead of "at different radial distances" but this will not address the essence of this difficulty where we can legitimately ask about problems that involve physical variables (which what we usually have and where the questions about them is usually more natural).

2. What is the fractional change in the time interval that is experienced by a clock as it moves[229] from the surface of the Earth to very far away from the Earth (i.e. infinity)?

   Answer: Assuming a Schwarzschild spacetime (see the upcoming note 3), the fractional change is given by:

$$\begin{aligned}
\frac{d\tau_2 - d\tau_1}{d\tau_2} &= 1 - \frac{d\tau_1}{d\tau_2} \\
&= 1 - \left(1 - \frac{2GM}{c^2 r_1}\right)^{1/2} \\
&\simeq 1 - \left(1 - \frac{2 \times 6.674 \times 10^{-11} \times 5.972 \times 10^{24}}{(3 \times 10^8)^2 \times 6.371 \times 10^6}\right)^{1/2} \\
&\simeq 6.951 \times 10^{-10}
\end{aligned}$$

where in line 2 we use Eq. 303, and in line 3 we use the mass of the Earth for M and the radius of the Earth for r_1.

Note 1: in this answer we used the radius of the Earth for r_1. This seems inconsistent with the fact that the radius of the Earth is a physical radius while r_1 is a coordinate radius. The justification of this according to some general relativists is that the radius of a circle is its circumference divided by 2π. So, let use this definition of radius to find the coordinate radius r that corresponds to the physical radius of the Earth R in the Schwarzschild metric. Now, the radius of a great circle of a sphere (which represents the Earth) is the same as the radius of the sphere. We can use (with no loss of generality) the equator circle (which corresponds to $\theta = \pi/2$ in the Schwarzschild coordinates) in this calculation. On the equator circle the ct, r and θ coordinates are constants (because in measuring length ct is fixed, r is the radius of a circle and $\theta = \pi/2$) and hence $dt = dr = d\theta = 0$. Accordingly, on the equator circle the Schwarzschild line element becomes:

$$ds = r \sin \frac{\pi}{2} d\phi = r\, d\phi$$

where ds is an infinitesimal line element on the physical circumference and r is the coordinate radius. The physical circumference s is then obtained by integrating ds from $\phi = 0$ to $\phi = 2\pi$, that is:

$$s = \int_0^{2\pi} r\, d\phi = 2\pi r$$

---

[228] For finite r and $\frac{2GM}{c^2 r} < 1$ we have $\left(1 - \frac{2GM}{c^2 r}\right)^{-1/2} > 1$.

[229] We note that "moves" in this question and its alike means change of its location with no consideration of any kinematical effects.

8.3 Gravitational Time Dilation

Now, if we note that the physical circumference is 2π times the physical radius (i.e. $s = 2\pi R$) then from the last equation we get $2\pi R = 2\pi r$ and hence $R = r$, i.e. the physical radius is equal to the coordinate radius.

We note that this general relativistic argument is crucially based on the premise: "the physical circumference is 2π times the physical radius (i.e. $s = 2\pi R$)" which is valid in flat space but not in curved space. As discussed earlier (see § 2.3), in curved space the relation between the circumference C of a circle and its diameter D is $C \ne \pi D$ while in the above argument we have $C = \pi D$ (where C and D refer to the metrical quantities). So in brief, the above definition of radius (i.e. the radius of a circle is its circumference divided by 2π) is incorrect (if radius means physical radius) since it holds only in flat space. The physical radius should therefore be obtained by integrating the line element ds along the radial direction (i.e. where only dr changes in the metric).[230] In fact, the above general relativistic argument can be easily refuted (with no need for any detailed considerations) by the obvious fact that in curved space the radial coordinate differential dr is not equal to the radial proper length differential ds (as can be seen from Eq. 256) and hence the radial distance (or physical radius) cannot be equal to the radial coordinate (or coordinate radius).

We should note that in the literature of general relativity there are similar arguments to the above argument; some of which are more elaborate and may appear technically more sound than the above argument but none of these are sufficiently convincing and can overcome this difficulty. Moreover, this problem is not restricted to the radial coordinate r or to the Schwarzschild geometry and hence even if we find a solution to this problem for r in the Schwarzschild geometry there is no guarantee that we can find a solution for other coordinate variables and in other spacetime geometries in all cases and circumstances.

We should finally remark that in the literature of general relativity some authors distinguish between radial coordinate and radial distance quantitatively (i.e. they treat them as unequal) while other authors treat them as quantitatively equal. Moreover, some authors claim that the radial distance of a point from the origin is not defined.[231] So, we have at least three opinions about this issue.

Note 2: the above practical problem about solving problems involving coordinate variables (which are usually unknown) may be eased in the case of weak gravity (as it is the case in classical systems) where the difference between the physical and coordinate variables is usually negligible (since the spacetime is essentially flat) and hence we can use the known physical variables in place of the unknown coordinate variables. However, this does not address this problem completely because this is limited to weak-gravity systems and hence we still face this problem in strong-gravity systems where the physical and coordinate variables differ significantly. Moreover, even in the weak-gravity systems the effect of the difference between the physical and coordinate variables can become significant by accumulation over extended space or/and time period due to the violation of locality as seen in the case of perihelion precession (see the note of problem 1 of § 8.1 and refer to § 9.1). Furthermore, the convergence of coordinate variables to physical variables in weak-gravity systems may apply in the Schwarzschild coordinates but not necessarily in other coordinates where the relation between the two sets of variables can be more complicated. Anyway, this may be acceptable as a practical solution but it cannot be regarded as a valid theoretical solution that addresses this problem fundamentally and from its roots.

Note 3: the use of the Schwarzschild metric in this question and its alike may be justified by being a good approximation or by being an idealization. Otherwise, the Earth is not a Schwarzschild object due for example to its rotation around its axis (as well as revolution around the Sun) and its lack of spherical symmetry. Moreover, the spacetime surrounding the Earth is not really a Schwarzschild spacetime due for example to the existence of other gravitating objects like the Sun and other planets. Also, the surrounding space is not really empty (as required by the Schwarzschild metric as a vacuum

---

[230] We note that calculating such an integration is not viable in practical situations where mass varies along the radial direction especially if the body is not spherically symmetric. Yes, in the case of black holes the situation is eased by the assumption of singularity.

[231] Although this seems to be an attempt to address this problem, it actually worsens it because if it is not defined then we cannot solve problems that involve radial distance from the origin.

solution) due to the existence of matter and electromagnetic energy (although in very tiny quantities).

8.4 Gravitational Frequency Shift

Gravitational frequency shift means that a light signal ascending in a gravitational well will be red shifted while a light signal descending in a gravitational well will be blue shifted. Referring to our discussion in § 6.3.4, this effect according to general relativity is based on the effect of gravitational time dilation due to the reciprocal relation between periodic time and frequency. Gravitational frequency shift may also be explained classically (at least qualitatively) by the principle of energy conservation because a photon of light ascending in a gravitational well will increase its potential energy and hence this gain in energy should be compensated by a red shift, while a photon descending into a gravitational well will decrease its potential energy and hence this loss in energy should be compensated by a blue shift.[232] Accordingly, this effect can be predicted in principle even by classical physics. As indicated above, gravitational frequency shift was investigated earlier (from a general relativistic viewpoint assuming a Schwarzschild spacetime) in § 6.3.4 and hence we do not repeat here.

We should remark that if gravitational frequency shift is based on gravitational time dilation (as it is the case in general relativity) then a single explanation (i.e. gravitational time dilation) applies to both time dilation and frequency shift. However, we may need two different explanations (i.e. one for time dilation and one for frequency shift) if we used for example the conservation of energy (following a classical approach) to explain gravitational frequency shift because the conservation of energy may not be able to explain the observations that are specifically related to time dilation, e.g. when we compare two clocks at two different levels in a gravitational potential well and hence there is no frequency shift of radiation (because there is no descent or ascent of radiation in a gravitational well) to be explained by energy conservation. This issue will be investigated further in the following paragraphs.

Let have two observers A and B who have identical sources of monochromatic radiation, i.e. when A and B are at the same location in the spacetime the frequency of the source of A is identical to the frequency of the source of B. Also, let assume that A is located at a higher potential in a gravitational well than B. Now, we have two physical situations that need to be investigated. The first is when each one of A and B observes his own source without sending signal to the other although they can compare the frequencies of their sources (e.g. by using the frequency as a basis for a clock that can count and store time). The second is when A and B exchange signals from their sources (i.e. A sends signal to B and B sends signal to A) and hence they can compare the frequencies of their sources directly each at his location. Now, instead of starting from theoretical analysis let start from observations where it is claimed that in the first situation the clock of A runs faster than the clock of B (and hence the frequency of A source is higher than the frequency of B source), while in the second situation A will observe the signal of B red shifted and B will observe the signal of A blue shifted. In the following paragraphs we will try to analyze and explain these observations general relativistically and classically.

According to general relativity, time runs faster at the location of A and hence both situations can be explained by gravitational time dilation. As for the first situation, since time runs faster at the location of A then the clock of A should run faster and this means that his source of radiation (which is the basis of his clock) should have higher frequency compared to the frequency of the source of B. As for the second situation, since time runs faster at the location of A then the unit of time of A should be shorter than the unit of time of B and hence if the frequency of the identical sources at A and B is calibrated by the units of time at a given location (i.e. at A or at B) then it will be higher at B. For example, if we have 10 cycles of radiation that descended from A to B where the size of the unit of time at B is twice the size of the unit of time at A (due to time dilation) then these 10 cycles (which represent 10 units of time at their location of emission at A) will be seen at B to occur in a time interval that is half their time interval at A (i.e. they occur in 5 time units of B since they correspond to 5 cycles of B) and hence their frequency

---

[232] We note that according to Planck's relation $E = h\nu$ the photon kinetic energy E is proportional to its frequency ν and hence red shift means decrease in kinetic energy while blue shift means increase in kinetic energy. The involvement of Planck's relation should make the explanation partially non-classic although it is classic from a gravitational perspective.

8.4 Gravitational Frequency Shift

at B will be seen as twice their frequency at A. Accordingly, we can claim that gravitational time dilation (according to general relativity) provides a single and consistent explanation for both situations.

According to classical physics, although the second situation can be explained by energy conservation (see Exercises), it is not obvious that the first situation can also be explained by energy conservation unless we assume that the difference in energy (due to the difference in frequency) at the two locations represents an energy loss in or by the gravitational field (and hence if the two sources emit the same amount of energy at the two locations then there should be an amount of energy stored in the field; otherwise the amount of energy emitted by the sources should depend on the strength of the gravitational field at their locations and hence source A emits more energy than source B although this may not be based on energy conservation). To be more clear we can say: if we have to follow the reported observations then identical radiation sources will emit radiations at different frequencies at A and B. Hence, the energy of the radiation at B will be lower than that at A. As the radiation emitted at A descends to B its potential energy will decrease and hence its frequency will increase further (i.e. by energy conservation apart from the increase by the emission process at A as compared to the emission process at B). Accordingly, the upcoming derivation (see Exercises) that is based on the conservation of energy (of descending/ascending signal) should account only for the part of the change of frequency that is caused by the descent/ascent process. The significance of this is that we may find classical derivations that may agree with the general relativistic predictions in general when we account for all the causes of change of frequency and energy (i.e. it is possible that the quantitative difference between the predictions of general relativity and classical gravity in some cases is because the conservation of energy can partly explain gravitational frequency shift). The interpretation of the change of frequency at the emission process will then require the assumption that the emission process is affected by the gravitational field such that the energy at emission is a function of the gravitational field. Anyway, this classical explanation of the first situation (even if it is assumed to be rational) seems less convincing than the general relativistic explanation. Nevertheless, we will continue to consider the classical explanations (like energy conservation) of the phenomena that are commonly classified as gravitational time dilation as viable explanations at least in principle and in some cases and instances (e.g. second situation).[233]

We should finally note that considering the close relation between *energy* and *time* (e.g. energy conservation is a consequence of the homogeneity of time, and energy divided by c is the temporal component of 4-momentum), the classical explanation of gravitational frequency shift by *energy* conservation and the general relativistic explanation by *time* dilation may originate from the same principle and this could partly explain the closeness of the predictions of these two explanations at least in some physical systems and circumstances. In fact, even the Planck relation $E = h\nu$ can be added to the examples of the close relation between energy and time since the Planck relation is essentially a relation between energy and reciprocal time (i.e. frequency) and hence we may need to look for a more fundamental principle that may explain not only the close relation between *time* dilation and *energy* conservation but even physical relations, like the Planck relation, that correlate energy to time or frequency (also see Problems).

Problems

1. Discuss the above general relativistic analysis and its implication on energy conservation.
   Answer: As explained above, the source of A is at a higher level in the gravitational well than the source of B. Now, according to general relativity the time at B runs slower than the time at A and hence identical sources at A and B will be seen by A and B to have identical frequencies and identical energies when each frequency is observed in its location of emission because the units of time at A and B are different, i.e. each one calibrates his observations by his own time unit. So, to compare the frequencies of A and B meaningfully the signal of B should ascend to A (or the signal of A should descend to B), i.e. the signal at one location should be moved to the other location so that we can compare the two signals sensibly at the same location since the time unit is unique at any specific location. The logical scenario in this case is that the signal of B will be seen at A to have lower frequency and energy (or the signal of A will be seen at B to have higher frequency and energy) because the unit of time at B is

---

[233] The main reason for the insistence on the viability of the classical explanations is that in most cases they produce almost identical results to the results of general relativity.

larger than the unit of time at A. However, this does not violate the universality of time only but it also violates the universality of energy (since the units of energy are dependent on the units of time which are position dependent) and hence the conservation of energy at global level will lose its significance. In fact, the invalidity of the conservation of energy globally in general relativity can be seen as a logical consequence of the invalidity of global time in general relativity since time (which is a basis for the quantification of energy as seen for example in the Planck relation or in the correspondence of energy to the temporal component of the momentum 4-vector) has only local significance.

2. Investigate and assess some of the qualitative arguments for the gravitational frequency shift that circulate in the literature of general relativity.
Answer: One of the common arguments in the literature of general relativity is based on the use of the equivalence principle. In brief, a light signal in an accelerating frame moving along the orientation of the propagation of signal (i.e. in the same direction or in the opposite direction) should experience frequency shift, i.e. "expansion" of waves and hence red shift in the case of same direction and "compression" of waves and hence blue shift in the case of opposite direction. Hence, according to the equivalence principle the same should happen in an equivalent gravitational frame. However, this argument may be criticized by the following:

• The choice of "expansion" and "compression" in the two cases seems arbitrary and the reverse may also be claimed and this should spoil the whole argument. In fact, both claims require the presumption that light has a proper (or primary) frame and improper (or secondary) frame and this could lead to the necessity of assuming an absolute or privileged frame for the propagation of light. The analysis may also require considering the fact (which we found in B4 from analyzing the formalism of Lorentz mechanics) that the light signal obtains a velocity component from its source although the speed of light remains constant. So, if we extend the validity of this finding from inertial frames to non-inertial frames then we may say: if the source of light is in the accelerating frame then the compression and expansion should follow one of the above patterns while if the source of light is not in the accelerating frame then the compression and expansion should follow an opposite pattern (although other patterns may also be possible).

• We may claim that the opposite argument may be more appropriate and logical, that is the frequency shift in a gravitational frame is justified first (e.g. by energy conservation or something else), then the frequency shift in an equivalent accelerating frame is inferred from the equivalence principle. However, this course of action could defeat the original purpose of the equivalence principle as an accessory for establishing facts and obtaining results about gravity and its consequences.

Exercises

1. Derive a classical formula for the gravitational frequency shift and hence compare the classical and general relativistic formulae. Also, determine how the classical formula becomes when the observer is at infinity.
2. Show that the general relativistic formula for the gravitational frequency shift agrees with the classical formula (see exercise 1) by using power series approximation with reasonable conditions and assumptions.
3. Find the fractional change in the frequency of a light signal emitted on the surface of a star (say white dwarf) with one solar mass and one Earth radius when it reaches the Earth. Use in your answer both the general relativistic and classical formulae and compare the results.
4. Write the gravitational frequency shift formula of general relativity and classical physics in terms of wavelength.
5. Show that if $\frac{GM}{c^2 r_1} \ll 1$ then the gravitational frequency shift formula of general relativity converges to its classical counterpart at infinity (i.e. $r_2 \to \infty$).
6. Use the classical and general relativistic formulations to find the wavelength of a light signal emitted at the surface of the Sun with a wavelength $\lambda_1 = 500$ nm when it reaches the Earth.
7. Repeat the previous exercise but assume this time that the Sun has the same radius as the Earth.

8.5 Gravitational Length Contraction

The essence of gravitational length contraction in simple terms is that measuring sticks shrink in the radial direction as they ascend in a gravitational well and stretch in the radial direction as they descend in a gravitational well. In fact, gravitational length contraction (assuming a Schwarzschild spacetime) was investigated rather thoroughly in § 6.4.3 and hence we do not repeat the investigation here.[234]

Problems

1. There are some opposite physical interpretations in the literature of general relativity about length contraction (as if it is length dilation) and its meaning. Discuss this issue.
   Answer: Apart from the possibility in some cases of being misunderstanding or mistake, some of these interpretations are based on using different metrics. The above description of gravitational length contraction is based on the Schwarzschild metric. The origin of some of these conflicting interpretations may also be the difference in perspective (similar to what we find in special relativity about similar issues; refer to B4) and the language of presentation.

Exercises

1. Derive a general relativistic formula for the relation between proper infinitesimal lengths along the radial direction in a Schwarzschild spacetime as obtained in two local stationary frames that are at different radial distance from the source of gravity. How this formula becomes when one frame is at infinity, i.e. very far away from the source of gravity?
2. What is the fractional change in the infinitesimal length along the radial direction that is experienced by a stick as it moves from the surface of the Earth to very far away from the Earth (i.e. infinity)?
3. List some issues that can be a source of controversy and conflict in the gravitational length contraction.[235]
4. In § 6.3.4 we derived gravitational frequency shift (see Eq. 251) from gravitational time dilation. Try to derive (or rather deduce) gravitational frequency shift from gravitational length contraction.

8.6 Gravitational Waves

Gravitational waves is one of the consequences and predictions of general relativity where fluctuations or ripples in the geometry of spacetime (i.e. gravitational waves) are generated by the motion of massive objects. These ripples are like the ripples generated by the motion of electric charge in the form of electromagnetic waves. In brief, as accelerated electric charges emit electromagnetic waves that carry energy and momentum, accelerated masses should also emit gravitational waves that carry energy and momentum (noting that the Birkhoff theorem puts some restrictions related to symmetry). For example, two gravitating bodies (e.g. neutron stars) orbiting around their center of mass should emit gravitational waves continuously due to their accelerated motion and hence they continue to lose energy in the form of gravitational waves and this leads to gradual contraction of their orbit and potentially eventual collapse and merge.[236] We note that according to general relativity the characteristic speed of gravitational waves is the same as the characteristic speed of light c.

Anyway, the theory of gravitational waves is very lengthy and complex with very trivial practical value and outcome and hence it is beyond the scope and objectives of this book. In fact, large parts of the gravitational waves theory are beyond the capability of modern science to verify or falsify and hence they should be regarded more aptly as mathematical speculations and curiosities rather than physical theories

---

[234] We should note that spatial distance (or length) is ambiguous when the metric is time dependent because it will also be time dependent due to its dependency on the actual trajectory (or world line) that connects the initial and final points in the spacetime. However, this does not apply to the Schwarzschild geometry which is time independent. Anyway, the validity of length contraction should not depend on this issue because even in the case of time dependency there could still be length contraction (or dilation) as a general feature even if it is time dependent. The same may be said about time interval when the metric of spacetime is time dependent.

[235] This question is general and is not restricted to a specific metric like Schwarzschild.

[236] Although this scenario applies to any such orbiting system (like the solar system) the rate of energy emission is so low on astronomical scale that it takes very long time (cosmological time scale) for these systems to collapse. This should explain why these systems are generally stable within the window of observation.

(see § 1.9). Therefore, we refer the interested reader to the literature of general relativity for details on this subject (although we strongly advise against spending valuable time on this virtually useless theory unless there are purely theoretical justifications and motivations).

Problems

1. List some of the properties of gravitational waves.
   Answer: According to the literature of general relativity:
   • Gravitational waves are ripples in the geometry of spacetime rather than ripples contained in the spacetime (as it is the case for example in electromagnetic waves).
   • They are generated by accelerating massive objects.
   • Like ordinary waves, they are characterized by wavelength and frequency.
   • Like ordinary waves, they carry energy and momentum.[237]
   • Like electromagnetic waves, they propagate with the speed of light.
2. Give examples of the physical events that are expected to generate gravitational waves.
   Answer: Examples are:
   • Massive objects (e.g. two stars) orbiting around their center of mass.
   • Collision and merger of massive objects (e.g. two black holes or two galaxies).
   • Explosion in astronomical objects or collapse of stellar cores (like nebulae and supernovae). However, for such astronomical events to generate gravitational waves they must be spherically asymmetric because otherwise we will have a Schwarzschild spacetime (according to the Birkhoff theorem) which is time independent (see § 4.1.1).
   • Rotation of spherically asymmetric compact objects (like white dwarfs).
   • Violent events during the creation of the Universe (i.e. Big Bang).
3. Should a radially pulsating star emit gravitational waves?
   Answer: No, because according to the Birkhoff theorem as long as the matter distribution of the gravitating body is spherically symmetric the geometry of spacetime in the vacuum region is described by the Schwarzschild metric even if the matter distribution varies in time. Now, since Schwarzschild metric is time independent there should be no disturbance in the spacetime, i.e. there is no emission of gravitational waves.

Exercises

1. Assess the sensibility and rationality of the existence of gravitational waves.
2. Assess the commonly accepted claim that the idea of gravitational waves is the brainchild of Einstein.
3. Why a spherically symmetric gravitating object cannot generate gravitational waves even if it varies in time (i.e. while remaining spherically symmetric)?
4. Justify the fact that gravitational waves are much weaker than electromagnetic waves (where weaker and stronger in this context may be quantified generically by the ease of detection or by the amount of energy they carry).
5. Why we might have "gravity of gravity" (according to general relativity or general relativists) but not "electromagnetism of electromagnetism"?
6. Outline the formulation of gravitational waves in general relativity using the linearized form of the Field Equation.

8.7 Black Holes

We should note first that this section about black holes is very brief (compared to similar sections or chapters in other books of general relativity) because the scope of our book is general relativity and hence its applications (in black holes and cosmology for example) are not of major interest to us. Moreover, we are skeptical about the physical reality of many of those mathematically oriented investigations and models of black holes that are based on the formulation of general relativity since any authentic physical

---

[237] Whether gravitational energy and momentum can be a source term in the Field Equations seems to be a controversial issue (which is a typical state in the relativity theories) although the common opinion among general relativists seems to deny this (where it may be justified by their implicit inclusion through the non-linearity of the Field Equations).

8.7 Black Holes

theory should be primarily driven by observations and experiments rather than by mathematical models. However, many important issues about black holes are addressed in the problems and exercises (which the reader can skip with no regret and with no loss of continuity).[238]

Black holes, which are regarded as one of the dramatic and exotic predictions of general relativity, are massive compact objects whose matter is highly concentrated in space (i.e. they have very high density) to such a limit that even light cannot escape from the grip of their strong gravitational field.[239] This means that any light emitted (hypothetically) from a black hole toward outside will bend and return back to the black hole and hence no light from the black hole can reach an outside observer and this should explain why they are "black". This is a direct consequence of the distortion of the spacetime caused by the dense concentration of matter according to general relativity because the distortion of spacetime in the vicinity of a black hole is so dramatic that all the geodesics originating from the black hole bend and turn back to the black hole. However, as we will see black holes can also be obtained (at least in principle regardless of any particular derivation) from a classical gravitational formulation with no need for this geometric paradigm.

Although it is impossible to observe black holes directly (because they are "black") black holes can be observed in principle indirectly by observing the effect of their gravitational field on neighboring objects (such as stars) or by observing energetic emissions (such as X-rays) generated by the accretion of gas, dust and debris onto the black holes from their surrounding. Black holes are allegedly detected (firmly or tentatively) or contemplated in certain binary orbiting systems and in the center of some galaxies. There are also theoretical speculations that they might exist randomly in space as residues from the alleged Big Bang and the subsequent processes in the evolution of the Universe.

It is generally claimed that there are three main types of black hole (although several other types, such as intermediate black holes, can also be found in the literature):
• Super-massive black holes which are supposed to exist in the center of most galaxies especially the large ones.
• Stellar black holes which are supposed to be created by the collapse of massive stars or the merger of binary objects.
• Miniature (or primordial) black holes which are supposed to be created by violent events and inhomogeneities during the Big Bang.

According to the theoretical models, black holes have very simple "structure" and hence (or because) they are characterized by only three mutually independent and externally observable intrinsic physical properties: mass, electric charge and angular momentum.[240] This may be stated colloquially as: "black holes have no hair" (which is commonly known as the "no hair theorem"). Black holes are commonly seen as physical singularities in the spacetime where mass accumulates in a single point of space and hence they are infinitely dense.[241] The singularity of black holes may also be interpreted differently and correlated to other parameters (e.g. curvature of spacetime).

Black holes are also supposed to be surrounded by an imaginary sphere (centered on the singularity) called the "event horizon" which is where the escape speed becomes equal to the characteristic speed of light c. Accordingly, nothing (including light) can escape from within the event horizon and hence what falls inside the event horizon will be trapped there forever (although black holes "evaporate" quantum-mechanically but not classically according to some theories; see § 8.7.3). The radius of the event horizon is known as the Schwarzschild radius which is given by $R_S = \frac{2GM}{c^2}$ where M is the mass of the black hole.

---

[238] Because the general relativistic literature on black holes and related topics is congested with nonsense, fantasies, absurdities and paradoxes we strongly advise against spending valuable time on these topics. In fact, our investigation is mainly for the purpose of completeness and to meet the demand of some readers (and that is why it is kept brief).

[239] The compactness and high density of black holes will be discussed later where we will see that they may not apply equally to all types of black hole according to certain criteria.

[240] In fact, angular momentum is not entirely independent of mass although it is independent (due to its dependency on other parameters). Also, the condition "intrinsic" is meant to exclude observer-dependent (i.e. extrinsic) properties such as velocity and kinetic energy.

[241] Before the discovery of neutron and neutron degeneracy pressure, it was believed that "neutron star" should collapse to a point (and hence become a singularity). We may currently be in a similar situation with regard to black holes where some discovery in the future may prevent the collapse of black hole to a point and becoming a singularity.

8.7 Black Holes

A rotating black hole has also an ergo-region (or ergo-sphere) around its event horizon where spacetime is dragged forcibly by the rotation of the black hole and hence no object within the ergo-region can remain stationary (relative to inertial observer at infinity). Another claim about black holes is that falling toward a black hole is an endless journey as observed by an external observer (see Exercises).

We should remark that many of the alleged "facts" about black holes are based on highly abstract mathematical models with no physical basis and hence they should not be taken seriously as scientific facts although they are commonly presented as such (in fact we are skeptical even about the mere existence of black holes despite the recent affirmative claims about the detection of these objects). This is one of the main problems in modern physics where the theory leads, rather than follows, the experiment and observation and hence in many cases we find the role of experiment and observation is to search for and harvest evidence in support of a predetermined theory. Accordingly, many bizarre illusions (such as wormholes, time machines, white holes,[242] and dark energy) have emerged in recent times thanks to this approach (which is largely based on general relativity and its methodology and framework) where they are presented as respected scientific theories or even as scientific facts. In fact, the literature of black holes (which is almost entirely based on general relativity) is full of these mathematical fantasies and illusions. Therefore, the audience of modern physics should be vigilant about these claims to avoid being fooled by these theories and "facts". Anyway, we believe that the existence of black holes (if proved observationally) can be explained even by classical gravity with no need for these illusions and fantasies (see the Exercises of this section and refer to § 9.7).

Problems
1. What we mean by "singularity" when we talk about black holes?
   Answer: Singularity in general relativity means a point in spacetime with an infinite property which (in the case of black holes) is the mass density (or curvature of spacetime) because the mass is supposed to be accumulated in a single point of space and hence the mass density at that point is infinite. However, some may view black holes as confined regions in spacetime with certain properties without talking explicitly about infinite density (or infinite curvature) to escape potential criticism or to allow possible intervention of other physical theories (namely quantum mechanics) that may modify (and hence prevent) the general relativistic nature of black holes as singularities in spacetime.
2. Analyze the meaning of "radius" in "Schwarzschild radius".
   Answer: Based on the general relativistic derivation of the expression of Schwarzschild radius (using Schwarzschild metric as explained in § 4.1.1 and § 5.1) it should mean coordinate radius rather than metrical radius although they may be seen equal according to some opinions (see problem 2 of § 8.3). This should have an impact on other aspects related to black holes like "average density" and "volume inside event horizon". However, we generally do not make such restrictions and clarifications to avoid unnecessary complications where we rely on this understanding.
3. Black holes are commonly described as compact objects. Does this apply to all types of black hole?
   Answer: If "compact" means high average density (where average density is defined as the mass divided by the volume inside the event horizon) then being compact objects applies only to some types of black hole. As we will see in the Exercises, the average density of some black holes (i.e. the very "heavy" ones) is very low. So, the characteristic property of black hole is being a concentration of a quantity of matter within its Schwarzschild radius which results in preventing light inside its event horizon from escaping to outside. Yes, if we accept the claim that black holes are physical singularities, then all black holes should have infinite density although their average density may be very low. However, we do not believe in the existence of physical singularities because there is no infinite quantities in the real world although such quantities may exist in abstract mathematical spaces. Nevertheless, the actual density of black holes (even the very "heavy" ones) could be much higher than their average density because the mass can be confined to a volume less than the volume inside the event horizon. We should also draw the attention to the peculiar meaning of "volume inside the event horizon" due to the presumed difference between coordinate variables and metrical variables as indicated in the previous exercise. In fact, it is more appropriate in this context to think general relativistically rather than classically.

---
[242] White hole may be defined as the time-reverse of black hole.

8.7 Black Holes

4. What is the mass of a black hole whose Schwarzschild radius is 3.5 km?
   Answer: We have $R_S \equiv \frac{2GM}{c^2} = 3500$ m. Hence:

   $$M = \frac{c^2 R_S}{2G} \simeq \frac{(3 \times 10^8)^2 \times 3500}{2 \times 6.674 \times 10^{-11}} \simeq 2.360 \times 10^{30} \text{ kg}$$

   which is more than the mass of the Sun.

Exercises

1. Give a succinct and simple definition of black hole.
2. Define event horizon and Schwarzschild radius of black hole.
3. In what sense black holes are "black" and "holes"?
4. How do we detect and observe black holes if they are "black"?
5. Give examples of the mechanisms by which black holes can release energy (regardless of the energy release being by the black hole on its own or through its interaction with its surrounding).
6. Make a rough estimate of the energy released by accretion of a certain amount of matter of mass m onto a non-rotating black hole of mass M. Comment on the result.
7. List some of the properties of black holes.
8. What we mean by the structure of black hole?
9. What are the main schemes for classifying black holes?
10. Derive a mathematical expression for the Schwarzschild radius using a classical argument. What is the significance of this classical derivation?
11. Assess the classical derivation of the previous exercise.
12. What is the Schwarzschild radius of a black hole with a mass (**a**) equal to the mass of the Earth and (**b**) equal to the mass of the Sun?
13. Use some of the previously developed general relativistic formulae related to gravitational time dilation and gravitational frequency shift to infer some of the physical properties and consequences of black holes.
14. Discuss the mathematical singularities of black holes assuming a Schwarzschild metric. Do these singularities mean that black holes are physical singularities?
15. Can we ascribe Schwarzschild radius to objects other than black holes?
16. What characterizes the event horizon of black holes?
17. Develop a formula for the average mass density of black holes and analyze it.
18. Calculate the average density of a black hole whose mass is (**a**) equal to the mass of the Earth, (**b**) equal to the mass of the Sun and (**c**) equal to 10^9 solar masses. Compare these densities to the density of the atomic nucleus and comment.
19. What can you conclude from the results of the last two questions with regard to the likelihood of the formation of black holes?
20. Derive a classical formula for the gravitational field gradient of black hole along the radial direction and analyze its classical implications.
21. Considering that black holes are fully characterized by three properties: mass, angular momentum and electric charge, what are the main types of black hole from this perspective?
22. Considering the mass range of black holes, what are the main categories of black hole from this perspective? Provide some details about these categories.
23. Describe briefly the free fall journey of an astronaut A toward a non-rotating black hole as experienced by him and as observed by a stationary observer B located at infinity. Assume that A is initially at infinity[243] and he is falling from rest and hence he follows a radial path toward the black hole. Also, assume that A and B have identical light sources and identical and initially-synchronized clocks.
24. State the hypothesis (or conjecture) of cosmic censorship and comment.
25. What is the implication of the "no hair theorem" (i.e. black holes are characterized by only three observable independent physical properties: mass, angular momentum and electric charge) with regard

---

[243] Infinity here should mean very far where the gravitational field is very small but not zero.

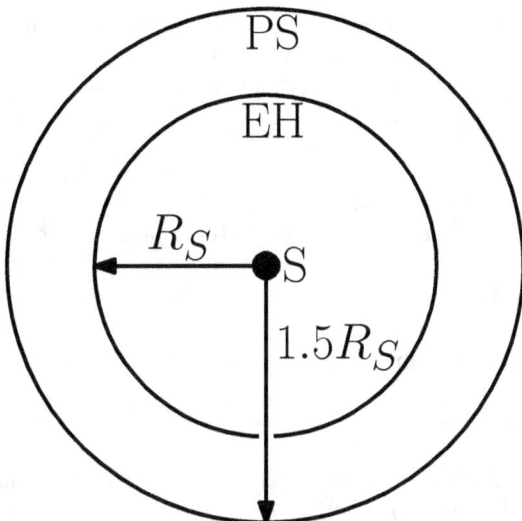

Figure 7: A 2D schematic of the structure of a Schwarzschild black hole (and the surrounding space) where the singularity S, event horizon EH, Schwarzschild radius R_S and photon sphere PS are indicated.

to the issue of absolute frame? What is the implication with regard to the issue of being physical singularities?

8.7.1 Schwarzschild Black Holes

These are static black holes characterized by mass $M > 0$, zero angular momentum $J = 0$, and zero electric charge $Q = 0$. Accordingly, the spacetime that associates these black holes in the absence of other sources of gravity is described by the Schwarzschild metric as their name suggests (see § 4.1). The "structure" of the Schwarzschild black hole (and the surrounding space) is very simple as it consists of a singularity surrounded by an imaginary sphere (i.e. event horizon) of "radius" R_S centered on the singularity (see Figure 7). A photon sphere, which is a sphere like event horizon but with a "radius" of $1.5R_S$ (see Figure 7 and Exercises), may also be added to this structure. Schwarzschild black holes should be considered the simplest objects in the physical world since they are characterized by mass only and hence two Schwarzschild black holes of the same mass are identical. So, in a sense a Schwarzschild black hole of a given mass has no identity as it is no more than a chunk of matter in spacetime.

We should remark that describing this category of black holes as Schwarzschild black holes can be misleading since it may suggest that the entire spacetime surrounding these black holes is necessarily described by the Schwarzschild metric. In fact, "Schwarzschild black hole" should only mean a black hole with mass, zero angular momentum and zero electric charge even if the spacetime contains other sources of gravitation and hence the metric of the extended spacetime is more complex than the Schwarzschild metric. So, the "Schwarzschild" label originates from the fact that such a black hole is the same as the object in the Schwarzschild solution and hence if this black hole is the only object in the spacetime then it will generate the Schwarzschild metric. The "Schwarzschild" label should also indicate that the structure of the black hole (i.e. the features of the spacetime in the immediate neighborhood of the singularity such as event horizon and photon sphere) is described by the Schwarzschild metric although the extended spacetime may not be described by this metric.

Problems

1. What we mean by saying "two Schwarzschild black holes of the same mass are identical"? Can you find an analogy in more familiar and ordinary objects?
   Answer: We mean they are indistinguishable from each other by any intrinsic property although they could be distinguished by an extrinsic property such as being at two different points in spacetime. We

may make an analogy with two electrons or two protons which are indistinguishable from each other by any known intrinsic property such as mass or charge (within our observational limits) although they may be distinguished by extrinsic properties like their position or velocity or kinetic energy.

Exercises
1. Define Schwarzschild black hole in a few words.
2. What is the photon sphere of Schwarzschild black hole?
3. Derive the general relativistic formula for the "radius" of the photon sphere (i.e. $r = 1.5R_S$) of Schwarzschild black hole.
4. Derive a classical formula for the radius of the "photon sphere of Schwarzschild black hole" and comment on its significance.
5. Investigate the singularities of Schwarzschild black hole.

8.7.2 Kerr Black Holes

These are rotating black holes characterized by mass $M > 0$, angular momentum $J \neq 0$, and zero electric charge $Q = 0$. Accordingly, the spacetime that associates these black holes is described by the Kerr solution as their name suggests (see § 4.2). Because these black holes are rotating, they have more complex structure than the simple structure of the static Schwarzschild black holes (refer to Figure 8). According to the Kerr solution of the Field Equations, rotating black holes are surrounded by a zone called the ergo-region (or ergo-sphere)[244] where no object within this zone can stay at rest relative to an inertial frame at infinity because of frame dragging and hence any object in this region is forced to rotate in the sense of rotation of the black hole. However, objects within the ergo-region (and outside the event horizon) can still escape the gravitational grip of the black hole and hence matter and energy can be extracted from this region. The ergo-region is defined externally by an ellipsoidal surface (which has axial symmetry around the axis of rotation) called the stationary limit (or static limit) which is a surface of infinite red shift. The stationary limit may be described by some as the locus in spacetime where spacetime flows at the speed of light.

Exercises
1. What is the implication of having rotating (or Kerr) black holes?
2. Why should Kerr black holes be the most common type of stellar black holes in the Universe (assuming they do exist)?
3. What is the ergo-region of a rotating black hole?
4. Mention some of the characteristic features of the ergo-region of a rotating black hole.
5. Give some of the characteristic features of the stationary limit of a rotating black hole.
6. Compare the event horizon and the surface of infinite red shift in the Schwarzschild and Kerr black holes.
7. Classify the regions surrounding Kerr black holes according to the ability of a physical object (as seen by an inertial observer at infinity) to stand still and rotate in one sense or another.
8. Try to use the terms "stationary" and "static" (as used commonly in the literature of general relativity) to describe Schwarzschild and Kerr black holes.

8.7.3 Hawking Radiation

This is a hypothetical effect predicted by Stephen Hawking as a result of applying quantum mechanics to general relativity and hence it is a quantum gravitational effect. In fact, the theory of Hawking radiation may be regarded as the first quantum gravitation theory. Hawking radiation may also be called Hawking effect or black hole evaporation. According to this theoretical prediction, black holes can emit black body radiation as a result of quantum mechanical processes in the vicinity of the event horizon and hence they "evaporate". It should be obvious that because of the equivalence between mass and energy, emission of radiation means losing mass and hence "evaporation". Accordingly, if these quantum losses are not

---

[244] In our view, "ergo-region" is more appropriate than "ergo-sphere" (unless sphere means something like zone or field).

8.7.3 Hawking Radiation

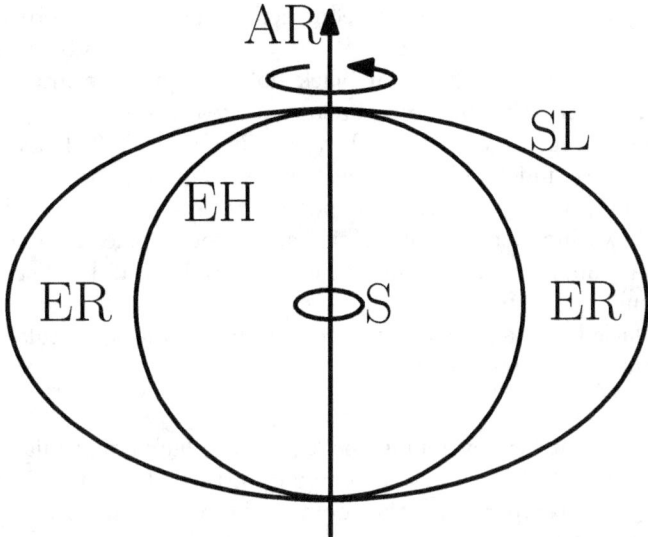

Figure 8: A 2D schematic of the structure of a Kerr black hole where the singularity S, event horizon EH, ergo-region ER, stationary limit SL and the axis of rotation AR are indicated. We note that the singularity at the center is not a point but a ring of coordinate radius a in the equatorial plane (see § 4.2). We also note that some delicate details of the structure of Kerr black hole (e.g. inner event horizon) are not demonstrated in this simplistic illustration.

compensated by gaining matter and energy from the surrounding of the black hole then the black hole can in theory evaporate totally and hence disappear.

Hawking effect may seem contradictory to the premise that nothing (including radiation) can escape from the gravitational grip of black holes. However, this is addressed in the theory by a strange hypothetical scenario in which virtual particle-antiparticle pairs participate and hence there is no violation to this premise.[245] In fact, this is one of the masterpieces of wild imagination in modern physics and hence we do not go through these useless details (although we will touch on the issue mildly in the questions). Anyway, it should be understood that the above premise is about classical emission processes related to classical black holes while Hawking radiation is a quantum mechanical process related to quantum mechanical black holes.[246] In fact, Hawking radiation (according to the alleged mechanism of emission) is not a radiation originating from inside the black hole to violate the aforementioned premise.

Hawking radiation is expressed quantitatively by the following relation which links the "temperature" of the black hole (which may be called the Hawking temperature) as a black body radiator to its mass:

$$T = \frac{\hbar c^3}{8\pi k G M} \tag{304}$$

where T and M are the temperature and mass of the black hole, $\hbar$ is the reduced Planck constant ($= h/2\pi$), c is the speed of light, k is the Boltzmann constant and G is the gravitational constant. As we see, the temperature of black hole as a thermal radiator solely depends on its mass in a reciprocal relation and hence "light" black holes are hotter than "heavy" black holes.

Apart from being an entirely theoretical effect with no observational foundation or verification, Hawking radiation has been criticized and challenged in its theoretical foundation, e.g. by the so called "information loss paradox" (which is a fantasy built on a fantasy). Anyway, if we accept Hawking radiation (and hence

---

[245] Virtual particles and antiparticles are paradigms representing physical entities within the theoretical framework of relativistic quantum mechanics.

[246] "Classical" in such contexts is opposite to quantum mechanical and hence even the relativity theories and their implications are classical in this sense. Also, "quantum mechanical black holes" means black holes that follow the rules and laws of quantum physics.

accept the above formula) then we may need to consider renaming these objects (e.g. by calling them "dark holes") since they are not really "black". We may also need to adjust the "no hair theorem" by adding the temperature as a fourth property of black holes (even if we interpret the temperature in a non-conventional sense) although this may be challenged by the dependency of temperature on mass. However, the quantitative dependency of T on M according to Eq. 304 does not affect the fact that mass and temperature are two independent properties even though they are linked through a certain quantitative relation. In fact, we have a similar dependency between mass and angular momentum and we still consider these as two independent properties according to the "no hair theorem". However, we should admit that angular momentum depends on other variables and hence overall it is independent of mass; therefore the challenge could be valid. Also, the "no hair theorem" seems to be formulated specifically to classical black holes and hence the restriction to mass, angular momentum and electric charge is justified since temperature is quantum mechanical.

Problems

1. Discuss the issue of the credit for developing the theory of black hole radiation.
   Answer: Although the name of Hawking is commonly attached to the theory of black hole radiation and black hole entropy and temperature, the roots of these ideas seem to originate from the work of Bekenstein. However, the detailed quantum mechanical application and development of formalism of these ideas seem to belong mainly to Hawking (although Hawking was initially against the idea of black hole radiation). Anyway, we think the credit for the theory of black hole radiation should be shared to include Bekenstein (and possibly others).
2. What is the temperature of a black hole with (**a**) 10 Earth masses and (**b**) 0.1 Earth mass?
   Answer:
   (**a**) We have $M \simeq 5.972 \times 10^{25}$ kg and hence:

$$T = \frac{\hbar c^3}{8\pi k G M} \simeq \frac{6.62607 \times 10^{-34} \times \left(3 \times 10^8\right)^3}{16\pi^2 \times 1.38065 \times 10^{-23} \times 6.674 \times 10^{-11} \times 5.972 \times 10^{25}} \simeq 0.00206 \, \text{K}$$

   (**b**) We have $M = 0.01$ of the mass in part (a) and hence the temperature should be scaled by a factor of $1/0.01 = 100$. Therefore, $T \simeq 0.206$ K.

Exercises

1. What is the essence of Hawking radiation and its basic mechanism?
2. Analyze Hawking radiation formula for the temperature of black hole.
3. Derive a formula for the power radiated by a black hole assuming it is a black body and analyze it.
4. It is claimed that classical black holes violate the third law of thermodynamics. Try to justify.

8.8 Geodetic Effect

Geodetic effect (which is also known as geodetic precession or de Sitter effect or geodesic gyroscope precession) is another characteristic prediction of general relativity that has no parallel in classical gravity. The effect is simply based on the fact that the spacetime of general relativity is curved Riemannian, unlike the space of classical gravity which is flat Euclidean, and hence when vectors are parallel-transported in the spacetime they will be affected by the curvature of spacetime. To be more clear, let have a gyroscope (see Exercises) in a flat Euclidean space that moves freely around a closed path in the space, and assume that the gyroscope starts its journey from an initial position P_1 where its spin (or angular momentum) vector points into a fixed direction (say the direction of a given vector $\mathbf{V}_1$). Due to the conservation of angular momentum in the absence of external torques the gyroscope spin vector will keep pointing into the same initial direction $\mathbf{V}_1$ during its entire journey and hence when it returns to its initial position P_1 the final direction of its spin vector will be the same as its initial direction $\mathbf{V}_1$. In other words, the angular momentum vector of the gyroscope is parallel-transported along its closed trajectory in an absolute Euclidean sense of parallelism.

Now, let the gyroscope be moving freely in a curved Riemannian space (i.e. the spacetime of general relativity) where it performs a similar journey around a similar closed path. Since the space is curved

8.9 Frame Dragging

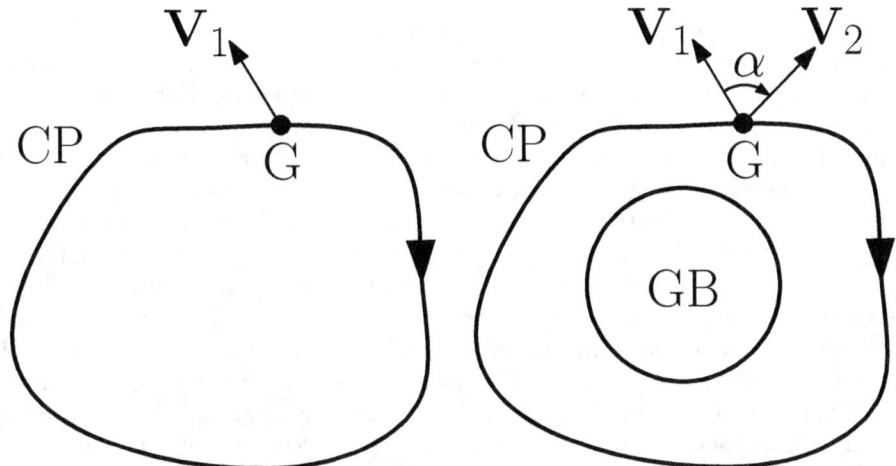

Figure 9: A gyroscope G moving freely around a closed path CP where on the left the gyroscope is in a flat spacetime and hence it returns to its starting point without change of its spin vector $\mathbf{V}_1$ while on the right the gyroscope is in a curved spacetime due to the presence of a gravitating body GB and hence it returns to its starting point with change of its spin vector $\mathbf{V}_1$ to $\mathbf{V}_2$ (where α is the precession angle).

then by parallel-transporting[247] the angular momentum vector along the closed trajectory the direction of the gyroscope spin will not be preserved in a Euclidean sense. In other words, the final direction of the gyroscope spin vector is not the same as its initial direction (refer to § 2.8 and see Figure 9). Accordingly, if we have for example a gyroscope on board a satellite orbiting the Earth then after each revolution the gyroscope spin vector will deviate slightly from its previous initial direction. In fact, the gravitational field of the Earth is very weak and hence the curvature of the spacetime is very small and therefore the deviation in each revolution is very tiny. However, after many revolutions the effect of these individual deviations should accumulate and become measurable (if this effect does exist).

It is shown in the literature that the angle of precession per revolution (assuming Schwarzschild geometry) is given by:

$$\alpha = 2\pi \left[1 - \left(1 - \frac{3GM}{c^2 r}\right)^{1/2}\right] \qquad (305)$$

where α is the angle of precession in radians, i.e. the angle through which the spin vector of the gyroscope is rotated in the direction of the orbital motion in the plane of the orbit in one revolution around a gravitating body of mass M following a circular orbit of radius (or radial coordinate) r.

Exercises
1. Give a brief definition of gyroscope.
2. Find the precession angle experienced in one revolution by the spin vector of a satellite gyroscope skimming the Earth in a circular orbit at very low altitude. What is the precession angle in one year?

8.9 Frame Dragging

This effect (which may also be called inertial frame dragging or dragging of inertial frames) is a purely general relativistic effect where a rotating massive body[248] is predicted to distort the metric of the spacetime in its surrounding by its rotation resulting in modifying the geodesic trajectories (or world lines) by rotation. So, unlike classical gravity where gravitation depends on the mass of the gravitating

---

[247] In fact, parallel-transporting is the requirement of the conservation of angular momentum in the absence of external torques.
[248] In fact, rotation is an instance of motion of the gravitating object which drags inertial frames in its vicinity.

object but not on its rotation, in general relativity the rotation of the gravitating object has direct impact on the geometry of spacetime which leads to gravitational effects. For example, a massive body falling from rest along a radial direction toward a massive rotating object (e.g. Kerr black hole) is expected to be dragged by the rotation and hence its motion will gain a non-radial (or azimuthal) component.

In more technical terms, a freely falling particle in the field of a rotating gravitating object at the center of a standard spherical coordinate system[249] (with the $\theta = 0$ axis representing the axis of rotation and ϕ representing the azimuthal coordinate that quantifies the rotation) is expected to experience a variation of its azimuthal coordinate ϕ by a variation in its radial coordinate r (i.e. $\frac{d\phi}{dr} \neq 0$) and hence the particle is dragged by an angle $\Delta\phi \neq 0$ as it falls radially toward the gravitating object. This is obviously in contrast to the classical gravity where no such variation is predicted and hence $\frac{d\phi}{dr} = 0$ and $\Delta\phi = 0$.

The prominent example of frame dragging is the so-called Lense-Thirring effect (or Lense-Thirring precession) where gyroscopic precession is predicted to occur due to frame dragging caused by a rotating gravitating object like the Earth (see § 9.9.2). This may be demonstrated by a gyroscope in a satellite that orbits the Earth in a polar plane (i.e. a plane that embeds the axis of rotation of the Earth) where frame dragging will force the orbital plane to rotate around the Earth causing the gyroscope to precess (see Exercises).

Exercises
1. Explain frame dragging in a few simple words.
2. Why is frame dragging called *inertial* frame dragging?
3. Explain, with a simple sketch, gyroscopic precession caused by frame dragging in a satellite orbiting the Earth in a polar plane.
4. What feature of the Kerr metric relates to frame dragging?

8.10 Wormholes and Other Fantasies

Wormholes in spacetime are supposed to be one of the consequences and predictions of general relativity. In fact, wormholes is just an example of many bizarre ideas in modern physics whose roots originate from general relativity. We believe that wormholes and their alike are fantasies and utter illusion (in fact they are typical examples of how modern physics can become absurd). This sort of ridiculous science should point out to potential defects in the theory of general relativity itself since it leads to such nonsensical consequences and predictions. We believe no one should waste his time in these futile investigations and hence the interested reader in wormholes and their alike should refer to the literature of general relativity.

8.11 Cosmological Predictions

There are many cosmological consequences and predictions that are based on general relativity (as part of the so-called relativistic cosmology). In fact, modern cosmology is almost entirely based on general relativity. However, almost all these consequences and predictions are theoretical in nature and utterly hypothetical with no observational basis since they represent speculations and contemplations about the Universe and its origin and destiny and hence they are a strange mix of philosophy and mathematics with some science (mainly physics and chemistry). Moreover, large parts of these consequences and predictions are based on the alleged Big Bang and hence the result is not expected to be more credible than its genesis. In fact, modern cosmology cannot be regarded as a scientific discipline in a strict sense due to the involvement of many non-scientific elements and the contamination with metaphysical ideas and wild speculations which have no observational ground. Furthermore, most of its predictions and results are scientifically unverifiable and hence they cannot be classified as science (see § 1.9). Therefore, we should not waste our time in these useless investigations and speculations. The interested reader should consult the literature of general relativity and relativistic cosmology about these nonsensical details.

---

[249] We are considering only the angular part of the system (i.e. the θ and ϕ coordinates).

Chapter 9
Tests of General Relativity

The subject of this chapter is the experimental verification and observational evidence in support of the theoretical consequences and predictions of general relativity that were investigated in the previous chapters (and the last chapter in particular). We should note that the tests of general relativity are commonly classified into several categories. The tests are initially classified as classical and non-classical where the classical refer to those tests that were proposed by Einstein (i.e. perihelion precession of Mercury, light bending by the Sun and gravitational red shift of light). Some may also add to the list of classical tests the tests that are similar in their working principle to these three tests such as gravitational time dilation and Shapiro time delay tests. The list of classical tests may also be extended by some to include tests of special relativity since according to the equivalence principle of general relativity special relativity holds locally in freely falling non-rotating frames and hence any valid test to the theory of special relativity should be a valid test to general relativity.

The non-classical tests are also classified into several categories where different authors generally adopt different categorization plans. For example, some authors classify non-classical tests into modern tests (such as frame dragging tests), strong field tests (such as black holes observations) and cosmological tests (such as gravitational lensing) while other authors classify non-classical tests into satellite based tests (such as geodetic precession tests), astronomical tests (such as black holes observations) and gravitational waves tests (such as LIGO observations).

In fact, many of these categories are not mutually exclusive (or disjoint) and hence we may find a single test listed (or can be listed) into two categories or more. Moreover, some of these categorization plans do not seem comprehensive. Anyway, most of these categorization plans are based on somewhat arbitrary criteria (e.g. practical or historical criteria or personal preference and convenience) rather than fundamental criteria. In fact, some of these categorization plans are so arbitrary and uninformed that they can be described as nonsensical and illogical. Accordingly, we do not see real benefit or value in adopting any one of these categorization plans or creating a new categorization plan. Therefore, in the following sections we include the main tests (as well as alleged tests) of general relativity in a single list (or category) considering in the sectioning and ordering a broad range of reasons such as clarity, chronology and convenience. In fact, we are generally guided by the structure of the previous chapter although that structure is partly based on the intended structure of the present chapter.

Problems
1. Discuss the factors that should be considered in determining the value and significance of any evidence in support of a theory about a particular physical phenomenon such as gravity.
   Answer: We think the following factors (among other factors) should be considered when assessing the value and significance of any evidence in support of a scientific theory:
   (**a**) Whether there is or there is not another theory that can explain the phenomenon. Accordingly, the value of the evidence should diminish if there is another theory because the evidence is not exclusive to the theory and conclusive about its validity since it can be explained by the other theory. In other words, even if the theory is wrong the evidence can still be explained by the other theory and hence the evidence does not necessarily imply the correctness of the theory.
   (**b**) Whether the evidence does exist before or after the creation of the theory. Accordingly, if the evidence does exist before then its value should diminish because the theory has been (or at least could have been) created and tuned to match the evidence.
   (**c**) Whether the evidence is correlated to other evidence. For example, time dilation and frequency shift are correlated effects (see § 6.3.4) and hence the value of the evidence on each one of these effects should diminish since they both relate to the same aspect of the theory and hence in this sense they

are not independent of each other. In other words, they should be regarded as a single type of evidence and not as two different types of evidence.

(**d**) Whether there is bias for or against the theory or not. Accordingly, the value of the evidence should diminish (or increase) if there is bias in favor of (or against) the theory because the likelihood of prejudice and error should increase (or decrease) in this case. In fact, general relativity is a typical example of a theory with bias in its favor.

(**e**) Whether the evidence is based on trust or not. For example, a theoretical evidence that any one (with appropriate qualification) can check does not depend on trust while the result of an exotic experiment or observation depends on trust because no one (except the experimenter or observer) has access to the experiment or observation to verify its authenticity and validity.

(**f**) Whether the evidence requires the full formulation of the theory or not. Accordingly, an evidence that can be explained by an independent rule or principle within the framework of the theory without need to the full formulation is less significant in its endorsing value because the full formulation of the theory is not needed to justify the evidence and hence even if the full formulation of the theory was wrong we can still explain and justify the evidence.

(**g**) Whether the physical setting from which the evidence is extracted is sufficiently simple or too complicated and hence how many factors are involved. This is because when the system is sufficiently simple then the sources of error and uncertainty are less and the result will usually lend itself to a rather simple, clear and conclusive analysis while the situation will be different in the opposite case. Accordingly, the value of scientific evidence should decrease as the complexity of the physical system and the experimental and analytical tools increases due to the increased probability of uncertainty and error in the results as well as in the analysis of the results. In fact, the physical systems from which the general relativistic evidence are extracted are generally very complex and they require very complex experimental and theoretical tools and hence their value should diminish from this perspective.

(**h**) Whether the results belong to a single theory or they represent combined effect of two (or more) theories. Accordingly, the value of any evidence should diminish when the results represent combined effect of more than one theory because we do not know what is the actual contribution of each theory to the observed combined effect and hence even if the observed combined effect agrees with the theoretically predicted combined effect the results should still be ambiguous about the endorsement of the theories individually and separately. For example, if the theoretical prediction attributes 50% of the combined effect to theory 1 and 50% to theory 2, it is still possible that the actual contribution of the effect of theory 1 is 25% while the actual contribution of the effect of theory 2 is 75% and hence the evidence is still inconclusive about endorsing the theories. In fact, when the results represent combined effect of more than one theory we may need the applicability of the superposition principle (i.e. the effects of the individual theories are additive) and this may not be evident or well established.

(**i**) Whether the evidence is obtained from a physical system whose state is described and modeled by tentative theoretical models and subject to hypothetical conditions and assumptions. Accordingly, the value of any evidence obtained from such a system should not exceed the value of its origin and hence its value should diminish substantially. This is particularly true with regard to the evidence obtained from most astronomical and cosmological systems where these systems are subject to highly hypothetical models and tentative presumptions. So, the value of astronomical and cosmological evidence is generally less due to many uncertainties and sources of error (see exercise 6 of § 9.2). In fact, considerable amount of the claimed evidence in support of general relativity is obtained from such systems.

(**j**) Whether there is a conflict in the results or/and in the opinions about the interpretation and significance of the results. For example, if a specific experiment or observation was conducted more than once (say by two scientific teams) and the two disagree about the results then the significance of the evidence (even if both results endorse the theory in some aspects) should diminish because the possibility of error (which should have already occurred somewhere) should increase due to the conflict. Similarly, if the result of a specific experiment or observation is (or can be) interpreted and evaluated differently by different opinions then the value of the evidence should also diminish due to the lack of clarity about the implication and significance of the result.

It is noteworthy that there are many other factors that contribute to the determination of the value of scientific evidence (some of which may be specific to each individual evidence) and hence the above list represents a sample of general factors that could serve as guiding principles in assessing the value of evidence in science.

Exercises
1. What is the significance of endorsing a theory by evidence that belongs to a particular aspect of the theory? What about having evidence against a particular aspect of the theory?
2. Discuss the relation between the tests of general relativity and the tests of special relativity.
3. What correct/incorrect theory means?
4. Can we have correct predictions from wrong theories, and is this consistent with the rules of logic?

9.1 Perihelion Precession of Mercury

Historically, this may be seen as the first "evidence" in support of general relativity because it emerged during the formulation of the theory. In fact, the formulation of the theory was tuned to match the observation and hence it should not be considered as "evidence" in the technical sense. Yes, the correctness of the theoretical "prediction" of general relativity to this effect means that the theory is compliant with the precession observations and hence it is correct in this regard. As discussed in § 8.1, the perihelion precession of Mercury is just an instance of the periastron precession that is predicted by general relativity to occur in orbital motion and hence this evidence is more general than Mercury although historically the precession of Mercury was the focus of this prediction. Accordingly, this evidence extends to include many instances where there is no "tuning to match observation" and hence it has predictive value. As we will see, there are claims of strong experimental evidence in support of the predicted relativistic precession in the orbital motion of binary star systems. Also, the precession of other planets in the solar system generally agrees with the predictions of general relativity although the agreement may not be as good as in Mercury.

In our view, the precession of Mercury is seemingly the most convincing evidence in support of general relativity as a gravity theory. However, we may note the following about the significance of this evidence:
• As indicated above, the theory was tuned to match this observation and hence the significance of this prediction should be diminished. It should be obvious that the significance of this evidence will be more important if this agreement came without knowing what we should expect. However, the subsequent agreement with other precession observations (i.e. of binary systems and of other planets of the solar system) should strengthen the significance of this evidence (although the agreement in most cases is not as good as in the case of Mercury).
• It may also be claimed that the derivation of the perihelion precession formula from a classically-based method (i.e. the retarded gravitational potential theory of Gerber) should diminish the value of the endorsement of this evidence to general relativity. However, it is claimed that the retarded potential theory is an *ad hoc* fix rather than a complete and consistent theory. Unfortunately, we do not have first hand information about this theory to assess its validity and significance and if the theory is sufficiently general to predict other precession observations (i.e. other than Mercury). Anyway, being an *ad hoc* fix (even if this claim is correct) should not diminish its value as a potential competitor to general relativity in explaining the precession observation(s). It should also be noted that the precession formula was also obtained later from other non-classical gravitational theories (e.g. those of Whitehead and Birkhoff).
• We should also note that there are more recent Newtonian formulations of the precession phenomenon in orbital motion (although they generally do not agree with the general relativistic prediction quantitatively). Accordingly, the value of any evidence in this regard in endorsing general relativity specifically should be diminished due to the existence of these classically-based formulations which compete (at least qualitatively) with general relativity in explaining this phenomenon.
• There may also be a possibility of "tuning" the observations to make a "perfect match" with the prediction of general relativity. We may use the following quote from Clemence (which is taken from Synge) to justify our concern: "The observations cannot be made in a Newtonian frame of reference. They are referred to

9.1 Perihelion Precession of Mercury

the moving equinox, that is, they are affected by the precession of the equinoxes, and the determination of the precessional motion is one of the most difficult problems of positional astronomy, if not the most difficult. In the light of all these hazards it is not surprising that a difference of opinion could exist regarding the closeness of agreement between the observed and theoretical motions". This concern may at least be a basis for suspecting the existence of an exaggeration in the claimed accuracy of the agreement. In fact, there are other sources of concern about the claimed accuracy of the agreement such as the lack of relativistic treatment of the many-body problem and the lack of well determined reference for the rotation of perihelion (see for example Synge in the References).

• The precession of periastron is also observed in certain binary pulsar systems and it is claimed that the evidence is conclusive in its agreement with the general relativistic prediction.[250] This should undisputedly give more weight to this evidence in its support to general relativity. However, this should depend on the level of certainty in the parameters (like the mass of the orbiting objects) of the binary system and if these parameters are obtained independently of general relativity or not. We should also add the precession observations of other planets in the solar system which seem to endorse the general relativistic predictions although the agreement (in some cases at least) may not be as impressive as in the case of Mercury.

• As discussed briefly in § 8.1, the "excellent" agreement between the general relativistic prediction and observation is the result of questionable derivation and calculation which are partly based on using classical results whose validity is restricted to flat spacetime and hence they partly ignore the effect of spacetime curvature.[251] Although the effect of spacetime curvature is negligible in this classical case (as long as we respect the locality condition), the violation of the locality condition makes this effect considerable by accumulation. In fact, the large spatial extension of the orbit associated with the large time period of a century will make ignoring the effect of even a very tiny curvature unacceptable and will reduce the result useless (or even totally wrong). The reality is that the violation of the locality condition itself is what made this precession observable and considerable in this classical case, so we cannot consider the impact of the violation of locality in the effect itself but ignore it in the consideration of the curvature of spacetime. This is inline with our previous observation (see exercise 8 of § 8.1) that even with the consideration of the orbiting system as a classical system we still observe the extremely tiny precession of the orbit over extended space (of a planetary orbit) and long time period (of a century), so even the tiny effect of spacetime curvature should not be neglected over this extended space and long time period. In fact, qualitative and quantitative analysis should lead to the conclusion that the effect of the curvature of spacetime should be comparable to the predicted 43 arcseconds that we obtained from the implicit assumption of flat spacetime. This should put a big question mark on the validity of the perihelion precession in endorsing general relativity. This applies at least to the problem of the perihelion precession of Mercury although this should extend even to other precession problems, e.g. periastron precession of binary systems whose results seem to be obtained from a similar semi-classical method of derivation and calculation. In fact, some of these systems are not classical and hence the effect of the spacetime curvature and the impact of the violation of locality should be more serious and pronounced. To sum up, fully-relativistic analytical treatments and calculations for the perihelion/periastron precession problem are required for this evidence to be valid and conclusive in endorsing general relativity (or otherwise). The current treatments and calculations contain classical and flat-space elements and hence their validity and conclusivity are questionable.

• Due to the involvement of several causes of precession (see § 8.1), we may need to assume the validity of the superposition principle (although this seems logical in this case).

Exercises

1. Is the general relativistic prediction about extra precession of orbits verified in the case of planets other

---

[250] We note that in our derivation of the orbital precession formula we used some classical arguments and hence the derived formula may not be valid in non-classical gravitating systems. However, these arguments can potentially be generalized; moreover there are other methods for deriving the orbital precession formula in general (although they also contain questionable aspects).

[251] This is demonstrated for example by employing flat-space formulae in the derivation and using physical variables, in place of coordinate variables, in the calculations (refer to problem 1 of § 8.1).

than Mercury? What about other orbiting systems?

9.2 Light Bending by Gravity

This may be seen as one of the most established evidence in support of general relativity. However, there are several factors that diminish its value as supporting evidence to general relativity:

• Most observations (which are in the optical part of the spectrum) have very large uncertainty and hence they do not lend full support to the general relativistic prediction. Yes, there are recent claims of observations with high certainty that agree very well with the predictions of general relativity. These observations are related to the deflection of radio waves using baseline interferometric techniques. However, there are some sources of uncertainty about the source of emission as well as the detection technique and hence we wonder if the analysis of the data is sufficiently objective towards factors like these.

• Light bending by gravity can also be predicted by the classical theory of gravity (see § 8.2) and hence it is not a decisive evidence for general relativity. So, qualitatively classical physics and general relativity are equal in their relation to the prediction of light bending and hence both theories can claim to be endorsed by this evidence.[252] However, from a quantitative perspective the predictions of the two theories are different since the classical prediction is half the general relativistic prediction. Nevertheless, considering the large uncertainty in the observations (which we indicated in the previous point), the evidence can still be shared by classical physics since the quantitative prediction of general relativity is not confirmed and hence the amount of bending may well be explained by classical gravity (at least in some cases). Moreover, there are claims in the research literature of classical derivations of the light deflection formula that agrees with the general relativistic formula and this could cast more doubts over the validity of the light bending by gravity as an endorsement to general relativity specifically. We should also note that light bending by gravity can also be predicted by non-classical metric gravity theories other than general relativity.

• At least some of the proclaimed evidence for light bending by gravity (e.g. gravitational lensing) may be explained by non-gravitational causes. For example, some cases of alleged gravitational lensing may arise from purely optical causes such as refraction of light in the atmosphere of the lensing object or diffraction of light by the lensing object. This may be particularly true in the case of "Einstein ring" where the pattern is very similar to diffraction. In fact, some of the proclaimed evidence may even be explained (tentatively at least) by fortunate occurrences and accidents such as the accidental existence of very similar observed objects around the presumed lensing object. The reality is that many of these observational evidence rely on the validity of general relativity and hence they are not independent of the theory itself; in other words they are effectively circular. In brief, there is a huge bias in analyzing the alleged observational evidence since the observers (most of whom are staunch supporters of general relativity) assume the validity of this theory (implicitly at least) and hence when they analyze the evidence they already assume a gravitational cause and assume that the proper theory for conducting this analysis is general relativity. So, the purpose of any presumed analysis is to verify the quantitative agreement between the observation and the theory and in many cases this verification can be easily done by tuning the parameters (consciously or unconsciously) to achieve the match where the numerous sources of uncertainty and error in such observational evidence grant many degrees of freedom in manipulating the evidence (legitimately or illegitimately). In other words, when these evidence[253] are examined and analyzed individually and independently of each other and of the theory of general relativity most of them will not lend definite support to general relativity or favor this theory, but when they are investigated within a single context and under a single banner (i.e. *light bending by gravity* as predicted by *general relativity*) then the evidence will look convincing and even overwhelming.

---

[252] As we will see, the prediction of light bending is not specific to general relativity or one of the achievements of Einstein as it is claimed in the literature. In fact, the deflection of light by gravitational field was predicted and formulated even in classical physics where work and results related to this issue are attributed to Newton, Cavendish and Soldner as well as other classical physicists (see Problems and refer to the literature about light deflection by gravity).

[253] We use "these evidence" and its alike for clarity (which is of primary importance in scientific texts) even if it is not allowed linguistically.

9.2 Light Bending by Gravity

- Considerable part of the claimed evidence in support of light bending by gravity has only qualitative value at most. For example, most or all observations of gravitational lensing do not provide quantitative evidence independent of the theory used in the analysis because in gravitational lensing most parameters (such as the mass of the lensing object) are guessed rather than measured or observed. In fact, this guessing extends in some cases even to the existence of the presumed lensing object because there is no independent evidence for the existence of the lensing object (such as black hole whose existence cannot be observed and verified directly) apart from the presumed gravitational lensing. Accordingly, even if we assume that the claimed evidence supports light bending by gravity qualitatively, it does not provide a conclusive quantitative evidence in support of a particular theory unless we assume *a priori* the validity of a particular theory, and this is obviously circular. In brief, gravitational lensing (and its alike) is at the best a qualitative evidence and hence it can be easily explained by other theories (classical and non-classical) and cannot be a conclusive evidence in support of general relativity specifically.
- The general relativistic derivation of the formula of light bending by gravity does not necessarily require the full formulation of general relativity (as represented by the Field Equations)[254] and hence it cannot be seen as evidence for the full formulation of general relativity. In other words, the general relativistic formula can be obtained from simple arguments based on the weak equivalence principle and hence it does not require the full force of general relativity. Accordingly, even if we assume that this test is compliant with the prediction of general relativity it cannot be regarded as definite and specific evidence to the full formulation of general relativity although it can be regarded as support to some of the arguments and principles that general relativity rests upon. It should be remarked that the predictions of the derivation from the equivalence principle (according to some widely-accepted derivation methods) and from the full formulation differ quantitatively and hence from a quantitative perspective the agreement with the prediction of the derivation from the full formulation (or otherwise) should be a decisive factor. However, this does not apply to all the derivation methods from the equivalence principle. Moreover, many of the claimed evidence have only qualitative value and hence they should be indifferent with respect to the quantitative aspect.
- The derivation of the relativistic light bending formula (i.e. Eq. 293) is based on certain approximations (see the Problems of § 8.2) and hence its general applicability is questionable. This means that even the highly reliable observations (e.g. the observations related to the deflection of radio waves) that supposedly endorse general relativity with high confidence and certainty can in fact endorse general relativity only within the limits of validity of these approximations. Therefore, these observations cannot endorse the predictions of general relativity in general (i.e. when the above approximations do not apply).
- We should also mention that there are numerous sources of uncertainty and error in the light bending observations in general (especially those in the optical range) due for example to optical distortion by the instruments or to refraction by the Earth atmosphere. These factors should affect the certainty and reliability of this test (at least in some cases).

To conclude, light bending by gravity which is seen as one of the best established evidence in support of general relativity is questionable in many aspects and details and hence it cannot be classified as conclusive evidence in support of general relativity despite the common claims in the literature about its validity and significance.

Problems

1. Provide a short historical account on the investigation by classical physicists (prior to general relativity) of light bending by gravity.
   Answer: In the literature of general relativity the deflection of light by gravity is commonly presented as one of the unprecedented achievements of this theory. However, this issue has been investigated by several classical physicists (as early as Newton). The available historical records show that Newton has contemplated about the deflection of light by gravity, which is inline with his corpuscular theory about the nature of light. The records similarly show that this issue has also been examined qualitatively by Jean-Paul Marat in the 1780s. The first quantitative calculation (or estimation) of light bending

---

[254] We note that the geodesic equations are based on the space metric and hence they are ultimately a product of the Field Equations which produce the metric.

by gravity seems to belong to Cavendish in the second half of the 18th century. The first confirmed derivation of the classical formula of light bending by gravity is attributed to Johann Soldner who published a paper in the early years of 1800s in which he did not only derive the classical formula but he also estimated correctly (according to the classical prediction and within the uncertainties in the physical constants at that time) the amount of deflection that should be caused by the Sun. Moreover, he suggested corrections to the observations of the positions of the fixed stars (if observed too close to the Sun) based on the errors in the apparent position due to the effect of bending.

In brief, the issue of light deflection by gravitational fields has been investigated qualitatively and quantitatively rather thoroughly by classical physicists and hence it is not an invention or discovery or precedence of general relativity. Yes, the general relativistic prediction is quantitatively different from the classical prediction and hence if general relativity should be accredited it should be for this and only for this. However, in the light of the uncertainties about the evidence in support of the general relativistic prediction even this credit should be provisional and require more endorsement from observations. Moreover, there are non-relativistic predictions that agree quantitatively with the relativistic prediction and hence any credit will not belong to general relativity specifically and exclusively.

Exercises

1. Investigate bending of star light by the Sun and assess its value as one of the important evidence in support of general relativity.
2. Discuss the historical development of the derivation of light bending formula within the framework of general relativity and the significance of this on the value of this evidence.
3. Examine the possible reasons for the premature endorsement of general relativity by the results of the solar eclipse of 1919.
4. Briefly discuss and assess gravitational lensing as one of the claimed evidence in support of general relativity.
5. Talk about the tests of light bending by gravity in the radio band (rather than the optical band) of the electromagnetic spectrum.
6. Investigate the sources of error and uncertainty in the results of astronomical and cosmological investigations in general.

9.3 Gravitational Time Dilation

Gravitational time dilation may be regarded as one of the most compelling evidence in support of general relativity. In the following subsections we briefly investigate a number of experiments and applications that are based on the gravitational time dilation and hence they can be considered as direct evidence in support of this aspect of general relativity. We should remark that because of the correlation between gravitational time dilation and gravitational frequency shift (see § 6.3.4) most (if not all) of the claimed evidence of gravitational time dilation can be explained indirectly by classical gravity (and hence they are not conclusive evidence in support of general relativity) although time dilation does not exist classically. In other words, although theoretically and conceptually gravitational time dilation may not be possible to explain classically, it is possible to explain indirectly as gravitational frequency shift due to the correlation between time dilation and frequency shift at least in some cases. This is because in § 8.4 we derived a classical formula for gravitational frequency shift using the conservation of energy and it was shown there that the classical formula and the general relativistic formula produce virtually identical results in most cases since their power series approximations are identical under certain practical conditions and assumptions. So, at least some of the instances of gravitational time dilation evidence (which meet the conditions of the power series approximation and involve frequency as timing mechanism)[255] can be

---

[255] Referring to our investigation in § 8.4, we may need to add another condition that is for the gravitational time dilation to be possible to explain classically it must be based on gravitational frequency shift by sending a signal from one gravitational potential level to another potential level and not by generating two independent signals at two potential levels. In fact, by adopting this condition we essentially make a distinction between gravitational time dilation and gravitational frequency shift as two different phenomena and hence the significance of the evidence as time dilation

explained classically and hence they are not conclusive as general relativistic evidence. In fact, they can also be explained non-classically by modern non-relativistic gravity theories. We should finally remark that in most of the time dilation (and frequency shift) experiments the frames of occurrence or/and observation are not fixed in space (e.g. they are freely falling as in the case of global positioning system or they follow certain spatial trajectory as in the case of Hafele-Keating experiment) and this should require generalization of the argument and derivation of the gravitational time dilation (and frequency shift) that we presented earlier.[256]

Exercises

1. What are the cases in which gravitational time dilation cannot be explained classically?

9.3.1 Hafele-Keating Experiment

This experiment, which was conducted in 1971 by Joseph Hafele and Richard Keating, is based on measuring the combined effect of time dilation caused by motion (i.e. special relativistic) and by gravity (i.e. general relativistic) where precise cesium atomic clocks carried on board airplanes were used to detect time dilation by performing round trips around the world eastward and westward and comparing their time with the time recorded by identical clocks that stayed stationary on the Earth. The principle of the experiment is very simple that is the traveling clocks on board the airplanes should experience slowing down (or speeding up depending on the direction of travel) due to kinematical (or special relativistic) time dilation effect caused by the motion, and speeding up due to gravitational (or general relativistic) time *dilation* effect caused by the increase in altitude. Accordingly, the experiment is a simultaneous test to both relativity theories in one go as it is based on testing the combined effect of time dilation predicted by these theories and not on testing the individual effects of these theories.

However, the validity and significance of the results of this experiment may be questioned for several reasons. For example:
• Since this experiment is based on testing the combined effect of the kinematical and gravitational time dilation it is not clear about the effect of each individual cause. Therefore, the result is not conclusive about the endorsement of each individual theory even if the observed combined effect agrees with the theoretically predicted combined effect (assuming the validity of superposition as will be discussed in the next point) because the actual contribution of the effect of each theory to the combined effect is unknown.
• Since this experiment is based on testing the combined effect of the kinematical and gravitational time dilation, we need to assume the validity of superposition (or additivity of effects, i.e. the total effect is the sum of the special and general relativistic effects) which is not obvious (see Problems).[257]
• There are question marks about the validity of the involvement of special relativity and its application in this experiment and its analysis. For example, the analysis seems to suggest the need for an external (or privileged) frame to make sense of the claimed special relativistic effects where the change in time is positive in one direction and negative in the other direction.
• The marginal error in the reported data is considerable and hence the results are not conclusive.
• The results may also be explained partially by classical gravity as discussed earlier although this can be challenged since gravitational time dilation in this experiment is not based on gravitational frequency shift. The results may also be explained partially by other gravity theories.

In fact, there are many sources of challenge and refutation for this experiment and its rationale and analysis as well as its procedural aspects (see Exercises). In brief, the results of this experiment are neither specific nor reliable and hence this test is not conclusive in its endorsement to special relativity or general relativity (let alone both). An improved version of this experiment was conducted later where it is reported that the marginal error was reduced substantially. However, this may have eliminated some

---

evidence should vanish or diminish.

[256] We note that the inclusion of special relativistic effects should not be sufficient to account for the required theoretical treatment of the gravitational effects in non-stationary frames unless we assume superposition (which is not obvious).

[257] The reader should note that the validity of superposition (even if it is established) is not sufficient to address the concern in the previous point about the combined effect.

causes of concern in the previous experiment but not all causes and hence in our view this test is still inconclusive and potentially wrong.

Problems

1. Assess the validity of the commonly-accepted analysis of the observations and experiments (such as the Hafele-Keating experiment) that involve a combination of special and general relativistic effects (and potentially other effects).
Answer: In our view, this analysis is questionable (at least in some cases) because it is implicitly based on the assumption of the validity of superposition[258] which is not obvious in these situations. For example, the formulation of gravitational time dilation is based on employing stationary gravitational frames and hence this formulation may not be valid in dynamic gravitational frames. Similarly, the applicability of the special relativistic formulation in the presence of gravity may not be valid. In fact, the uncertainty about the validity of superposition should introduce more sources of uncertainty on the validity of the tests that involve combinations of more than one effect (particularly special and general relativistic effects), i.e. apart from the uncertainty in endorsing the individual theories.

Exercises

1. Make a list of potential criticisms to the Hafele-Keating experiment and its analysis.
2. Discuss briefly the "general relativistic version" of the twin paradox.

9.3.2 Shapiro Time Delay Test

The theoretical foundation of this test, which was theoreticized in 1964 by Irwin Shapiro, is based on the claim that due to the gravitational time dilation[259] an electromagnetic signal passing by a massive object should slow down due to its descent and ascent in a gravitational potential well. Accordingly, a radar signal sent from the Earth and reflected by another planet (e.g. Mercury or Venus) should experience a delay if the signal passed by the Sun in its onward-return journey. Such a test has been carried out in a number of experiments where it is claimed that the results match well with the predictions of general relativity.

We note that the theoretical foundation of this test may seem inconsistent with the alleged constancy of the speed of light. However, this can be refuted by the claim that the constancy of the speed of light in general relativity is local. In fact, even the claim that this test is based on the gravitational time dilation effect may be challenged because it is inconsistent with the meaning of time dilation as used in other contexts and the controversy about the theoretical basis of this time delay (see the upcoming discussion). In our view, this test and its claimed theoretical foundation and formulation are shrouded with many question marks and the literature in this regard is rather messy and unreliable. Moreover, large part of the literature in this regard (especially on the experimental aspects) is based on trust with no scrutinized analysis or examination. So, even if such a delay is actually observed as reported, its theoretical foundation and formal justification require further assessment and analysis. The experienced delay may be explained by a general relativistic aspect (and potentially non general relativistic aspect) other than the proposed cause (e.g. gravitational time dilation) and the two are not necessarily equivalent in their significance and implications (see next paragraph).

In this context, we note that there is confusion in the literature about the cause of this delay and if it is gravitational time dilation (as classified above) or gravitational length "dilation" (or what is described as spacetime dilation which may be stated by some as lengthening of the light path due to the curvature of spacetime) or dependency of the speed of light on the strength of the gravitational potential or gravitational light bending.[260] Although some of these causes may provide equivalent explanations, the validity of

---

[258] We may also call it "additivity of effects".

[259] As we will see, this is one potential explanation of the Shapiro time delay. We think the effect of gravitational length contraction (and possibly other effects) should also be considered in the estimation of the time delay if gravitational time dilation is supposed to be involved in this effect. However, a correct analysis based on the spacetime metric should automatically consider both temporal and spatial effects (as embedded in the spacetime metric).

[260] These alleged causes are what we found in the literature and there could be more. Although some of these explanations may reduce to others, it is not always the case. We note that the original derivation of the formula of this time delay

9.3.2 Shapiro Time Delay Test

each one of these causes should be demonstrated explicitly and formally to identify which aspect of general relativity is allegedly endorsed by this test and assess the value of this claimed evidence. We should also note that the confusion about the cause of this delay seems to be reflected in the confusion about its formulation, as will be seen in the questions.

Problems

1. Discuss the reliability of the Shapiro time delay test in its endorsement to general relativity.
   Answer: We believe this test is not conclusive. Some reasons for this belief are:
   • The theoretical foundation and formulation of this test are not well established. In fact, there are several methods and formulae that claim to represent Shapiro time delay and these methods and formulae do not produce the same predictions. For example: **(a)** some use coordinate variables while others use metrical variables, **(b)** some employ isotropic coordinates while others employ Schwarzschild coordinates, **(c)** some use straight path while others use deflected path, **(d)** some employ correction for the Earth movement while others do not. There are also disputes about its theoretical foundation (noting that the theoretical foundation should affect the formalism); some of these disputes are indicated in the text.[261]
   • There are question marks about the consideration of the speed of light and how it enters in the theoretical foundation of this time delay. We also question the potential role of gravitational length contraction (which is a result of having coordinate variables and metrical variables; see § 6.2 and § 6.4) and if it should be considered in the analysis. These issues are not addressed properly and consistently in the literature and this gives many degrees of freedom in manipulating the theoretical foundation to suit general relativity and its supporters. In brief, among this lack of clarity it is easy to select or tailor a formulation or explanation or method of calculation that suits the endorsement of general relativity even if this is inconsistent with the framework of general relativity or other established facts and principles.
   • There are many sources of uncertainty and error in the technique of radar ranging which is used in this sort of time delay experiments. This technique is complicated and challenging and hence the experimental results are not as straightforward as might be imagined. This criticism may be less severe with regard to other versions of the Shapiro time delay experiments where signals emitted from a satellite or spacecraft (e.g. Cassini probe) are used in these experiments and hence the time delay of the grazing signals can allegedly be obtained more directly than in the case of radar ranging experiments. However, there are still significant sources of error and uncertainty even in these versions.
   • The time delay caused by gravity is a very tiny fraction of the total journey time of the signal in the round trip and hence the effect of any error or noise will be so significant that the result will not be conclusive (especially when considering the above theoretical and practical uncertainties).
   • The formulae and calculations of this test are generally based on the use of physical variables for what is supposed to be coordinate variables and this could introduce significant error considering the extended spacetime interval of the onward-return journey even though the curvature of spacetime is small.
   • The reported (high) accuracy of the experiments of this test requires unrealistically high certainty about the distances involved and this is almost impossible to achieve in such dynamic orbiting systems.
   • Other effects (e.g. optical such as the refraction of the signal by the solar corona) are involved in these experiments and this should increase the uncertainty and error in the results (especially when considering the tininess of this time delay within the overall trip time).

Exercises

1. Give two formulae that allegedly represent Shapiro time delay and compare their quantitative predictions.

---

by Shapiro is based on assuming the dependency of the speed of light on the strength of the gravitational potential in a Schwarzschild spacetime.

[261] We should also note that there are technical reservations on the derivation methods of some of the time delay formulae (the details are out of scope).

2. Give some examples of expected sources of uncertainty and error in the Shapiro time delay test and assess their effect.

9.3.3 Gravity Probe A

This is a space mission conducted by the Smithsonian Astrophysical Observatory (SAO) in collaboration with the National Aeronautics and Space Administration (NASA) in 1976 to test gravitational time dilation where a probe carrying a hydrogen maser of very precise frequency was launched to a height of about 10^4 km[262] above the surface of the Earth. The working principle of this experiment is that due to the difference in the gravitational potential at the surface of the Earth and at 10^4 km above the surface of the Earth the rate of time flow should differ (i.e. it is faster at 10^4 km above the surface of the Earth since it is higher in the potential well of the Earth) and this should be reflected in a change of frequency due to the reciprocal relation between the periodic time and frequency (see § 6.3.4). Accordingly, this may be seen as a test to the gravitational frequency shift as well as to the gravitational time dilation due to the fact that these two phenomena are based on the same physical principle (as explained in § 6.3.4) and hence they are essentially the same.

It is claimed that the experimental results confirmed the predictions of general relativity to a high level of accuracy. However, even if we accept this claim in principle the evidence may not be conclusive in its endorsement to general relativity specifically since the results may be explained classically by gravitational frequency shift as a consequence of energy conservation (see § 9.4.3) although this explanation can be challenged because there is no actual frequency shift in this experiment but a generation of two independent signals at two gravitational potential levels (see § 8.4). We should also note that this experiment may be explained by non-classical gravity theories other than general relativity. In fact, it may be explained even by the equivalence principle without need for the full formulation of general relativity. Hence, it is not conclusive in its endorsement to the formulation of general relativity specifically and definitely. The reader is also referred to § 9.4.3 for further discussion and details about this experiment and related issues.

Exercises

1. What is the fractional change in the time interval between a clock on the surface of the Earth and a clock at an altitude of 10^4 km above the surface of the Earth (i.e. as in the Gravity Probe A experiment)?[263]

9.3.4 Global Positioning System

Global positioning system (GPS) consists of a number of satellites positioned at high altitude ($\simeq 20200$ km) around the Earth. The principle of operation of the GPS is based on satellite ranging where precise position and time signals are sent from the satellites and received by the users on the Earth to determine the position of the user (as well as other information like time and velocity). Accordingly, clocks on the satellites are used in conjunction with a triangulation-like method to determine the position on the surface of the Earth and in its proximity (e.g. in marine and aeronautical navigation). For the system to work reliably and determine positions accurately the clocks must be very precise to provide accurate and synchronized timing (and hence positioning) information.[264] Now, according to the gravitational time dilation of general relativity the clocks at high altitude run faster and hence regular corrections are needed to keep the timing accurate (i.e. synchronized with the timing on the surface of the Earth) and the system functioning. It is claimed that without the gravitational time dilation corrections of general relativity the GPS will become useless within a few minutes or hours due to the accumulation of timing errors and hence positioning errors. Accordingly, the GPS is regarded as a live evidence for the validity

---

[262] Some sources in the literature report different height. However, the above figure seems more credible.

[263] The question is restricted to the change caused by gravitational time dilation due to difference in height (or potential level) and hence it does not include other possible causes.

[264] In talking about things like preciseness, accuracy, correction and synchronization we are essentially considering the timing on the surface of the Earth and the intended function of the GPS which is designed and deployed to serve its users on the Earth. Otherwise the satellite clocks are working properly within their physical environment without error or malfunctioning.

9.3.4 Global Positioning System

of general relativity. It should be remarked that special relativistic time corrections (due to kinematical time dilation) are also needed to keep the GPS working (see the upcoming discussion and Exercises). We should also expect other corrections related for example to the Shapiro time delay due to the descent of the satellite signals in the gravitational potential well of the Earth.

Anyway, we have a number of reservations about the validity and significance of this test. For example, the test is supposed to involve both special and general relativistic effects (i.e. kinematical and gravitational time dilation) and hence it cannot be conclusive in endorsing any particular theory because we do not know what is the actual contribution of each theory (assuming that both theories have contribution) to the observed time difference between the satellite clock and the ground clock. In fact, the situation is more complicated due to the involvement of other effects and required corrections as indicated above. We also need to assume the applicability of superposition, i.e. the resultant effect of time change is the algebraic sum of the individual contributions of the effects of each theory (as if the other theory is not in action). The least that we can say about the validity of this assumption is that it is not obvious and hence it requires justification. Another reservation is that the applicability of special relativity is questionable because according to special relativity in its Einsteinian interpretation each clock (i.e. satellite clock and ground clock) should experience kinematical time dilation relative to the other clock. In fact, this should lead us to the controversy of twin paradox which was thoroughly investigated in B4 and hence we do not repeat the investigation here although we should mention that some of the alleged refutes of the twin paradox do not apply here (as indicated earlier in a similar context within the answer of some exercises). Moreover, the inertiality of the satellite and ground frames is questionable. We should also repeat what we stated before about the possibility of classical explanation for the gravitational time dilation based on the frequency shift and the conservation of energy although this can be challenged since what we actually have is two signals generated at two potential levels by two separate clocks. Anyway, alternative explanation may be provided by non-classical gravity theories other than general relativity even if we reject the classical explanation. The explanation may also be provided by the equivalence principle without need for the full formulation of general relativity.

So in brief, we believe that this test is invalid or at least it is inconclusive. We can even claim that this test can be an evidence against at least one of the relativity theories, i.e. it can be an evidence against special relativity since time dilation takes place in a privileged frame (i.e. satellite frame) and not in the other frame (i.e. ground frame) which is a clear violation of the relativity postulate of special relativity in its Einsteinian sense. In fact, the collapse of this evidence in its special relativistic aspect should lead to its collapse in its general relativistic aspect because the observed combined time dilation effect (which is claimed to match the combined predictions of these theories) depends on the involvement of both theories in this observed combined effect and hence the collapse of the contribution of one theory to this combined effect should lead to the collapse of the contribution of the other theory.

Finally, we should remark that there are claims in the literature that the time corrections in the GPS are based on empirical (rather than relativistic) formulae and the two are not identical. If so, then the global positioning system could lose its status in endorsing the relativity theories and could even be an evidence against the relativity theories. Unfortunately, we are not in a position to confirm any one of these conflicting claims because we have no detailed knowledge about the design and operation of the global positioning system. Hence, at the best the validity of this test (like most other tests) is a matter of trust and faith (assuming that its theoretical basis is sound).

Problems

1. Give a brief account of how the global positioning system works.
   Answer: The satellites of the global positioning system regularly transmit electromagnetic signals carrying information about their time and position in space. A receiver on the surface of the Earth (or in its surrounding such as airplane) uses the information from a number of these satellites. Hence, by knowing the location of the transmitting satellites and the time required by the signals to reach the receiver it can work out its position relative to the satellites and from this the position of the receiver on the Earth can be determined since the position of the satellites relative to the Earth is known. In brief, the satellites of the global positioning system work as reference points (or markers) in the space

9.4 Gravitational Frequency Shift 197

to facilitate the determination of the position of the receiver on the Earth through the information about their position and the time required by the transmitted signals to reach the receiver.

Exercises
1. Calculate the general relativistic time correction that is required to keep the GPS functioning.[265]
2. Calculate the special relativistic time correction that is required to keep the GPS functioning. Comment on the method of calculation.
3. Considering the results of the previous two exercises, find the required combined correction due to relativistic effects.[266]

9.4 Gravitational Frequency Shift

Gravitational frequency shift may also be considered as one of the compelling evidence in support of general relativity. In the following subsections we briefly investigate a number of experiments and observations that allegedly verify the gravitational frequency shift of general relativity. However, before that we should draw the attention to the following remarks:

• Gravitational frequency shift in general relativity is based on gravitational time dilation (see § 6.3.4) and hence it is not an independent evidence. This should diminish the value of this evidence since in essence it does not support an aspect of general relativity other than gravitational time dilation. In brief, all the claimed evidence in support of gravitational frequency shift are no more than instances of gravitational time dilation (and hence they do not represent a different type of evidence). The reverse is also true in general (or at least in most cases) due to the general use of frequency for timing because timing is usually based on using repetitive frequent physical phenomena like pendulum swing and atomic emissions.[267]

• As we saw in § 8.4, the classical and general relativistic formulations of gravitational frequency shift are different but their quantitative predictions are very similar in most cases[268] and this should cast a shadow over at least some of the claimed gravitational frequency shift evidence in support of general relativity because these predictions cannot be distinguished practically from the classical predictions and hence the evidence is not conclusive in its endorsement to general relativity specifically since it can be justified classically (noting also that it may be justified by other modern gravity theories). In fact, this should also cast a shadow over at least some of the claimed gravitational time dilation evidence in support of general relativity because of the close connection between gravitational time dilation and gravitational frequency shift (as explained in the previous remark and in numerous places in the previous sections and chapters) and hence most (if not all) of the time dilation evidence may be regarded as frequency shift evidence and vice versa (although this may be disputed at least in some cases).

• Based on the previous remark and referring to our discussion in § 8.4 (see exercise 5 of § 8.4), if $\frac{GM}{r_1 c^2} \ll 1$ then the gravitational frequency shift formula of general relativity at infinity (i.e. $r_2 \to \infty$) is practically indistinguishable from its classical counterpart. In fact, this is the case in most astronomical observations of gravitational frequency shift. Therefore, for the gravitational frequency shift evidence in astronomical observations to be potentially conclusive in its endorsement to general relativity specifically the evidence should belong to a situation where the condition $\frac{GM}{r_1 c^2} \ll 1$ does not apply (by considering black holes and possibly neutron stars).

• We should also note that the gravitational frequency shift is essentially a test for the equivalence principle rather than the Field Equation and this view is expressed by some general relativists. For example, according to Rindler (see the References) "The well-established gravitational Doppler shift, occasionally

---

[265] In fact, the question is about the main general relativistic correction caused by gravitational time dilation due to difference in height or potential level (noting that other general relativistic corrections are required).
[266] In fact, the question is about the main (not all) relativistic effects (i.e. the effects investigated in the previous exercises).
[267] We note that timing may be achieved by non-repetitive processes such as assigning (according to Newton's first law) equal time periods to equal distances traversed by free particles (although the non-repetitivity of this process may be questioned as well as the axiomatic foundation of this timing technique). In fact, there are many details and potential debates and controversies about these issues (which we do not investigate).
[268] In fact, it was shown in § 8.4 that power series approximations of the classical and general relativistic formulations (plus reasonable conditions and assumptions which are valid in most cases) produce identical formulae.

also referred to as a crucial effect, is, in fact, not a test of the field equation but merely of the equivalence principle - at least in lowest order, which is all that can be observed at present". In fact, there are many interesting details about this issue which the inquisitive reader may wish to investigate (see for Example Weinberg and Rindler in the References).

Exercises

1. Justify the fact that the correlation between gravitational time dilation and gravitational frequency shift should diminish the value of the evidence related to these effects in endorsing general relativity.
2. What is the significance of having a classical formulation for the gravitational frequency shift that is identical to the general relativistic formulation under a power series approximation that is acceptable in most cases?
3. If gravitational time dilation and gravitational frequency shift are equivalent then why should we investigate these effects (in this chapter and in the previous chapters) separately?
4. Discuss the conventional nature of the physical definition and determination of time.

9.4.1 Gravitational Red Shift from Astronomical Observations

Red shift of electromagnetic emissions from astronomical objects, such as white dwarfs, has been observed regularly in astronomical investigations and it is commonly attributed in many cases to gravitational cause in accord with the prediction of general relativity. Some of these claimed instances of gravitational red shift go back to the 1920s where a spectral line from the companion of Sirius was observed red shifted and this was explained gravitationally and hence it was regarded as endorsement to general relativity (although this was discredited later). There are many other subsequent observations that are similarly explained and hence they are alleged to endorse general relativity.

However, in many cases this evidence is not conclusive due to the many uncertainties in these observations and in the physical characteristics of the emitting objects where these characteristics (such as mass and radius) enter in the estimation of the gravitational red shift. In fact, some of these claimed instances can be circular in their indication since the observed emissions are used to infer the physical characteristics of the emitting objects and hence these characteristics (even though they may be obtained from physically-similar but different objects) cannot be used in the estimation of the gravitational red shift.

In brief, we have several reservations on the validity and conclusivity of the gravitational red shift in astronomical observations as an endorsing evidence to general relativity; these reservations include:

• There are many observational and theoretical sources of uncertainty and error in these observations and their theoretical foundation and methods of analysis (as it is the case in most astronomical observations). One demonstration of this fact is the contradiction between some observations and other similar observations (refer for example to the huge difference in the Sirius observations and the large uncertainties in the results as reported in the literature).
• Some of these observations may be explained by non-gravitational causes (e.g. kinematical such as Doppler shift or moderation by interstellar medium). In fact, such a possibility should ruin the significance of many claimed instances of this evidence even if we accept in principle the theoretical foundation of the general relativistic gravitational red shift.
• Most of these observations (assuming they are certain and are of gravitational origin) can also be explained by the classical formulation of the gravitational red shift (as discussed earlier) and hence they are not definite in their endorsement to general relativity. These observations can also be explained by metric gravity theories other than general relativity (and possibly even by the equivalence principle with no need for the formalism of general relativity).

9.4.2 Pound-Rebka Experiment

The essence of this experiment, which was conducted in 1959 by Robert Pound and Glen Rebka, is to measure the frequency shift experienced by gamma rays traveling between the top and bottom of a tower

9.4.3 Gravity Probe A

at Harvard University over a distance of about 22.5 meters in the gravitational field of the Earth.[269] It is reported that this experiment verified the prediction of general relativity (i.e. blue shift in the descending rays and red shift in the ascending rays) to high level of accuracy.[270] Hence, this experiment is regarded as one of the most conclusive evidence in support of the gravitational frequency shift and the theory of general relativity.

However, as we will see in the Exercises the classical formulation of the gravitational frequency shift provides very similar quantitative prediction to the prediction of general relativistic formulation and hence the results of the Pound-Rebka experiment are not conclusive in endorsing general relativity specifically since the results can be similarly explained by the classical formulation. Moreover, the results can also be explained by other metric gravity theories (and possibly even by the equivalence principle without need for the full formulation of general relativity).

Exercises

1. Derive a simple classical formula for the expected frequency shift in the Pound-Rebka experiment.
2. Referring to the answer of the previous exercise, justify why high-frequency radiation (like gamma ray) should be desirable to use in experiments like the Pound-Rebka experiment.
3. Use the available data about the Pound-Rebka experiment with the classical formula that was derived in exercise 1 to estimate the frequency shift and the fractional frequency shift in the Pound-Rebka experiment.
4. Repeat exercise 3 using this time the general relativistic formula for the gravitational frequency shift that was derived in § 6.3.4. Comment on the results.
5. Repeat exercise 3 using this time the classical formula for the gravitational frequency shift that was derived in § 8.4. Comment on the results.

9.4.3 Gravity Probe A

This test was investigated earlier in § 9.3.3 from a gravitational time dilation perspective. However, we briefly re-discuss this test here from a gravitational frequency shift perspective to show that the results of this test may not be as significant in endorsing general relativity as it might be thought because similar quantitative results can be obtained from classical formulation (see Problems). In fact, this test may not be really a frequency shift test and hence the classical formulation (which may be restricted to the case of actual frequency shift) may not apply because as explained earlier we have in this experiment two independent signals generated at two different gravitational potential levels (i.e. we do not have a single signal that descends or ascends in a gravitational well) and hence the classical argument of frequency shift (which is basically based on the conservation of energy during the transition of a signal between two potential levels) may not be valid. This issue has already been investigated rather thoroughly (see for example § 8.4) and hence we will not discuss it anymore.

Problems

1. Estimate the gravitational frequency shift in the Gravity Probe A experiment using classical and general relativistic formulations. Comment on the results.
   Answer: We subscript the variables on the surface of the Earth with 1 and the variables at 10^4 km above the surface of the Earth with 2. In the following we use the fractional frequency shift in our estimation.
   According to the classical formula that was derived in § 8.4 we have:

$$\nu_2 = \nu_1 \left(1 - \frac{GM}{c^2 r_1}\right)\left(1 - \frac{GM}{c^2 r_2}\right)^{-1}$$

---

[269] The Pound-Rebka experiment was more elaborate than what our description might suggest. The details (which are of little value to our purpose) can be found in the literature of general relativity. We should remark that there are some reservations on some aspects of the Pound-Rebka experiment other than being justifiable by other theories (and classically in particular). However, we see no necessity in going through these lengthy details.

[270] In fact, the accuracy of the original Pound-Rebka experiment was rather poor. But it was later improved substantially by Pound and Snider according to the reports.

$$\nu_1 - \nu_2 = \nu_1 - \nu_1 \left(1 - \frac{GM}{c^2 r_1}\right)\left(1 - \frac{GM}{c^2 r_2}\right)^{-1}$$

$$\frac{\nu_1 - \nu_2}{\nu_1} = 1 - \left(1 - \frac{GM}{c^2 r_1}\right)\left(1 - \frac{GM}{c^2 r_2}\right)^{-1}$$

$$\frac{\Delta\nu}{\nu_1} \simeq 1 - \left(1 - \frac{6.674 \times 10^{-11} \times 5.972 \times 10^{24}}{(3.0 \times 10^8)^2 \times 6.371 \times 10^6}\right) \times$$

$$\left(1 - \frac{6.674 \times 10^{-11} \times 5.972 \times 10^{24}}{(3.0 \times 10^8)^2 \times (6.371 \times 10^6 + 10^7)}\right)^{-1}$$

$$\frac{\Delta\nu}{\nu_1} \simeq 4.2460057692 \times 10^{-10}$$

According to the general relativistic formula that was derived in § 6.3.4 we have (noting that we use physical quantities for what is supposed to be coordinate quantities):

$$\nu_2 = \nu_1 \left(1 - \frac{2GM}{c^2 r_1}\right)^{1/2}\left(1 - \frac{2GM}{c^2 r_2}\right)^{-1/2}$$

$$\nu_1 - \nu_2 = \nu_1 - \nu_1 \left(1 - \frac{2GM}{c^2 r_1}\right)^{1/2}\left(1 - \frac{2GM}{c^2 r_2}\right)^{-1/2}$$

$$\frac{\nu_1 - \nu_2}{\nu_1} = 1 - \left(1 - \frac{2GM}{c^2 r_1}\right)^{1/2}\left(1 - \frac{2GM}{c^2 r_2}\right)^{-1/2}$$

$$\frac{\Delta\nu}{\nu_1} \simeq 1 - \left(1 - \frac{2 \times 6.674 \times 10^{-11} \times 5.972 \times 10^{24}}{(3.0 \times 10^8)^2 \times 6.371 \times 10^6}\right)^{1/2} \times$$

$$\left(1 - \frac{2 \times 6.674 \times 10^{-11} \times 5.972 \times 10^{24}}{(3.0 \times 10^8)^2 \times (6.371 \times 10^6 + 10^7)}\right)^{-1/2}$$

$$\frac{\Delta\nu}{\nu_1} \simeq 4.2460057692 \times 10^{-10}$$

Comment: as we see, the classical and general relativistic results are identical (to the quoted accuracy).[271] So, from a gravitational frequency shift perspective the results of this test are not as significant in endorsing general relativity as it might be thought because from this perspective the quantitative prediction of the general relativistic formulation can be obtained from the classical formulation. However, the applicability of the classical formulation in this experiment may be questioned (as explained earlier).

9.5 Gravitational Length Contraction

We are not aware of any alleged evidence in support of gravitational length contraction of general relativity. One reason could be the alleged impossibility of observing length contraction (assuming it does exist). However, the proposed reasons and arguments in this regard (or at least their generality) are largely questionable and the investigations and discussions about this important issue are rather terse and superficial. In fact, there is even some mess and confusion in this part of the literature of general relativity which may indicate a crack in the theory. Anyway, even if this does not indicate a crack in the theory it may indicate that there are some aspects in the theory that are not well developed or may not be properly examined and assessed and this should put a question mark on the endorsement of the theory. Accordingly, it should be premature to assert that the theory of general relativity is endorsed even if all

---

[271] In fact, they are also identical with the result of exercise 1 of § 9.3.3 which is about time dilation (as it should be).

the claimed evidence are valid and conclusive because these evidence are limited to certain aspects of the theory and hence many important aspects of the theory (such as length contraction) are still waiting verification.

We may also claim that the existence of evidence in support of gravitational time dilation, but not gravitational length contraction, in addition to the existence of a classical formulation that can explain gravitational time dilation (i.e. indirectly through frequency shift) at least in some cases and circumstances may give the classical interpretation of these results more weight. In other words, we may not actually have gravitational time dilation (because if we have then we should also have gravitational length contraction) and hence all the effects that are attributed to time dilation may be explained by other causes such as frequency shift and conservation of energy.

Exercises
1. Try to justify the claim that observing gravitational length contraction is not as easy as observing gravitational time dilation (and it may even be impossible).
2. What is the important issue that is highlighted by the previous exercise?

9.6 Gravitational Waves

There are claims of direct and indirect observation and detection of gravitational waves and the research in this field is going on vigorously. However, unequivocal confirmation is still sought despite the recent claims that definitely confirm the detection of gravitational waves from cosmic cataclysms (e.g. the collision of two black holes or the coalescence of two neutron stars) using laser interferometric techniques. We should remark that in principle gravitational waves may also be deduced from other gravitational theories (including classically-based theories) and hence they do not make a decisive evidence in support of general relativity unless the observations quantitatively agree with the predictions of general relativity and disagree with the predictions of other theories. So, up to the present time gravitational waves (even if they are really detected) cannot be regarded as a conclusive evidence in support of general relativity. In fact, gravitational waves should be a natural consequence of any full field theory with temporal and spatial dependency and hence gravitational waves, even if confirmed experimentally, should not be considered a decisive evidence in endorsing general relativity specifically but they should be seen as evidence for the tempo-spatial field model of gravitation.

We should also note that there is a more fundamental challenge to the significance and value of gravitational waves as evidence for general relativity because the current general relativistic theory of gravitational waves is largely based on the linearized form of general relativity (see § 3.4) in the weak field approximation and hence there are theoretical objections to the endorsement of gravitational waves to the full non-linearized form of the theory.[272] In fact, on inspecting the literature of gravitational waves we can conclude that the general relativistic formulation of gravitational waves is theoretically problematic from multiple aspects (whether linear or non-linear). Hence, the available approaches for developing formulations for this phenomenon cannot lend full support to general relativity (even if these formulations are vindicated by observations) because these formulations are not entirely, genuinely and surely general relativistic since they are based on many approximations, twists and controversies.

There are two main methods for verifying gravitational waves: indirect and direct. These methods are investigated and assessed in the following two subsections.

9.6.1 Indirect Observation of Gravitational Waves

Until recently, this was the only method for the *detection* of gravitational waves. Indirect *observation* is based on the principle of conservation of energy in conjunction with the relation between the periodic time of orbiting system and its total energy. According to the basic theory of this test, a binary orbiting

---

[272] We should note that the recent research literature on the theory of gravitational waves includes claims of non-linear solutions of the relativistic gravitational waves problem. However, there are many ambiguities, obscurities and controversies about this issue and hence we can say that the currently-accepted theory of gravitational waves is based on the linearized general relativity.

system[273] of two massive objects (e.g. neutron stars) which loses energy by the emission of gravitational waves, should change its orbital period continuously (see Exercises). The famous (and possibly the only) indirect test of gravitational waves by this method is related to the PSR B1913+16 binary pulsar system which consists of two neutron stars of approximately equal mass of about 1.4 solar masses. This binary system was discovered in 1974 by Hulse and Taylor and was subsequently investigated by them thoroughly. According to the results of this investigation, which is largely based on a general relativistic framework, the orbital period of the system should decrease by about 70 μs per year[274] due to the loss of energy by the emission of gravitational waves in excellent agreement with the observations (the reported uncertainty is less than 1%).

We think the significance of this test in endorsing general relativity is highly exaggerated in the literature of general relativity. For example, the current general relativistic theory of gravitational waves is based on the linearized form in the weak field approximation and the applicability of this approximation in this case as well as its endorsement to the non-linearized form can be strongly challenged. We should also mention that there are many sources of theoretical and observational uncertainties (as well as potential circularity) in quantifying this binary system and extracting the reported results. Moreover, these results are partly based on some classical formulations and presumptions where the applicability of some of these should be questioned (because of the strong gravity of this system and/or because of the accumulation of relativistic effects due to violation of tempo-spatial locality). We should also consider the fact that being an indirect test in itself should diminish the value of this test as a decisive evidence for gravitational waves due to the possibility of the involvement of mechanisms other than gravitational waves in the shrinkage of the orbit and the decrease of the orbital period. In fact, this alleged indirect *observation* should at the best be regarded as circumstantial evidence. We therefore believe that this test is at least not as conclusive as it is depicted by general relativists.

Exercises

1. Why should the emission of gravitational waves by a binary orbiting system shorten the orbital period?
2. Assess the claimed evidence of indirect observation of gravitational waves through the detection of changes in the orbital period of binary system.

9.6.2 Direct Observation of Gravitational Waves

Direct observation of gravitational waves is supposedly achieved (or expected to be achieved) by using gravitational wave detectors like LIGO (which is an acronym for: Laser Interferometer Gravitational-wave Observatory) and LISA (which is an acronym for: Laser Interferometer Space Antenna). In this context, we should also mention Virgo which is an Italian interferometer (based near the city of Pisa) that is dedicated to gravitational wave detection and it collaborates with LIGO in gravitational wave observations. These detectors employ Michelson-like laser interferometric techniques. The principle of operation of these detectors is that as the gravitational waves pass by the detector they cause changes in the geometry of spacetime in its vicinity and hence these waves introduce differential changes on the length of the arms of the laser interferometer which cause a measurable change in the interference pattern and its detectable effects.

We should remark that the above-described method of detection by laser interferometric techniques is one of the two main methods for direct observation of gravitational waves. This technique belongs to what can be generically described as "free particle" method of detection. The other main direct detection technique (which belongs to what can be generically described as "elastic body" method of detection) is based on the use of hefty metal bars equipped with very sensitive sensors to detect minute changes in the bar dimensions caused by the passage of gravitational waves. However, the latter detection technique is not sensitive enough to be of significant practical value (at least for the time being) in detecting gravitational

---

[273] Being binary is not necessary, but it is the case in the system(s) used in this test. Anyway, it is a legitimate example and hence it is sufficient to demonstrate the principles.
[274] We note that the rate of change should change with time.

9.7 Black Holes

waves of common origin and hence no credible claim of detection of gravitational waves by this method has been made yet.

Problems

1. Assess gravitational waves as evidence in support of general relativity.

    Answer: In assessing gravitational waves as evidence in support of general relativity we note that there are two aspects of this test: a qualitative aspect which is the mere existence of gravitational waves, and a quantitative aspect which is the measurement of the properties of these waves such as their amplitude and frequency.

    Regarding the qualitative aspect, we note that the endorsing value of this aspect is very limited because as we pointed out earlier the existence of gravitational waves can be accommodated within theories other than general relativity and it was a subject of contemplation and investigation even in classical physics (see § 8.6). So, it is not specific to general relativity and hence it is not conclusive in its endorsement to the theory. In fact, any tempo-spatial gravity theory that accepts the finity of the speed of propagation of gravitational interactions should imply the existence of gravitational waves (in some sense).

    Regarding the quantitative aspect, we note that although the endorsing value of this aspect is significant and could be decisive in endorsing one theory or another, there is no conclusive evidence in support of general relativity from this aspect. The reason is that indirect observations have no quantitative aspect from this perspective[275] (since there is no actual detection of waves to measure their properties and see if they agree with the predictions of general relativity or not) while direct observations have many uncertainties from a quantitative perspective and hence they cannot determine beyond any doubt if the observations agree quantitatively with the predictions of general relativity or not. Moreover, some metric gravity theories can also provide similar quantitative predictions.

    In brief, gravitational waves (whether allegedly observed directly or indirectly) is at the best a qualitative evidence, moreover it is not specific in endorsing general relativity since it can also be explained by other theories. Furthermore, its endorsement to the full general relativistic formulation (rather than linear or semi-classical formulation) is questionable.

Exercises

1. Provide a brief description of LIGO. Also, list some of the alleged observations of gravitational waves made by LIGO.
2. Provide a brief description of LISA.
3. Assess the recent claims of detecting gravitational waves by LIGO.
4. Assess the value of the experimental projects (such as LIGO and LISA) about the detection and observation of gravitational waves.
5. Discuss briefly the "elastic body" method for the direct detection of gravitational waves.

9.7 Black Holes

In our view, black holes are not a conclusive evidence in support of general relativity. In the following points we outline some of our reservations and observations about this alleged evidence:

• Although the existence of black holes is advocated by some physicists as a scientific fact and may be regarded as one of the strong evidence in support of general relativity, their existence is not confirmed beyond any doubt. Moreover, even if the existence of black holes is confirmed this does not apply to many details about their existence and their physical structure and properties. For example, there may be evidence for the existence of stellar black holes and super-massive black holes but there is no evidence for the existence of primordial black holes. Also, there is no evidence for the theoretical models of their structure (e.g. singularity surrounded by event horizon and photon-sphere, etc.). In fact, there are many question marks on many details about black holes in general, the claimed evidence on their existence, their types and their alleged physical structure and properties. Most parts of the physics of black holes

---

[275] In fact, even the qualitative aspect of indirect observations is generally contemplative and has no decisive value.

are based on pure theoretical models (which are generally based on general relativity) and the existence of black holes does not imply that their detailed physics follows these models. Moreover, some of the claimed evidence is motivated by the search for evidence in support of these theoretical models and predictions and hence there are many sources of bias and lack of objectivity and impartiality. Therefore, we should be cautious about the physics of black holes and the claimed evidence on their existence and the alleged circumstantial evidence in support of some details of their physics.

• If there is a conclusive evidence about the existence of black holes then their existence can be explained even by classical gravity, as well as by other gravity theories, and hence they are not conclusive evidence for general relativity. Yes, there are certain consequences and implications about black holes that are specifically derived from general relativity and hence they may provide supporting evidence for general relativity. However, none of these consequences and implications have been proved definitely by experiment or observation and therefore we see no evidence from the physics of black holes to support general relativity specifically and definitely.

Exercises

1. Assess black holes as evidence for general relativity.
2. Assuming Hawking radiation, calculate the power radiated by a black hole of one Earth mass and comment on the result.

9.8 Geodetic Effect

In this section we investigate two claimed evidence in support of geodetic effect of general relativity (see the subsections of this section). Apart from the feebleness of these claimed evidence from a practical perspective, the claimed evidence may also be challenged theoretically on the basis of lack of meaningful physical significance of momentum (and hence its conservation) on global scale within the framework of general relativity. This should cast a shadow on the rationale and significance of the implications of these claimed evidence. However, the challenge may be questioned by the reliance of this effect on its local significance as demanded by parallel transport (although even this can be challenged by questioning the value and significance of parallel transport). We may also add another theoretical reservation that is according to general relativity spinning objects deviate in their trajectories from the geodesic paths of non-spinning objects and this should introduce an error into the general relativistic formulation of geodetic effect which is based on assuming geodesic trajectories. However, this deviation (and hence the error) is negligible when the spinning is not excessive (which presumably is the case in the claimed evidence). We should also note that the calculations in the experiments of this test (see the subsections of this section) seem to be based on ignoring the difference between physical variables and coordinate variables which is important in this case even though the gravitational field is weak and the spacetime is almost flat. This is due to the violation of locality and the accumulation of tiny relativistic effects over extended tempo-spatial interval (similar to the situation in the case of perihelion precession; see § 8.1 and § 9.1). In fact, the violation of locality should magnify even the effect of the aforementioned "deviation from the geodesic paths due to spinning" and make it significant.

9.8.1 Precession of "Moon-Earth Gyroscope" in Motion around Sun

This is an effect predicted by de Sitter in his analysis of the revolution of the "Moon-Earth gyroscope" around the Sun. It is claimed in the literature that this effect is verified in a number of experiments using radio interferometry and laser ranging techniques and the accuracy of the agreement between the experimental results and the predictions of general relativity ranges between 10% and 1%. However, there are many question marks and uncertainties about these experiments and associated claims as well as the insufficiency of the reported accuracy (at least in some cases) to establish the claimed proposition firmly and definitely. Anyway, the general challenges to geodetic effect (as outlined in § 9.8) should be sufficient to question the significance of this test in endorsing general relativity.

9.8.2 Gravity Probe B

This is a space experiment in which 4 gyroscopes on board a satellite orbiting the Earth in a polar orbit are used to test the general relativistic prediction of geodetic effect (as well as frame dragging which will be investigated in § 9.9.2). Despite the common claims that the experiment confirmed the general relativistic prediction of geodetic effect to high level of accuracy (about 1% and later reduced to about 0.5%), there are serious suspicions and challenges to these claims where some respected physicists disputed the accuracy of this experiment (and perhaps even some of its theoretical aspects). We refer the interested reader to the literature about this issue. Anyway, the general challenges (as outlined in § 9.8) should be enough to question the endorsing value of this test.

9.9 Frame Dragging

Before we go through the claimed evidence in support of frame dragging, we should note that there are conceptual and theoretical difficulties about the significance of frame dragging and the interpretation of its formalism as well as its relation to the Mach principle and geodesic motion. Some of these issues (at least) should cast a shadow on the significance and value of some of the claimed evidence in support of frame dragging and its general relativistic foundations. The interested reader should refer to the literature of frame dragging for details. We should also note that the calculations in the experiments of this test (see the subsections of this section) are apparently based on ignoring the difference between physical variables and coordinate variables which is important in this case even though the gravitational field is weak and the spacetime is virtually flat. This is due to the violation of locality and the accumulation of tiny relativistic effects over extended tempo-spatial interval (similar to the situation in the case of perihelion precession; see § 8.1 and § 9.1). Anyway, as we will see there is no credible evidence in support of frame dragging.

9.9.1 LAGEOS Satellites

LAGEOS satellites (which stands for LAser GEOdynamics Satellites) are simple spheres covered in retro-reflectors. These satellites are used for laser ranging from Earth-based stations. It is claimed that the orbital planes of these satellites are observed to change in accordance with the predictions of frame dragging of general relativity. However, these claims are highly disputed because even the lowest estimates of the errors in these observations are too high to be conclusive and hence this test was dismissed even by loyal general relativists.

9.9.2 Gravity Probe B

Apart from the suspicions about the results of the Gravity Probe B experiment in general, the frame dragging effect in this experiment is reported to be below the noise level and hence the result of frame dragging is inconclusive even according to the supporters of this experiment.[276] Also, it may be claimed that if the two effects (i.e. geodetic effect and frame dragging) in the Gravity Probe B experiment are mixed together (as we expect) then the value of the entire experiment could become questionable in endorsing any particular effect. However, we should note that the total gyroscopic shift is made of two components: a component in the polar orbital plane due to geodetic effect, and a component in the equatorial plane (assuming the gyroscope spin points initially in a direction parallel to this plane) due to frame dragging effect (or Lense-Thirring effect) and hence these components are perpendicular and should be distinguishable from each other. Nevertheless, the uncertainty in one of these components could still affect the uncertainty in the other component. In fact, we may require the validity of superposition (or additivity of effects) as well to justify the analysis of the results of this experiment (although this requirement may be questioned for the same reason, i.e. being perpendicular and distinct). Nevertheless, the potential requirement of superposition could cast a shadow on the combined effect and hence on the confidence in the results of the individual effects.

---

[276] To the best of our knowledge, the best estimate of error in the frame dragging effect of this experiment is 15%. The interested reader should refer to the literature for details.

9.10 Wormholes and Other Fantasies

So far, no discovery of a wormhole (or one of the other fantasies) has been announced. However, we should not be surprised if such an announcement is made in the future thanks to the enthusiastic race in endorsing general relativity and the general attitude among physicists toward its validity.

9.11 Cosmological Predictions

As indicated earlier (see for example § 8.11), the cosmological consequences and predictions that are extracted and formulated according to the general relativistic framework are largely based on speculations and contemplations as well as philosophical and metaphysical ideas and hence we do not expect any genuine confirmation of the majority of these consequences and predictions. In fact, some of these consequences and predictions are outside the realm of physical science and hence they are not scientifically verifiable (e.g. many details related to the development of the alleged Big Bang). However, no one should be surprised if alleged confirmations to general relativity from these cosmological predictions emerged in the future. In fact, even now we can find some claims of cosmological evidence in support of general relativity by some enthusiastic relativists.

9.12 Tests of the Equivalence Principle

Experimental tests of the weak equivalence principle may be considered as indirect evidence in support of general relativity. However, most (if not all) these tests are actually tests to the classical equivalence principle (which is undisputed) and hence they have no significance or value in endorsing general relativity. In fact, from a practical perspective the classical equivalence principle contains all the actual physical content of the weak equivalence principle and hence the classical equivalence principle should be able to explain any alleged physical consequences of the weak equivalence principle. In other words, the value added by the weak equivalence principle to the classical equivalence principle is no more than an unverifiable theoretical claim. This should be supported by what we saw earlier (refer to § 1.8.2) that even some general relativists challenge the validity of the weak equivalence principle and this has no meaning other than challenging its theoretical implication and content because no one can question the validity of the classical equivalence principle.

Anyway, even if we accept that these tests are evidence for the weak equivalence principle they are not specific to general relativity as there are many gravitational theories that rest on this principle. In other words, the validity of the weak equivalence principle is not an evidence for the validity of the formalism of general relativity specifically (although it may be claimed to be an evidence for the validity of the family of theories that rest on this principle). In fact, the validity of the weak equivalence principle is not evidence for the validity of a specific theory that is based on this principle (whether this theory is general relativity or another theory in that family) or even the validity of that family in general because wrong theories can rest on correct and valid principles, e.g. a wrong maximization (rather than minimization) theory may rest on a valid variational principle.

In any case, there is no evidence in support of the strong equivalence principle (and possibly there is evidence against it as claimed by some and as will be seen later) despite some unreliable claims, which are mostly based on theoretical arguments, about its validity. In fact, what is critically important and significant to general relativity is the strong equivalence principle (which alleviates the theory from being just a gravity theory to become a "General Theory") because the weak equivalence principle is essentially similar to the classical equivalence principle (which is unquestionable) with some unverifiable and arguable theoretical extensions as we explained earlier.

Exercises

1. Assess the value of the equivalence principle tests as endorsing evidence to general relativity.

9.13 Tests of Special Relativity

As we noted earlier in the preamble of this chapter, the tests of special relativity are considered by some to be tests of general relativity. This may be justified by the claim that "special relativity holds true locally in the spacetime of general relativity" or by the claim that "special relativity is a special case of general relativity or contained in it". However, this may be challenged by the following:

• The locality condition deprives these tests from any value to general relativity as such because as long as we are assuming that we are within the domain of special relativity (even if this is only at local level) then the spacetime is flat and hence there is no significance of such evidence to general relativity as a theory for the entirety of the curved spacetime. This means that any implication or effect that is specific to general relativity cannot be obtained from such tests and hence it cannot be endorsed by these tests. In other words, these tests have no general relativistic content and hence they cannot be supporting evidence to general relativity as such, so they are only tests to special relativity (whether independently or within the general relativity at local level).

• Based on the previous point, any claim of endorsing general relativity by these tests implies duplication since this means endorsing special relativity by these tests once independently and once as a local application to general relativity. In other words, we are using the same evidence to prove the same premise twice and this should have no added value because it is mere repetition.

• The premise that "special relativity holds true locally in the spacetime of general relativity" has only theorized value with no real physical content. In other words, the originator of the theory of general relativity arranged the theory in such a way that special relativity holds true locally without deriving such a premise from the formalism of general relativity as represented by the Field Equations. To be more clear, we have a general-relativity-specific formalism represented by the Field Equations (and their direct consequences) and a special-relativity-specific formalism represented by the Lorentz transformations (and their direct consequences) and these two formalisms are independent of each other apart from the theorized connection (which is rather arbitrary) that is established by the originator of the theory of general relativity. The situation will obviously be different if this connection was a natural connection (i.e. by the convergence of the general relativistic formalism to the special relativistic formalism). Accordingly, even if the above premise is true it does not prove anything about general relativity as a gravity theory as represented by the Field Equations.

• The claim that "special relativity is a special case of general relativity or contained in it" is disputed (see § 1.7 and refer to B4). In fact, this claim can be challenged and rebutted. Moreover, the implications of this premise should depend on the meaning of "special case" or "contained" and this is not obvious or agreed upon. In fact, we may even claim that, according to some opinions and interpretations about the relation between special relativity and general relativity, the use of the special relativistic tests to endorse general relativity is circular since their validity as general relativistic tests depends on their validity as special relativistic tests while their validity as special relativistic tests is supposed to be a requirement (or special case) of general relativity.

In brief, there is no relation between general relativity as a gravity theory and special relativity and hence no test for one theory can be seen as a test for the other. Yes, there is a relation between general relativity as a "General Theory" and special relativity (due to the claim of local application of special relativity in the spacetime of general relativity). However, the physical setting in the two theories is different (i.e. global inertial frames of flat spacetime in special relativity, and local non-inertial frames of curved spacetime which are allegedly equivalent to inertial frames in general relativity) and hence the test for each theory should belong to that theory alone and cannot be considered a test for the other theory. Yes, if we ignore the difference in the physical setting (or abolish this difference by a theoretical rationale like the equivalence principle) then these tests have no general relativistic content and hence they cannot be supporting evidence to general relativity as such (as indicated earlier).

Anyway, even if we interpret the evidence of special relativity as evidence for the validity of special relativity locally in the spacetime of general relativity (as required by the equivalence principle) and hence it is evidence for general relativity as a "General Theory", the generality of this cannot be established (noting that without this generality the evidence is useless). In fact, the claimed evidence against the strong equivalence principle (or at least against its generality) should invalidate this generality and hence the special relativistic evidence cannot endorse general relativity as a "General Theory" due to the lack of generality.

9.14 Circumstantial Evidence

In the literature of general relativity there are claims of many circumstantial evidence in support of general relativity such as the luminosity of quasars (which indicates accretion of matter onto black holes as a viable mechanism for energy release) or the distortion of some spectral lines in certain astronomical observations (which indicates the existence of black holes with certain characteristics). However, all these alleged circumstantial evidence and indications are neither specific to general relativity nor conclusive and definite in their value and significance and hence they do not provide any credible support to this theory.

9.15 Evidence for Newtonian Gravity

The evidence in support of Newtonian gravity in its classical domain is overwhelming and this may be seen as evidence for general relativity as well due to the convergence of general relativity to the Newtonian gravity in the classical limit (see § 5.1). However, this type of evidence is in fact an endorsement to the classical formulation (whether in the Newtonian gravity form or in the converged general relativity form which is a special case of the general formulation of general relativity) and hence it cannot be regarded as evidence for the general formulation of general relativity as represented by the Field Equation. In other words, the evidence in this special case is not a valid evidence for the general case (which is the important thing to establish since the special case of classical gravity is already established regardless of general relativity). In fact, what is significant for the endorsement of general relativity is the evidence for the general formulation which distinguishes general relativity from Newtonian gravity rather than the evidence for this special formulation because the classical formulation is sufficiently endorsed and hence it does not need more evidence. Yes, the convergence of general relativity to the Newtonian gravity (which is a well established theory considering its limitations and restrictions) in the classical limit shows that general relativity is compliant with the requirement of the correspondence principle (see § 1.8.3) and this is not the same as being an evidence in support of the theory of general relativity.

We should also add two other factors that diminish the value of this type of evidence in endorsing general relativity:

• The convergence of general relativity to the Newtonian gravity is not a naturally occurring convergence but it is a planned convergence since the formulation of general relativity is tuned to converge to the Newtonian gravity in the classical limit and hence this convergence has no value from the perspective of endorsing general relativity. In other words, general relativity is formulated in such a way that it converges to the Newtonian gravity in the classical limit (i.e. it is forced to converge). In fact, this planned (or premeditated) convergence can be easily seen from the demonstration (in this book as well as in other books) of how general relativity converges to the Newtonian gravity in the classical limit, e.g. how κ is defined to fulfill this convergence as seen in § 5.1. Yes, if the theory of general relativity was formulated without awareness or regard to the Newtonian gravity and it appeared later that general relativity converges to the Newtonian gravity then this convergence could be significant in endorsing general relativity (conditionally and in this case).

• The formal arguments used to establish the convergence of general relativity (as represented by the Field Equation) to the Newtonian gravity in the classical limit are questionable in many details. In fact, these arguments (whether those given in this book, as presented in § 5.1, or their alike in the literature) which supposedly establish this convergence are based on many questionable assumptions, approximations, twisting, etc. and hence they can be challenged rather comprehensively and systematically.[277]

---

[277] Despite this, we did our best in this book to present the arguments and derivations of general relativity (whether related to the above convergence or to something else) in the most logical and sensible form (although they are usually assessed subsequently in a more realistic style that exposes their potential weaknesses).

9.16 Final Assessment

The claimed evidence in support of general relativity can be classified into several main types (which are not mutually exclusive):

1. Evidence that is not valid at all in its endorsement to general relativity. An example of this is the tests of the equivalence principle since these tests are related to the classical essence of this principle rather than to the general relativistic essence; moreover the evidence for the principle is not an evidence for the theory. Another example is the tests of special relativity since these tests have no endorsing value to general relativity as such. A third example is the evidence for Newtonian gravity as discussed in the last section.

2. Evidence that the theory was initially tuned to match (i.e. the theory was formulated in a way that yields the anticipated prediction). The obvious representative of this type is the precession of the perihelion of Mercury. As this type of evidence does not represent an actual prediction of the theory its value is rather limited, i.e. all the evidence can imply is that the theory is consistent with some observations. In fact, even the evidence for Newtonian gravity may be classified into this type (if it is regraded as a valid evidence for general relativity in this special case).

3. Evidence that is based on simple principles and arguments without the need for the full formulation of general relativity. An example of this type is the gravitational frequency shift or gravitational time dilation which can be obtained from the equivalence principle. The value of this type of evidence is also limited because it is not an evidence for the theory as a whole but it is an evidence for the principles (e.g. equivalence principle) which the evidence is based on. Although some of these principles may constitute essential components or foundations for the formalism of the theory the evidence is not conclusive in endorsing the theory itself. In fact, a survey of the literature of general relativity shows that a number of the consequences and predictions that form the basis of numerous alleged evidence are not necessarily based on the full formalism of general relativity, but they can be obtained from general and simple arguments (some of which are not related at all to the framework of general relativity). In this context, it is important to note that even from a historical perspective several of these consequences have been formulated and predicted (at least qualitatively) before the full formulation of general relativity and the emergence of the Field Equation.

4. Evidence that is questionable in its significance and value since it can be explained by other theories. An example of this type is the evidence that can be explained and interpreted by alternative gravitational theories, e.g. retarded gravitational potential theory whose prediction about the precession of Mercury perihelion is similar to the prediction of general relativity. Also, many instances of gravitational frequency shift (and possibly even gravitational time dilation) fall in this category since they can be explained classically. In fact, we can include in this category almost all the qualitative evidence that are not specific to general relativity such as the existence of black holes since it can be predicted in principle by any valid gravitation theory including classical gravity because there should be no restriction in any valid and sensible gravitation theory on the existence of high concentrations of matter with certain gravitational properties (such as being "black"). This similarly applies to gravitational lensing and gravitational waves. In fact, if we include modern metric gravity theories (which emerged before and after general relativity) then almost all the predictions (whether qualitative or quantitative) of general relativity can be obtained from other theories and hence they are not specific and definite in their endorsement to general relativity. This is unlike, for example, quantum mechanics (and potentially even Lorentz mechanics) where no credible alternative theory can replace it and provide similar or identical predictions and explanations.

5. Evidence that is based on trust. Such evidence cannot be accessed and assessed independently and hence its value is questionable because fraud, bias, mistakes and illusions are commonplace in science. In fact, they could be more common in science than in daily life due to the complexity of science and its special position and prestigious status.

6. Evidence that is questionable in its authenticity as there are indications of falsification or exaggeration. An obvious example of this sort of evidence is the light deflection by the Sun in the 1919 solar eclipse expedition of Eddington and his team where the evidence was sexed up. It is wrong to believe that

this historical incident is a one-off event and it is a rare exception. In fact, there are other examples (or at least potential examples) of this type of alleged evidence (especially in the exotic experiments and observations) but we do not go through the details. As indicated earlier, fabrication, prejudice and exaggeration are as common (if not more common) in science as in daily life.

7. Evidence that is based on simplifications and restricting assumptions in the formulation of general relativity and hence it is based on special conditions and compromising assumptions for the application of the theory. An example of this is some claimed astronomical observations whose implications are based on the use of a particular metric (e.g. Schwarzschild) or a conditional solution or a special case. Such evidence cannot be used to endorse the theory in its entirety and generality. We may also include in this category the evidence from certain gravitational regimes (which are mostly Newtonian or quasi-Newtonian) since the value of such evidence is rather limited with regard to the general validity and applicability of general relativity and its distinctive features against other theories. We may also include in this category the evidence from the linearized form of general relativity (e.g. the common formulation of gravitational waves) since the evidence on the validity of the linearized general relativity is not sufficient to establish the non-linearized general relativity (which is the real general relativity that critically needs evidence).

8. Evidence that is not conclusive due to many uncertainties and large error margins. Examples of this type of evidence include most light bending observations as well as at least some claimed observations of gravitational waves and frame dragging.

9. Evidence that is not independent of other types of evidence such as gravitational frequency shift which is correlated to gravitational time dilation. Such evidence has no value as independent evidence since it targets the same or similar aspect of the theory.

10. We should also note that almost all the claimed evidence in support of general relativity are related to the theory as a gravity theory and not as a "General Theory". Therefore, even if all these evidence are valid and conclusive they just establish the theory (or rather part of it) as a gravity theory and not as a "General Theory".

11. The overwhelming majority of the claimed evidence is based on the Schwarzschild metric and this should diminish the value of the evidence due to the Schwarzschild restriction. For example, the obtained results could be specific to this metric. Moreover, if this metric (or its equivalent) is obtained from another theory or argument then the evidence will not be conclusive in its endorsement to general relativity specifically.

To sum up, the claim that the theory of general relativity is supported by overwhelming and conclusive scientific evidence or it is proved beyond any doubt should be rejected. All the claimed evidence (even with the exclusion of the possibility of fabrication, exaggeration, error, etc.) do not provide full endorsement to the theory in its entirety and over its whole domain of validity. In fact, we may even claim that there is evidence against at least certain aspects and implications of general relativity such as the validity of the strong equivalence principle (and hence the validity of the "General Theory"). In brief, if we compare the evidence in support of general relativity to the evidence in support of other respected theories of modern physics, such as quantum mechanics and Lorentz mechanics, we can conclude that the evidence for general relativity is feeble and limited.

Problems

1. What is the strongest evidence in support of general relativity? Assess its significance.
   Answer: Such a question has no definite answer since it depends on personal views and many questionable and controversial factors. However, in our view the strongest evidence in support of general relativity is seemingly the perihelion precession of Mercury. This may also apply to other claimed evidence of precession in orbital motion, e.g. the evidence of periastron precession in some binary pulsar systems.
   As we saw, this evidence can also be explained by other theories (e.g. retarded gravitational potential as well as other metric gravity theories) and hence it is not definite in its endorsement to general relativity specifically. Moreover, the theory of general relativity was tailored to match this prediction and hence the value of this evidence should be diminished by this factor (although this may apply only

to the precession of Mercury). We should also mention that there are technical challenges to the theoretical prediction of general relativity where the effect of the curvature of spacetime is partly ignored in the formulation and calculation (e.g. by ignoring the difference between physical and coordinate variables which should be important due to the violation of locality and accumulation of tiny effects; see § 8.1 and § 9.1). So in brief, even this "well established" evidence is not really well established as claimed.

Exercises

1. Discuss the main types of evidence of general relativity and assess their value and significance.
2. List some general factors that could diminish the value of the evidence in support of general relativity.
3. Our assertion that the predictions that are not specific to general relativity (by being classically explicable for instance) have limited endorsing value may be challenged. Discuss this issue.
4. Why factors like fraud and error should be considered more seriously in assessing the evidence in support of general relativity?

9.17 Evidence against General Relativity

The investigation in the previous sections of this chapter was largely about the evidence in support of general relativity. Regarding the opposite evidence, we believe there are evidence against general relativity especially in its "General Theory" side. For example, the indications about violations of the strong equivalence principle and its general validity or the indications about violations of the relativistic paradigm of speed limit (i.e. c is the specific speed for massless objects and it is the ultimate speed for any physical object) should undermine general relativity and put serious question marks on its general validity. Some of these evidence and indications against general relativity will be investigated in the next chapter.

Chapter 10
Challenges and Assessment

In this chapter we investigate a number of challenges and criticisms to the theory of general relativity. We will also try to assess general relativity from a general perspective. In fact, we discussed throughout the book many detailed challenges and criticisms to general relativity and assessed the theory from different aspects and perspectives. However, most of those were specific and of technical nature. So, in this chapter we outline some general challenges and criticisms to this theory, its epistemology and methodology with an assessment of a general and non-technical nature (noting that the contents of this chapter are generally based on the contents of the previous chapters).

10.1 Challenges and Criticisms

There are many theoretical and practical challenges and criticisms to general relativity; most of which are by its opponents and some are even by its proponents. In the following subsections we list and briefly investigate a number of examples of these challenges and criticisms (noting that some of these challenges and criticisms are specific to this book and hence they represent the view of the author specifically).

10.1.1 Limitations and Failures of the Equivalence Principle

Because the equivalence principle is at the heart of general relativity and one of its most fundamental propositions, any limitation or failure related to this principle is a major blow to the theory and its theoretical and epistemological framework. In this regard we note the following:
• There are fundamental theoretical challenges to the equivalence principle. In fact, the equivalence principle is challenged and rejected even by some of the followers of general relativity (see for example Synge in the References) due to the lack of rationality as well as physical evidence. For example, the equivalence principle (at least in its strong form) can be challenged by the possibility of the dependence of physical laws on the existence (and non-existence) of gravitational field gradients and tidal forces (regardless of being detectable by our observation or not). Moreover, even if we assume that the apparently-intuitive theoretical arguments in support of the equivalence principle can establish the weak form, they cannot establish the strong form. In fact, using the locality as a basis for having good quantitative approximation may be valid with respect to the weak form but it cannot be used to establish the strong form since the existence (and non-existence) of gravitational field gradients and tidal forces may result in a qualitative difference between gravity and acceleration and this qualitative difference may invalidate the strong form altogether, i.e. the physical system in which field gradients and tidal forces exist can be totally different in its behavior from a similar physical system in which no field gradients and tidal forces exist.[278]
• There is no real evidence for the validity of the equivalence principle (neither the weak nor the strong). This is because all the claimed evidence for the weak equivalence principle are actually for the classical equivalence principle with no evidence that can be claimed to support the theoretical content of the weak equivalence principle (see § 9.12). The lack of evidence for the strong equivalence principle is more obvious. As we indicated earlier, there is no evidence in support of the strong equivalence principle even according to the view of some general relativists. In fact, we may even claim (with some general relativists) that there is evidence against the strong equivalence principle (some examples will be given later).[279]

---

[278] In fact, the difference between gravity and acceleration (in the context of the equivalence principle in general) could be in their nature and hence it is more fundamental than by being having or not having field gradients and tidal forces.

[279] In this context, it is useful to be aware of the opinion of some general relativists about the equivalence principle as expressed in the following quote (which is taken, with some modifications, from lecture notes by Sean Carroll): "the principle of equivalence serves as a useful guideline but it is not a fundamental principle of nature and hence this

10.1.1 Limitations and Failures of the Equivalence Principle

- Apart from the above physical challenges to the equivalence principle (whether weak or strong), we may also mention some epistemological challenges to this principle. Some of these challenges are discussed in the Exercises.
- There are several sources of limitation and ambiguity in the equivalence principle. For example, to what type of gravitation rotating and decelerating frames correspond? And what role should be attributed in this context to absolute frame (see for example § 6.5)?
- The huge mess in the literature of general relativity about the equivalence principle (definition, classification, implication, interpretation, validation, etc.) discredits many of the claimed aspects and significance of this principle and degrades its integrity and credibility. In fact, the equivalence principle is one of the best examples in modern physics for controversies, contradictions, arbitrariness, and chaos. Most of these differences are based on lack of clear vision and consistent logic in dealing with this thorny issue where inherent ambiguities in the theory itself and numerous complexities in its theoretical framework allow many to dictate their personal views and validate their own results by alleged tests and verifications many of which are scientifically baseless.
- The locality condition should mean (and it is actually interpreted by some relativists as such) that the domain of the physical experiments should be restricted to the "local" region of the space, i.e. the experiments must be isolated from any influence or effect or consequence related to objects and fields outside that region. This restriction should further degrade the value and significance of the equivalence principle because if we remove this artificial restriction then the outcome of an experiment in a gravitational field could be totally different from its outcome in an equivalent accelerating frame. For example, an electric charge at rest in a gravitational field does not radiate while an identical electric charge in an equivalent accelerating frame should radiate.[280] In fact, examples like this (and there are many) should invalidate the strong equivalence principle altogether (or at least some of its implications and significance).
- As indicated earlier, it is quite possible that there is a qualitative difference between gravitation and acceleration due for example to the existence of field gradients and tidal forces in the former but not the latter.[281] This should invalidate any argument in support of the equivalence principle as a practically useful guiding rule (as some general relativists claim despite their doubts about the principle as a rigorous rule).
- As indicated in a previous point and can be understood from previous and upcoming examples, the equivalence principle is a local rule. This does not only mean that the equivalence principle is a local rule of limited validity but it also means that even if we extend the locality of the rule slightly it could fail (i.e. it could fail even in a sufficiently extended local neighborhood). The implication of this "confined" or "restricted" locality is that this alleged principle is an entirely artificial principle and hence it has no real physical value or significance. In other words, to make this principle valid and functioning we should create or determine or define our physical system in such a way that makes this principle valid and functioning (and this is really ridiculous).

Exercises

1. Discuss the failure of the alleged equivalence between inertial and gravitational mass in achieving this equivalence.[282]
2. According to the equivalence principle of general relativity, a stationary observer relative to a source of gravity (e.g. a terrestrial lab) is equivalent to an accelerating observer. Analyze and discuss this proposition (which should lead to another failure of the equivalence principle).
3. As part of the extension of special relativity to include the domain of general relativity, an accelerating frame should be represented by a series of instantaneous inertial rest frames (IIRF) where special

---

principle is not rigorously true".

[280] We should note that even the locality restriction may not be able to address this challenge.

[281] As indicated, the existence of field gradients and tidal forces is just a possibility and example; otherwise the difference between gravitation and acceleration could be more fundamental and hence it does not need even the existence of field gradients and tidal forces (or any other reason) for its existence.

[282] As discussed earlier (see § 1.8.2), the alleged duality of inertial mass and gravitational mass is a general relativistic issue and not a classical issue since in classical physics mass is mass whether it is in Newton's second law or in Newton's gravity law. Accordingly, this exercise is about a challenge to general relativity specifically.

relativity applies in these frames. Discuss and assess this.
4. Analyze and assess the local application of special relativity in the spacetime of general relativity according to the strong equivalence principle.
5. Give an example of a limitation of the equivalence principle related to the local application of special relativity in the curved spacetime of general relativity.

10.1.2 Necessity of Metaphysical Elements

The necessity of hypothesizing the existence of dark matter and dark energy to explain certain physical formulations and observations should be considered as one of the failures of general relativity. In fact, the requirement of dark matter and dark energy is a failure not only to general relativity but to any theory that requires such dark objects in its theoretical framework or in its practical applications. Generally, dark objects are metaphysical objects even if they are incorporated in an ostensible physical theory. We should note that although dark matter and dark energy are not built in the basic theoretical structure of general relativity,[283] they are strongly associated with it and its interpretations and applications (especially in cosmology) and for this reason general relativity can be criticized for its potential and aptitude to allow and accommodate such non-scientific concepts and entities. We should also note that dark matter may be legitimate in some cases where there is an indirect evidence for the existence of such matter independent of the theoretical necessity. In fact, we briefly discussed this issue earlier (see for example exercise 4 of § 3.3).

We may also include (by extending the meaning of metaphysics) the incorporation of non-physical concepts in physical theories such as the concept of singularity (e.g. of black holes and Big Bang) or the creation of spacetime in no-time and no-where or the expansion of the Universe (during and following the Big Bang) with no center, no-time and no-where. Although these concepts and ideas may be legitimate and meaningful in mathematics or philosophy or religion they are not legitimate in physical sciences because they have no scientifically verifiable meaning or content (or rather they have no scientific meaning at all). In fact, most of these non-physical elements originate directly or indirectly from fundamental principles and paradigms in the theoretical framework and epistemology of general relativity such as merging and incorporating spacetime into the fabric of the physical phenomena (or the other way around) as expressed by the principle of metric gravity (see § 1.8.4) instead of treating spacetime as a container in which physical phenomena take place.

Exercises

1. Outline the problem of dark matter.
2. Argue against the existence of dark matter.
3. It may be claimed that the existence of dark energy can be justified by the "observed" expansion of the Universe and hence dark energy has physical justification and evidence. How do you address this claim?

10.1.3 Creation Theory

It may be claimed that general relativity is a creation theory (and hence it contains non-scientific elements) due for instance to the fact that it does not conserve energy. As indicated earlier (see § 1.9), a creation theory cannot be totally scientific theory because creation requires supernatural elements and agents and hence it belongs to religion or philosophy, for example, rather than science.[284] However, the failure of energy conservation at global scale does not require creation if it is related to the relativity of time and frequency for instance.[285] Yes, the creation of dark energy is required in some relativistic cosmological

---

[283] This should be more true with respect to dark matter because dark energy is related to the cosmological constant which is built (although not necessarily) in the formulation of general relativity as represented by the Field Equation (see § 3.3).
[284] We note that "creation" in this context is more general than "creation" and "annihilation" in their non-scientific implication and significance.
[285] As discussed earlier (see for example § 7.9), since time is relative the principle of conservation of energy should lose its significance at global level. This is because the relativity of time means there is no global significance to time or any

models and this could be considered as an implication of general relativity and hence general relativity can be classified on this basis as a creation theory because it (at least) allows creation. There are other creation aspects in the applications and interpretations of general relativity (rather than in the theory itself) such as the creation of spacetime in the Big Bang according to relativistic cosmology (or some of its accepted premises) and hence general relativity should take the responsibility for accommodating such creation aspects in its applications even if we assume that general relativity itself is not a creation theory. In fact, the entire Big Bang theory (which is a fundamental pillar in relativistic cosmology or at least it is strongly associated with it) is effectively a creation theory (at least because it implies creation of space and time although the Big Bang itself may not be necessarily so if we follow a certain interpretation which may be different to what is commonly accepted).

Exercises

1. Discuss the issue of conservation of energy in general relativity and its applications.
2. The necessity of the conservation of energy principle in physical theories may be challenged. What is your argument against this challenge?
3. Assess the assertion that the breakdown of the conservation of energy at global level in general relativity can be seen as a logical consequence of the breakdown of global time since time in the spacetime of general relativity is well defined only locally.

10.1.4 Dependence on Special Relativity

It may be claimed that the dependence of general relativity on special relativity makes general relativity susceptible to questioning if special relativity (or Lorentz mechanics) proved to be wrong or incomplete. So, the challenges to special relativity may also be seen as challenges to general relativity because special relativity is supposed to be at the foundation of general relativity. This can also be justified by the local application of special relativity in the spacetime of general relativity. However, this may not affect general relativity as a gravity theory because the gravitational formulation of general relativity (as represented by the Field Equation and its consequences) is not based on the formulation of special relativity (as represented by the Lorentz transformations and their consequences). In fact, there is no direct or formal link between the two formulations (see § 1.7 and § 9.13 and refer to B4). Moreover, the local application of special relativity in the spacetime of general relativity may not be entirely equivalent to the global application of special relativity in the Minkowski spacetime, and hence the failure of special relativity (as such) may not necessarily imply the failure of its local application in the spacetime of general relativity. However, this should depend on the nature and cause of the failure and hence the failure of special relativity could lead to the failure of its local application in the spacetime of general relativity and to the collapse of the "General Theory". So overall, the failure of special relativity should put a question mark on the validity of general relativity because special relativity (in one sense or another) is at the foundation of general relativity and is strongly related to it.[286]

Exercises

1. Give an example of the challenges to special relativity that potentially threaten general relativity.

10.1.5 Triviality of General Invariance

The requirement of invariance of physical laws (i.e. the necessity of expressing the physical laws in tensorial form) which is considered as one of the great achievements of general relativity is just a trivial claim or convention with no actual physical substance or implementation. In this regard, we note the following about the requirement of invariance (or general covariance) in general relativity:
• As we discussed earlier (see § 1.8.1) the principle of invariance is no more than an epistemological convention or demand with minor scientific content and significance.

---

physical quantity that depends on time and determines energy such as frequency.

[286] We should also distinguish in this context between the failure of special relativity (as an epistemological theory of mainly interpretative significance) and the failure of Lorentz mechanics (as a scientific theory based on the formalism of Lorentz transformations).

- Even if we accept the physical necessity (rather than the epistemological necessity) of the invariance of physical laws it should mean the necessity of the invariance of the essence and actual content of the physical laws and not the necessity of formulating the laws in a particular form (e.g. tensorial form) although such a formulation should be beneficial to the objectives of invariance.
- There is no actual implementation of this proclaimed general invariance in the theory of general relativity because this alleged invariance is not more than a demand. Also, the validity of this general invariance is not shown in the theory or proved by it in detail and as applied to actual laws (as done for example in Lorentz mechanics).
- The tensors in special relativity are Lorentz invariant, i.e. they are "Lorentzian tensors" and not general tensors. The "General Theory" is based on generalizing the formalism of special relativity to apply locally in curved spacetimes and this requires the existence of valid general tensorial (covariant) transformations which is not guaranteed (in its generality as a minimum). At least, this existence requires demonstration and verification.

Exercises

1. The principle of invariance is no more than an epistemological convention with minor scientific content and significance. Discuss this proposition.
2. Distinguish between the invariance of laws in essence and the invariance of laws in form.
3. Clarify the assertion about the triviality of the alleged general invariance (or general covariance) in general relativity.
4. Justify the triviality of the *Principle of General Covariance* (see Problems of § 1.8.2).

10.1.6 Interpretation of Coordinates

As we saw earlier (refer for example to § 6.2), the interpretation of coordinate variables as labels with no metrical significance is not intuitive. In fact, this may be intuitive in abstract mathematics and geometry but not in physics. Moreover, this leads to practical problems in application where data representing physical systems or extracted from experiment and observation are metrical and hence they cannot be used in the formulae whose variables are supposed to be coordinate symbols of mere labeling value. In fact, the problem of the interpretation of coordinate variables in general relativity extends to many other physical quantities and principles (which depend on coordinates in their definition and determination such as energy and energy conservation) where they lose their physical significance or require special or tentative interpretation (refer to § 7). This does not only complicate and hinder the interpretation and understanding of the physics but it also introduces traps, hurdles and complexities in the application of the formalism (see Exercises). The reader should finally note that "interpretation of coordinates" in this criticism does not only mean attaching a conceptual meaning to coordinates but it also means treating them in reality as such (i.e. stripping off their physical and metrical significance in the formalism and implementation).

Exercises

1. What is the root of the idea that coordinates are labels with no metrical value?
2. List some of the problems caused by the interpretation of coordinate variables as labels with no metrical significance (which originates from employing the paradigm of curved spacetime).
3. It may be claimed that "As long as the coordinates are not measurable quantities then the actually-measurable physical quantities cannot be correlated in reality to these coordinates. In other words, the actual link between coordinate quantities and physical quantities is missing in reality because this link is theoretical rather than practical". Discuss this claim.

10.1.7 Strong Dependency of Physical Results on Coordinate System

A distinctive feature of general relativity (due to its highly mathematical nature and its geometric approach) is the strong dependency of the obtained physical results and consequences on the employed coordinate system (as can be easily seen in the difference between the predictions based on the use of

Schwarzschild coordinates and those based on the use of other coordinates such as Kruskal-Szekeres coordinates) where this strong dependency can sometimes lead to contradictions and absurdities.[287] In fact, even if we accept the standard defence of general relativists of blaming the coordinates for obtaining conflicting and absurd results, we cannot trust such a theory that can abuse the coordinates and make them dysfunctional to such extent. This is because we cannot be sure if the results that look rational and are not in conflict with other results obtained from other coordinates are reliable since they can also be faulty mathematical artifacts of the employed coordinates even though we are unable to detect this by comparing these results to "more rational" results obtained from other coordinates (especially when results from other coordinates do not exist to compare with).

Exercises
1. Try to find a link between the issue in this subsection (i.e. strong dependency on coordinates) and the issue in the previous subsection (i.e. interpretation of coordinates).
2. Give an example from the literature of general relativity about the strong dependency of the physical results on the type of coordinates used in formulating and solving physical problems. Also, discuss possible challenges and defence to this criticism.

10.1.8 Local Relativity

In special relativity we have "global relativity" of physical quantities represented by the difference of these quantities (such as time and length) across frames of reference where each one of these frames is global since it covers the entire spacetime. Thus, although we have for example length contraction and time dilation (which make length and time relative across different frames) these physical quantities are still global in each frame and hence there is no problem in the application of the formalism in each frame since the physical quantities are uniquely defined and determined over the entire frame and spacetime. In other words, although "relativity" introduces some ambiguity on the physical quantities (since they differ between frames) the global nature of this relativity (i.e. being across global frames) makes this ambiguity harmless (at least in most cases) because what is largely needed is the application of the formalism in each individual frame (and the results can be transformed unambiguously across frames when this is needed). This means that physical quantities in the spacetime of special relativity are actually absolute as far as each frame in concerned since they are defined globally over the entire (global) frame and spacetime. For example, if we have a physical formula that involves length then when we apply this formula in a given frame the length is defined and determined uniquely and unambiguously over the entire frame and spacetime and hence we have no problem if this length extends over different patches of the spacetime (or even over the entire spacetime) because the length in any patch is the same as the length in any other patch. Similarly, we have no problem if the length in one patch of spacetime is correlated to the length (or length-related quantity) in another patch because length is uniquely defined and determined over the entire spacetime.

The situation in general relativity is quite different (thanks to the spacetime curvature and the non-metrical significance of coordinates which affects the significance of physical quantities and makes them dependent on the local metric properties) because what we have in general relativity is local frames where each one of these frames covers only certain patch of the spacetime (and hence the relativity in general relativity can be described as "local relativity" since it occurs across local frames coordinating limited patches of the spacetime). Now, due to the curvature of spacetime and the metrical difference between different patches, physical quantities like time and length have no global significance since they are relative across the different patches of spacetime and hence they cannot be defined and determined uniquely across these patches. So, when we have a physical formula that involves length for example and this length extends over different patches (e.g. distance between galaxies) and possibly across the entire spacetime then there is an inherent ambiguity in the application of this formula due to the inherent

---

[287] This may be exemplified by the previously-discussed issue of the time of fall of an object to the singularity or event horizon of a black hole (as seen by an observer at infinity) where this time can be finite or infinite depending on the employed coordinates. It should be obvious that we are talking above about the Schwarzschild solution in these coordinates (i.e. Schwarzschild and Kruskal-Szekeres).

ambiguity of global length in such curved spacetime. In fact, this could be one reason why in general relativity we have different definitions of physical quantities such as ruler distance and radar distance. This ambiguity should affect the uniqueness of any obtained result as well as its interpretation and physical significance. It should also introduce practical difficulties in applying the general relativistic formulae that involve this type of physical quantities (e.g. how to obtain the physical radius of the Earth or the Sun when needed in a formula where integral techniques for example are not feasible or sensible since integration in such situations means sum of quantities defined and determined differently over patches of different metrical properties).

In this context, we should remark that integration (which is based on the notion of "sum") in curved space loses its ordinary meaning (as understood in flat space) since physical quantities (whether scalars or non-scalars) at different locations cannot be added as in flat space because they are measured and metricized differently. In fact, this should explain why physical laws in general relativity are generally differential (rather than integral) relations. It should be obvious that associating (such as adding, subtracting, inner multiplying, equating, correlating by $>$ or $<$, etc.) physical quantities (whether scalar or non-scalar) at different locations is strictly sensible only in flat space where there is a global and universal reference for defining, metricizing and calibrating these quantities. Accordingly, associating physical quantities at different locations in curved space is actually a relaxed or sloppy form of association in which paradigms like tangent space, parallel transport and local standards of calibration are used.

Exercises

1. Can you give an example of a physical quantity in the spacetime of general relativity that is absolute and global? What is the significance and implication of this?

10.1.9 Nonsensical Consequences and Predictions

A brief inspection of the literature of modern physics in the last few decades reveals that general relativity is a major source and fertile ground for growing fantasies and illusions and breeding bizarre ideas such as wormholes, time machines and dark energy. The methodology of general relativity has also contributed to the incubation of more ridiculous theories such as the string theory in its various forms. In fact, general relativity should be given the main credit for transforming physics from its rational classical methodology to its irrational modern methodology. This should point out to a fundamental defect in this theory that leads to such disastrous consequences (or at least there should be something defective in its methodology, philosophy and epistemology).

Exercises

1. Give some examples of some nonsensical general relativistic results that violate certain physical rules and concepts.
2. It may be claimed that general relativity is not different from classical physics in having nonsensical consequences and predictions because even in classical physics we may obtain nonsensical results (in the form of non-physical solutions such as imaginary roots) from legitimate classical formulations. Address this challenge.

10.1.10 Incompatibility with Quantum Mechanics

It is well known that general relativity and quantum mechanics are not compatible with each other and this incompatibility hindered the development so far of a comprehensive and viable theory of quantum gravity. This is one example of the many inconsistencies in modern physics that justify a call for revision and change. The least that can be said about this incompatibility is that even if both theories are fundamentally correct and valid in their domain of application, there could be an intrinsic incompatibility in their formalism and methodology[288] that prevents merging them into a single consistent theory, and this is not a trivial defect that can be fixed by just introducing minor or *ad hoc* changes here and there.

---

[288] A likely example for incompatibility is the use of curved spacetime in general relativity and flat space-time (or spacetime) in quantum mechanics (and its derivatives).

Regardless of what is right and what is wrong (and regardless of even the existence of right and wrong), it seems necessary to discard at least one of these theories to sustain the natural development and healthy progress of modern physics. In our view, if we have to keep one of these theories we should keep quantum mechanics because it is the theory that is well supported by experiment, or at least it is the one that is more supported by experiment. A brief look around us and an abrupt inspection to our daily life will reveal that quantum physics is everywhere while general relativity is almost no where (at least because all the alleged practical applications of general relativity as a gravitational theory can be equally explained by classical gravity while the alleged non-gravitational applications are connected directly to Lorentz mechanics which is an independent theory).[289] In fact, a broad familiarity with general relativity is sufficient to reveal that this theory is largely a sort of intellectual luxury that is enjoyed by an elite of theoretical physicists, mathematicians, philosophers and cosmologists most of whom are completely detached from reality. This should be enough to make our choice and move forward.

However, we should remark that this does not mean that quantum mechanics is a perfect theory. In fact, there are many epistemological problems and difficulties in quantum mechanics. So, despite its huge empirical success quantum mechanics is not an ideal theory. In other words, quantum mechanics may provide very reliable rules from a practical perspective, but it is not ideal from a theoretical and epistemological perspective. Anyway, quantum mechanics is not the subject of the present book and hence we have no reason to discuss these issues.

As we noted earlier, quantum mechanics (and all its derivatives such as quantum field theory) does not need the paradigm of curved spacetime and hence it is in harmony from this perspective with other branches of physics and this should make any potential unification of quantum mechanics with other theories viable and consistent from this perspective. This is unlike general relativity which is based on the paradigm of curved spacetime and this paradigm (which distinguishes general relativity from other major physical theories) can be a potential cause for hindering the unification. We should also note that we have reservations about the current approaches about quantum gravity and the unification of physical theories and the nature of debates and suggested methods in dealing with quantum gravity in modern physics. However, these issues are not within the scope of this book.

10.1.11 Gaps, Ambiguities and Question Marks

General relativity contains gaps, ambiguities and questionable issues in its theoretical structure and hence the position of the theory about many important issues is problematic or not clear cut. For example:
• What is the position of the theory when the acceleration is not uniform (i.e. what rules should apply, e.g. Lorentz rules or something else) and to which gravity this situation corresponds? This mainly originates from the many restrictions and conditions that are embedded within the equivalence principle and this should propagate to the formalism of the theory and its foundation.
• What is the position of the theory about decelerating and rotating frames and to which gravity they should correspond according to the equivalence principle and hence what physical laws should apply in these frames (see exercises 23 and 24 of § 1.8.2)? In fact, this should also be linked to the more fundamental issue of absolute frame (which is denied by general relativity or at least by general relativists) where the theory is required to address this issue and all its consequences (such as the effect of the existence of absolute frame on the equivalence principle; see § 6.5) properly and seriously. To the best of our knowledge, there is no proper and profound investigation in the literature of general relativity to these key issues (let alone viable solutions).
• What is the position of the theory when the gravitated object is not a test particle and hence it modifies the gravitational field of the gravitating object substantially? This arises from the fact that in this case the distribution of matter depends on the curvature of spacetime while the curvature of spacetime depends on the distribution of matter and this is a form of circularity (rather than a form of non-linearity as presented in the literature although it could lead to non-linearity). Accordingly, in highly complex physical situations

---

[289] Even if global positioning system is really an application of general relativity (considering the many problematic issues about this as discussed in § 9.3.4), it cannot be compared (neither qualitatively nor quantitatively) to the applications of quantum mechanics which enter in every detail of our daily life in a spectacular manner.

10.1.11 Gaps, Ambiguities and Question Marks

(where the distribution of matter is complicated and where gravitating and gravitated objects cannot be defined and distinguished from each other) the application of the theory and its strategy is problematic (at least because the paradigm of geodesics is not feasible). This means that sensible application of the theory is limited to rather simple physical situations and cases (where test particles follow well defined geodesics of a curved spacetime created by a gravitating object or source of gravity). It is noteworthy that in the literature of general relativity there are some proposed fixes to this problem but they are not convincing or general; moreover, in most cases they are not practical.

- What is the position of the theory when the gravitational frame is not in a state of free fall and hence special relativity should not apply even locally (see for example exercise 3 of § 6.1 and the Problems of that section)?[290] Some general relativists suggest that fictitious forces should be assumed in such a frame in order to account for the effects of gravity (and hence we may still be in need for the classical paradigm of gravitational force although it is classified as fictitious).[291] Other general relativists seem to suggest that a law that holds true in the absence of gravity (i.e. it is a valid special relativistic law) should also hold true in the presence of gravity. But this on its face (apart from the lack of observational evidence in its support and hence it is a baseless claim) is inconsistent with the equivalence principle where special relativity should hold (locally) only in freely falling frames (because otherwise the effect of gravitational acceleration should be null). A more rational suggestion (which may be an elaborate version or clarification of the latter suggestion) claims that to account for the effect of gravity in stationary gravitational frames (or non freely falling gravitational frames to be more general) we should first apply special relativity in the local (freely falling) inertial frame and then we transform the laws to the stationary (or non freely falling) frame by a general transformation where this transformation is justified by the equivalence principle. In fact, this is inline with the *Principle of General Covariance* (see Problem 3 of § 1.8.2) and it seems to be the most logical and consistent stand (within the framework of general relativity) in addressing the issue of stationary (and non freely falling) gravitational frames. However, apart from the dependency of the validity of this on the validity of the strong equivalence principle which is questionable (due to the lack of evidence in its support or even the existence of evidence against it), the logical foundations of this can be questioned and challenged. Moreover, ambiguities and question marks can still be found in regard to this type of frames and their physical status.

- Why some forms of energy (e.g. electromagnetic) can be a source of gravity while other forms of energy (e.g. gravitational waves) cannot (at least according to the view of some general relativists). Apparently, this originates from the bizarre treatment of gravity in general relativity as a curvature of spacetime rather than a physical phenomenon in spacetime and hence gravitational waves (which are supposed to be distortions of the fabric of spacetime) cannot be a source of gravity although they are essentially a form of energy since they can carry energy and momentum like electromagnetic waves. So, we simply deprive gravitational waves of the qualification of being a proper source of gravity only because we modeled them in a bizarre way and not because of a real physical necessity or requirement or an intrinsic difference from other forms of energy. In brief, we need clear and rigorous criteria, as well as justification, for what forms of energy can be a source of gravity and why being a source of gravity is restricted to these forms and why other forms are excluded. An issue that is closely related to the issue of which form of energy can be a source of gravity is the problem of the contribution of the binding gravitational energy to the gravitational mass of gravitating objects. This problem is similarly confused in the literature and requires clarification by general relativists.

---

[290] Referring to exercise 8 of § 1.8.2 we may clarify this by saying: from a dynamic perspective freely falling frame in gravitational field is equivalent to inertial frame in free space and hence it is subject to the laws of special relativity, but from a static perspective stationary frame in gravitational field is equivalent to accelerating frame in free space and hence it is subject to what laws? Is it subject to the laws of special relativity in the instantaneous inertial rest frames? If so, then what is the evidence for this (noting that there is in fact evidence against it since the physics in an accelerating frame is not necessarily the same as the physics in the series of its instantaneous inertial rest frames; see for example exercise 3 of 10.1.1)?

[291] We should note in this context that Newton's second law for example is not valid in gravitating frames (i.e. frames not in free fall) unless we assume fictitious forces (see § 6.1). We should also note that we may accept the gravitational component of general relativity but we may not accept the generalizations of the equivalence principle where Lorentz mechanics is extended to apply locally in all frames as if they are inertial (even if the conventional procedure of general covariance is followed).

- There are also many ambiguities, question marks and theoretical disparities about many details and technicalities like gravitational length contraction (refer to § 8.5 and § 9.5). This is also true with regard to the invariance of physical laws and the type of the required transformations between different frames as discussed earlier (see for example § 10.1.5). In brief, there are many ambiguities and uncertainties about these issues and their alike.
- Whether energy conservation principle (as well as other similar conservation and non-conservation principles; see for example exercise 1 of § 10.1.3) is valid in general relativity or not and in what sense (e.g. local or global). There are different opinions and attitudes about this issue where all these opinions and attitudes claim to represent the stand of general relativity. We clarified our opinion (which is based essentially on analyzing the formalism, epistemology and applications of the theory itself) about this issue in several places (see for example § 7.9 and § 10.1.3), and we do not need to repeat.

It is noteworthy that the above are just a few examples and thus there are many other examples of the lack of clarity (and even consistency) in the theory about important issues and this is a main source of controversy about the theory and its application and interpretation even among its followers (see § 10.2.5). In fact, there is no field in physics that contains such chaos of conflicting opinions and interpretations as the relativity theories (and general relativity in particular) even among the top experts in this field and this should reflect the lack of clarity in the theory itself (and indeed even lack of consistency since a theory that is capable of accommodating and accepting all these contradicting views and interpretations should be inconsistent or at least its consistency should be questionable).

Exercises
1. Discuss and assess the issue of whether the gravitational field itself contributes to the source of gravity (as represented primarily by the distribution of matter and energy) or not and whether gravitational forms of energy (such as gravitational waves) make such a contribution or not.
2. According to the *Principle of General Covariance* the effect of gravity on a physical system can be assessed by the procedure described in the Problems of § 1.8.2 (also see exercise 3 of § 3.5). Accordingly, the problem of stationary local frame in a gravitational field (or non freely falling gravitational frame to be more general) is not a gap or ambiguity or questionable issue since the effect of gravity can be accounted for by this procedure. How do you respond to this?
3. Discuss briefly the gaps (and potential gaps) in the investigation of the principles of invariance and equivalence (and their combination as represented by the *Principle of General Covariance*) in the literature of general relativity.

10.1.12 Over-Mathematization of Physics

General relativity might be criticized for its over-mathematization of physics. This should be obvious because general relativity geometrizes gravity which is a physical phenomenon. The danger of over-mathematization is that it naturally leads to illusions and fantasies since mathematics is full of artifacts and abstract objects that do not correspond to physical reality (e.g. we have imaginary solutions, singularities, infinities, senselessly-negative entities like mass, etc.). So, even if general relativity is essentially correct it could become dangerous from a theoretical perspective since it can lead to illusory beliefs about the physical world. In fact, we can see this in reality where general relativity generated and incubated many fantasies and pseudo-scientific theories thanks to its heavily mathematized approach. Over-mathematization also leads to complexities which consume precious resources and hinder progress and development. For example, solutions (whether analytical or numerical) are very difficult (if not impossible) to obtain except in very simple and rather trivial cases.

10.1.13 Violation of Sacred Rules

General relativity may be criticized for its lack of respect to fundamental physical and epistemological principles and its violation (or potential violation) of sacred rules such as the conservation principles and the rules of causality and physical reality. In fact, there are many examples of these violations (or potential violations); some of which were discussed earlier (see for example § 10.1.3) while others will be discussed in the questions of this subsection as well as in the upcoming subsections. This could point out to a major defect in the theory and its theoretical framework because these rules and principles are generally well established (or supposedly so) and hence their violation based on a pure theoretical basis (e.g. as a demand of the theory of general relativity) should not be allowed. In other words, in the absence of experimental and observational evidence on their violation any theory that violates these rules and principles should be rejected (or at least questioned) because it is inconsistent with experiment and observation (at least so far and prior to the verification of the theory). In fact, even if we accept in principle the possibility of the violation of these rules and principles by new evidence that can emerge in the future, the theory should not be accepted unconditionally until such violation is verified (meanwhile the theory should be put on hold).

Exercises

1. Give an example of a general relativistic violation (or potential violation) of a sacred rule from special relativity, and assess in detail the significance of this violation and its impact on special relativity as well as on general relativity.
2. Give another example of a violation (or potential violation) of sacred rules by general relativity.
3. The call for formulating the physical theories (including gravity) using flat spacetime may be challenged by the possibility that the physical spacetime could be inherently curved. Discuss and assess this issue briefly.

10.1.14 Circularity

The methodology of general relativity is based on the philosophy that the state of matter (i.e. its distribution and motion) determines the state of spacetime (i.e. its geometric properties) while the state of spacetime determines the state of matter. This is eloquently expressed by the famous idiom "matter tells spacetime how to curve, spacetime tells matter how to move". In fact, this may be fine if this idiom (which really reflects the essence of general relativity as a gravity theory and its methodology) expresses the relation between a gravitating source of gravity and a gravitated test particle where the "gravitating matter" tells the spacetime how to curve while spacetime tells the "gravitated matter" how to move. But in more complicated situations where there is no distinction between gravitating and gravitated objects this methodology is problematic conceptually and practically since it is effectively circular.

The circularity problem also extends (directly or indirectly) to the problem of "gravity of gravity". This problem is reflected in the literature of general relativity within different contexts such as the issue of non-linearity and the energy density of gravitational field. Almost all (if not all) these problems and difficulties originate from the problematic central paradigm of general relativity that is based on distinguishing gravity from other physical phenomena by considering it a curvature attribute of spacetime rather than a physical phenomenon that takes place in spacetime. In this regard, we find in the literature complaints like "the task of assigning an energy density to a gravitational field is notoriously difficult" (see Hobson *et al.* in the References). What is really "difficult" (if not impossible) is the central paradigm of general relativity which leads to such dead ends and this paradigm should be revised or replaced if we have to avoid these notorious difficulties.

10.1.15 Using Einstein Tensor to Represent Curvature

The use of the Einstein tensor to represent and quantify the curvature of spacetime in the Field Equation is rather arbitrary[292] and questionable because the rank-2 Einstein tensor contains rather limited curvature information since the full curvature information is contained in the rank-4 Riemann-Christoffel curvature tensor and the contraction of the latter tensor to obtain the Ricci curvature tensor and scalar (from which the Einstein tensor is formed) could remove essential curvature information. So, if gravity is

---

[292] In fact, even the assumption that the spacetime of general relativity is Riemannian (or rather pseudo-Riemannian) is arbitrary.

really correlated to the curvature of spacetime (since it originates from this curvature) then this should compromise any gravity theory that uses the Einstein tensor to formulate gravitation.[293]

For example, in a curved spacetime whose Riemann-Christoffel curvature tensor is not zero but its Ricci curvature tensor (and hence its Einstein tensor) is zero we have $T_{\mu\nu} = 0$ (according to the Field Equation) and therefore the spacetime is curved without the presence of matter and energy. This should be a consequence of the fact that less curvature information are contained in the Ricci curvature tensor than in the Riemann-Christoffel curvature tensor and hence despite the curvature of the spacetime the Field Equation fails to represent the gravity of this missing part of the curvature since it predicts there is no gravity despite the presence of curvature (or it predicts gravity, represented by curvature, without energy-momentum source). In brief, gravity seems to be linked arbitrarily to a certain part of the spacetime curvature (in a certain way) and this requires justification.[294]

Exercises
1. Summarize the essence of the issue discussed in this subsection.
2. What is the implication of using the Einstein tensor (rather than the Riemann-Christoffel curvature tensor) to describe gravity?
3. Propose a challenge to the following statement: using the Einstein tensor to represent and describe gravity in general relativity is arbitrary and incomplete representation of gravity since the rank-2 Einstein tensor captures only part of the spacetime curvature due to the fact that this curvature is fully represented by the rank-4 Riemann-Christoffel curvature tensor.
4. A justification for the seemingly arbitrary choice of the Einstein tensor $G^{\mu\nu}$ to represent the curvature of spacetime is that it is the only rank-2 tensor that is second order in the derivatives of the metric for which the divergence vanishes (i.e. $G^{\mu\nu}_{;\nu} = 0$). Is this justification sufficient to address the issue of arbitrariness and partial representation of curvature?
5. Comment on the following quote from Rindler (see References): "The choice of $G_{\mu\nu}$ for the LHS of the field equations may at first seem somewhat arbitrary. But it can be shown [cf. D. Lovelock, J. Math. Phys. 13, 874 (1972)] that $G_{\mu\nu}$ is the only tensorial and divergence-free function of the $g_{\mu\nu}$ and at most their first and second partial derivatives, that, when put on the LHS of the field equations, allows Minkowski space as a solution in the absence of sources".
6. Comment on the following quote from Synge (see References): "In Einstein's theory, either there is a gravitational field or there is none, according as the Riemann tensor does not or does vanish. This is an absolute property; it has nothing to do with any observer's world-line".
7. It may be claimed that although the curvature of spacetime is not totally represented in the Field Equation and even if the representation by the Einstein tensor is arbitrary, the main empirical criterion for any "correct" theory is its compliance with experiment and observation and hence as long as general relativity is supported by experimental and observational evidence this should not be a defect in the theory. How to address this?

10.1.16 Ambiguity of Vacuum Equation

As indicated earlier (see Exercises of § 3.2) the vacuum equation $R_{\mu\nu} = 0$ of general relativity cannot be solved unless we start from a guess about the metric. Hence, it cannot be solved by the natural (or ideal) method of obtaining the metric from the mass-energy distribution (as explained in § 4) and this should introduce serious difficulties in most cases. The reason is that the vacuum equation $R_{\mu\nu} = 0$ is local (since it is restricted to the vacuum region) and hence it does not include in itself the source of gravitation. Therefore, the effect of the source of gravitation should be incorporated indirectly through the metric

---

[293] We may even argue that the use of the Riemannian geometry to describe the alleged gravity-related curvature of spacetime is arbitrary and it requires justification because even if we accept that gravity is a demonstration of spacetime curvature it is not necessary that this curvature is based on a Riemannian geometry since the geometry of spacetime can be non-Riemannian.

[294] In our view, this issue should be linked to the Weyl (or conformal) tensor which is thoroughly investigated in the literature. We should also note that another potential source of arbitrariness in the formulation of curvature in general relativity is the common use of Christoffel connection rather than other types of connection.

which effectively is the only entity present in $R_{\mu\nu}$. This can be revealed by a careful inspection of the solutions of the vacuum equation (e.g. Schwarzschild solution; see § 4.1.1) and how they are obtained where the source of gravity is usually inserted in the metric in a rather twitchy way (e.g. by comparison to the Newtonian solution in the classical limit). Such approach may apply in certain circumstances and cases but it is not general and this should represent a major difficulty in obtaining a solution in most cases. This should also aggravate the need for outside theories (e.g. classical gravity) and the dependence on external help and guidance to obtain solutions (refer for example to § 10.1.17). Moreover, it may even impose limits on the general validity of the obtained solution, e.g. by appealing to the classical limit.

We should note that the criticism of ambiguity (and related complications) may also be directed to the vacuum problems in classical gravity (and indeed to any gravity theory) where the gravitational potential takes the role of the metric and hence this criticism is not specific to general relativity since it originates from the nature of the vacuum problems. However, the situation in classical gravity is much simpler (due for example to the scalar nature of the Poisson equation) and hence this is not a major handicap or hurdle for finding vacuum solutions.

We should also remark that there is no guarantee that in the vacuum problems and solutions of general relativity (and indeed in any problem in general relativity; see § 10.1.15) full curvature information are accounted for and represented in the formulation due to the fact that in principle the rank-2 Einstein tensor (which represents the curvature of spacetime in the Field Equation) contains only part of the curvature information of spacetime which is contained fully in the rank-4 Riemann-Christoffel curvature tensor (see exercise 9 of § 2.13).[295] This may be seen as another potential source of ambiguity in the vacuum problems and solutions of general relativity (and indeed in any problem in general relativity).

10.1.17 Need for Classical Gravity

An important thing to note about general relativity is its need to classical gravity in many situations and contexts and this should compromise the integrity and independence of the theory. In fact, this can be easily seen in many places in this book as well as in other books where the guidance of classical gravity is essential in analyzing the situation and formulating the physics. Moreover, in many practical situations related to real physical problems the proposed and employed general relativistic formulations are not purely general relativistic and hence they are actually hybrid classical-relativistic formulations (which should reflect the need of general relativity to external help). In fact, even the language and method of presentation of many relativistic ideas and paradigms require the help of classical gravity. For example, it is common in the literature of general relativity to use classical paradigms and terminology like gravitational field, tidal force, gravitational potential energy, etc.

We may also add in this context the possible need of general relativity (or the need of some general relativists) for the classical concept of "gravitational force" in some situations, e.g. in the frames that are in a gravitational field but they are not in a state of free fall (i.e. frames of type c in the investigation of § 6.1). However, this may be refuted by the allegedly fictitious (and hence presumably non-gravitational) nature of the force; moreover, it may be interpreted as an effect of acceleration according to the equivalence principle (although this requires physical substantiation). In fact, this may also be refuted more elegantly by the rationale of the *Principle of General Covariance* although this may not be sufficient to address all the difficulties that arise from the disposal of "gravitational force". We should finally refer to the upcoming discussion of § 10.1.18 which may be regarded as another example of the persistent need for gravitational force.

10.1.18 Failure of Geodesic to Replace Force

The paradigm of "geodesic" is hailed in the literature of general relativity as an ingenious and innovative replacement of the old fashion paradigm of "force". Apart from the fact that this is not such an ingenious

---

[295] We are implicitly assuming that the curvature information in the Riemann-Christoffel curvature tensor can be classified into two parts: a part represented by the Einstein tensor and a part not represented by this tensor. Hence, if the latter part is void then the curvature information in the Riemann-Christoffel curvature tensor is fully represented by Einstein tensor; otherwise it is not.

10.1.18 Failure of Geodesic to Replace Force

and innovative idea (as explained earlier; see for example exercise 8 of § 2.9), there are many examples in which we find ourselves still in need for the paradigm of force to explain the physical situation satisfactorily. In fact, the paradigm of geodesic may explain the geometrical and kinematical aspects of the physical behavior of the gravitated object (i.e. its trajectory in spacetime) but it cannot explain the dynamical aspects of this behavior.[296] For example, let have a gravitated object that is released from rest in the "gravitational field" of a gravitating object. According to the old fashion paradigm the gravitated object is attracted by the force of gravity and hence it will start moving and accelerating toward the gravitating object along a certain trajectory that is defined by the gravitational field.[297] However, the paradigm of geodesic cannot provide a full explanation of the movement of the object. To be more clear, even though the paradigm of geodesic may describe how the object moves along a certain trajectory (i.e. determines the shape of trajectory in spacetime) it cannot explain why it should move at all (initially and along this trajectory) and follow this particular trajectory. In brief, while the paradigm of force is dynamic and lively (due to its physical nature) and hence it can provide a full physical explanation, the paradigm of geodesic is static and lethargic (due to its geometric nature) and hence it can provide only partial (and rather phenomenological) physical explanation (i.e. how but not why). In fact, this could have practical (and not only theoretical and conceptual) consequences.[298]

Another demonstration of the failure of the geodesic paradigm to replace the force paradigm is the lack of symmetry in the concept of geodesic (where we have action of one mass, i.e. gravitating, and response by another mass, i.e. gravitated) in contrast to the concept of attractive force where each mass attracts the other mass as it should be (for obvious logical and physical reasons). The lack of symmetry in the concept of geodesic may invalidate Newton's third law whose essence is the symmetry of interaction between physical objects which is a requirement for the equivalence between the objects in the absence of any discriminating factor (i.e. why should one object attract the other object but not the other way around?).[299] The situation will be more obvious when the two objects are identical (i.e. have the same mass, the same shape, etc.) where the symmetry in the interaction (as expressed aptly by the concept of force but not by the concept of geodesic) is a necessity. In fact, this should also be linked to the problem that we discussed earlier (see the third point in § 10.1.11; also see § 10.1.14) in regard to the situation where the gravitated object is not a test particle (and hence its gravitational field is not negligible).

In brief, "geodesic" is a geometric paradigm and hence when it is applied to a physical situation it may reflect the physical phenomenon kinematically but not dynamically, while "force" is a full dynamic-kinematic paradigm and hence it fully reflects the physical phenomenon dynamically as well as kinematically. Moreover, the paradigm of force is more general than the paradigm of geodesic in its application to various physical situations. So, the paradigm of force contains more physical content (i.e. the dynamics and kinematics and in all cases) of the physics of gravity than the paradigm of geodesic (which is restricted to the kinematics of gravity and only in some cases).

Exercises
1. Discuss the issue of geodesic deviation (see § 7.13) in the context of the failure of "geodesic" to replace "force".
2. Assess the paradigms of "geodesic" and "force" from the viewpoint of point-like and extended gravitated objects.

---

[296] We may also say: the paradigm of geodesic may provide a phenomenological explanation but not a fundamental or essential explanation.
[297] We should note that with the help of Newton's second law the tempo-spatial trajectory (and not only the spatial trajectory) can be determined.
[298] We note that other relativistic paradigms (such as geodesic deviation) may aid the geodesic paradigm to overcome some difficulties and shortcomings. However, this cannot address all the difficulties and shortcomings in a general and fundamental way. In fact, even some of these aiding paradigms (such as geodesic deviation) are essentially about how rather than why.
[299] In fact, Newton's third law could be seen as a demonstration of fundamental spacetime symmetries and hence it has a fundamental value from this perspective (regardless of the paradigm of force).

10.1.19 Limitation of Evidence

As we saw earlier (refer to § 9), there are many limitations and restrictions on the claimed evidence in support of general relativity. For example:
- There are specific limitations on the claimed evidence individually since almost all the claimed evidence can be challenged in their validity or value (in various ways) as explained in § 9.
- There are limitations on the nature of evidence as it is generally restricted to the gravitational aspects of the theory and hence it does not extend to the "General Theory" aspects.
- There are limitations on the geometry of the spacetime since most evidence are related to the Schwarzschild metric in the vacuum region.
- There are limitations on the endorsed (or rather supposedly-endorsed) aspects of the theory, e.g. time dilation but not length contraction.
- There are limitations on the range of validity as exemplified by the strength of the gravitational field since most of the claimed evidence are related to the weak regime of gravitational field (Newtonian and quasi-Newtonian).
- There are limitations on the formulation in some cases (such as the use of the linearized form of the Field Equation) and therefore any alleged evidence based on such formulations should be restricted by the limited (and potentially questionable) validity of these formulations.

Overall, there are many features and aspects of the theory that are not validated (and even investigated) sufficiently and properly. Hence, the validity of the theory in its entirety and generality is not established even if we accept all the claimed evidence in support of general relativity and ignore all the claimed evidence and challenges against it (such as the violation of the strong equivalence principle and the inconsistencies and absurdities in its theoretical structure and applications).

Exercises
1. Discuss the issue of the limitation of the claimed evidence in support of general relativity to the gravitational aspects.
2. What is the significance of the fact that most of the claimed evidence and tests in support of general relativity are related to the Schwarzschild metric in the vacuum region?

10.1.20 Absurdities and Paradoxes

We can find many inconsistencies and paradoxes in general relativity where most of these inconsistencies and paradoxes originate from the highly mathematized nature of the theory and its use of the geometric paradigm of spacetime curvature to model physical phenomena. For example, let analyze[300] the fall of an observer O_1 radially toward a black hole where he is observed by another observer O_2 who stays stationary at a certain location in the space far away from the black hole. It is shown in the literature that the time required for such a journey (i.e. the fall to the singularity) is infinite according to O_2 while it is finite according to O_1 (also see exercises 2 and 4 of § 7.12). Now, any time of any observer in the real world should be finite and hence the fall of O_1 to the singularity (and even reaching the event horizon) will never happen according to O_2 while it will certainly happen at some time according to O_1. In fact, many direct and indirect physical consequences are based on the finity of the fall time, e.g. the increase of the black hole mass or angular momentum by the falling matter according to the outside observers and hence if the fall time is infinite according to these observers many of the claimed physical consequences related to the interaction of black holes with their surroundings will never be observed (and in fact will never happen). Many other nonsensical consequences and implications can be drawn from the general relativistic formulation and depiction of this journey, e.g. the swap of t and r and the decrease of t with the increase of τ in the region $r < R_S$ (noting that the coordinate time t represents the proper time at infinity). Absurdities and paradoxes can also be drawn from other general relativistic formulations, applications and solutions. Of course, general relativists are smart enough to find technical fixes to these absurdities and paradoxes and their alike but none of these fixes can overcome the impossibility of

---
[300] We use in this analysis the Schwarzschild solution in the Schwarzschild coordinates.

envisaging this journey consistently and physically as depicted by the general relativistic formalism even if we can convince ourselves of its consistency and correctness from a formal and mathematical perspective. We note that the standard defence of general relativists (when they admit the nonsensical nature of these absurdities and paradoxes) is to blame the "inadequate" coordinates (e.g. Schwarzschild coordinates) for these absurdities and their alike. However, instead of blaming the coordinates for these absurdities, it is more appropriate to blame the theory that made the coordinates dysfunctional to such extent.

It should be remarked that some authors seem to suggest that the infinity of the fall time from the perspective of the external observer in the above-described journey is apparent. If so, then this will put a question mark on the reality of some alleged general relativistic consequences and if the theory could lead to misleading results. In fact, this should create more problems than it is supposed to solve. We should also note that although some of the absurd implications are based on the use of Schwarzschild coordinates specifically (which are usually described as inappropriate or inadequate) the nonsensical nature of these implications reveals the limitations of the mathematical conclusions from the application of general relativity. In other words, if this nonsense can happen (with our awareness) in the Schwarzschild coordinates (i.e. in the formalism obtained by legitimate application of general relativity in association with the Schwarzschild coordinates) it could happen (most likely without our awareness) even in the non-Schwarzschild coordinates (and even in the non-Schwarzschild geometries).[301]

Exercises

1. What criteria we use to assess a theory or a premise from the perspective of inconsistency and paradoxes?
2. According to one version of the journey of a freely falling observer A toward a black hole as observed by another observer B at infinity: "B will never see A crossing the event horizon and the image of A will remain frozen at the event horizon for eternity". Comment on this.
3. Give some examples of general aspects in the theory of general relativity that are common sources of inconsistency and paradoxes.
4. Give some examples of absurdities and paradoxes that originate from the mathematical formalism of general relativity.
5. The blame for paradoxes and absurdities in the applications of general relativity is commonly attached to the inadequacy of the employed coordinate systems (e.g. Schwarzschild coordinates). How do you challenge this?
6. Give some examples for the general relativistic absurdities from the physics of Schwarzschild black holes.

10.1.21 The Paradox of Absolute Frame and Reality of Spacetime

This, in our view, is one of the biggest epistemological paradoxes of general relativity. In fact, it is so big that no one seems to notice. In brief, the reality of the Newtonian space-time was denied in special relativity where the epistemological essence of this denial is represented by (or originates from) the denial of the absolute frame of spacetime (because the existence of absolute frame is a natural consequence of the reality of space-time). This may be understandable and acceptable within the framework of special relativity because the spacetime in special relativity is a bare skeleton that does not seem to embed or contain any physical essence or content (and hence the denial of its reality and thus the denial of the existence of absolute frame seems reasonable).[302] Now, in general relativity (which inherits the spacetime of special relativity) we have a spacetime that is so real that it is curved by the presence of matter and

---

[301] We should note that even in the mathematical basis of at least some of the "good coordinates" (e.g. Eddington-Finkelstein coordinates) we still have the problem of singularity at $r = R_S$ in the transformation from the Schwarzschild coordinates to the "good coordinates" but the problem is ignored, and hence the solution is still not ideal from this perspective because the singularity is ignored (and thus it is still there) rather than dealt with deeply and fundamentally. In fact, if the coordinates are really just labels (as claimed by general relativity) then the nonsense that we obtained in the Schwarzschild coordinates should not be resolved by changing the coordinates. We should also note that nonsensical results (e.g. traveling back in time) are also obtained from other spacetime geometries like Kerr geometry.

[302] In fact, even the spacetime in special relativity is not really a bare skeleton due to its possession of certain physical properties represented for example by "free space" constants like ε_0, μ_0 and c.

energy (as if it is a material object) and it even embeds real physical content since gravity (which in essence is a physical element) is part of the structure of this "very real" spacetime. In fact, the spacetime of general relativity is more real than even the space-time of the Newtonian theory (which was denied earlier by special relativity). Nevertheless, general relativity (or at least general relativists) continues to deny its absoluteness.

10.1.22 Constancy, Invariance and Ultimacy of the Speed of Light

One of the fundamental principles of the relativity theories is that the speed of light is a constant, invariant and ultimate speed for any physical motion (although the two theories differ in the global and local validity of this principle). However, there are many indications that this principle is not true in general since violations of this principle have been observed or deduced (at least seemingly and tentatively) from a number of experiments and observations. The most serious challenges to this principle seem to be from quantum entanglement and its consequences. If this principle collapsed, then the entire theoretical structure of the relativity theories will be under threat. The reader should be aware of the serious physical and epistemological consequences of any violation of the relativistic restrictions on the speed such as potential violation of causality and action at a distance within the relativistic framework.

Exercises
1. What makes the constancy of the speed of light in general relativity local and not global?
2. Discuss the significance of quantum entanglement and its potential impact on the relativity theories.

10.1.23 Lack of Practicality and Realism

General relativity and its applications (notably relativistic cosmology) contain many non-practical aspects and this should affect their status as viable and realistic scientific theories (see § 1.9). For example, solutions to general relativistic problems are very difficult (and in most cases impossible) to obtain. Also, the duality of coordinate and metrical variables leads to serious practical difficulties some of which are outlined earlier (see for instance § 6.2 and § 10.1.6). There are also many examples of non-verifiable consequences (e.g. dark energy or certain cosmological and astrophysical occurrences and developments) and non-obtainable quantities (e.g. some proper cosmological quantities). The theory and its applications also contain or imply many non-realistic fantasies and illusions. In brief, general relativity and its applications are not very practical or realistic theories because of their complexity, over-mathematization and highly theoretical nature (as well as other reasons).

10.1.24 Limited Usefulness

Despite its alleged elaborate structure and aesthetic nature, the usefulness of general relativity as a "General Theory" is very limited. This is mainly due to its excessive generality and vagueness (as well as other potential reasons discussed earlier in the book and in the previous subsections in particular such as triviality of invariance, gaps and ambiguities and lack of practicality and realism). Hence, we can hardly find any practical use or application to this alleged "General Theory". In fact, the actual usefulness of even the "gravity theory" is limited since almost all the practical applications of general relativity as a gravity theory can be obtained from classical gravity (and potentially from other theories as well). This issue will be investigated further later on.

10.2 Overall Assessment

In the following subsections we outline a number of general aspects in general relativity that can be used to assess the theory from practical and theoretical perspectives. We should note that these aspects represent a sample and not a comprehensive list. We should also note that many of the issues discussed and investigated in the subsections of the last section can be used for assessing general relativity.

10.2.1 Geometric Nature of General Relativity

General relativity is a geometric theory of physical phenomena, and for this reason (as well as for other reasons) it is not intuitive. In fact, this is one reason for our belief that general relativity is not an ideal physical theory (neither scientifically nor epistemologically) because its geometric nature diminishes its capability to capture the "spirit" of the physical laws behind these phenomena even if we assume that the theory is "correct" in essence and can provide precise description and prediction of the physical phenomena. In other words, the theory may describe the phenomena by the symptoms rather than by the physical content of the phenomena. Moreover, the geometric nature of the theory makes it susceptible to mathematical artifacts and fantasies as a result of over-mathematization of what is supposed to be a physical theory. Also, unnecessary complications and hurdles will be introduced to physics (whose main objective is to understand the physical world and reduce its complexity by using relatively simple physical and mathematical patterns) thanks to this excessive mathematical approach and methodology.

10.2.2 High Complexity

General relativity is notorious for its complexity and sophistication due largely to its total dependency on very complex mathematical machinery (mainly differential geometry and tensor calculus and their sophisticated concepts and techniques) which makes the theory difficult to understand and puts limits on its usability (e.g. analytical solutions are rarely available and only in very special cases and numerical solutions are very difficult to obtain if they are accessible at all). In fact, this is another reason for general relativity to be non-ideal. Therefore, developing a simpler theory will be a big advantage to science. The possibility of developing an alternative theory originates from our belief that science is a mix of discovery and invention and hence any physical phenomenon can be formulated in many different ways all of which are correct.[303] In other words, science is not unique and hence alien civilizations should develop different versions of science to the version that we have. In fact, even in our earthly science we can find many examples of multiple correct formulations related to a single physical phenomenon where some formulations are superior or more advantageous in comparison to other formulations (although this is usually conditional and depends on the contexts and circumstances). It is worth noting that the non-unique nature of science (and indeed any form of knowledge) should not affect the principles of physical reality and truth (specifically the uniqueness of reality and truth) since these different versions represent different demonstrations of the same reality and various expressions of the unique truth.

10.2.3 Highly Theoretical Nature

The highly theoretical nature of general relativity (which is an obvious feature of this theory thanks to its reliance on a very abstract scientific and epistemological framework) makes physics vulnerable to detachment from reality and susceptible to fantasies and illusions such as wormholes and travel in time. In fact, general relativity contributed to the departure of modern physics from the realism and rationality which characterized classical physics over its long history prior to the appearance of general relativity and other similar theories of modern physics. The highly theoretical nature of general relativity also encourages the emergence of many hypothetical and conjectural adventures as well as virtually useless experimental and observational projects. These adventures and projects do not only consume huge amounts of precious resources into futile lavish enterprises of negligible return and waste huge amounts of money, effort and brain power which are badly needed in many theoretically and practically important areas of science, but they also divert science from its real route of progress and development and take it in the wrong direction far away from its real and beneficial objectives.

10.2.4 Publicizing and Politicizing Science

The historical route of development of modern physics has changed dramatically since the endorsement of general relativity by Eddington and his team in 1919. Although the theory itself cannot be blamed for such

---

[303] We are implicitly assuming the correctness of general relativity although this is not necessarily the case.

historical developments and their subsequent consequences, they remain a decisive factor in determining the route and destiny of science and hence they should be seen as part of the legacy of the theory and its practical impact on science. Therefore, they should be taken into account when assessing the contribution of the theory to science and evaluating its potential impact (whether positive or negative) in the future.

In brief, since the endorsement of Eddington and his team to general relativity, strange elements and factors entered into science, and physics became publicized and politicized. This is a result of the huge prestige that Einstein gained by this endorsement where he became a highly revered public figure and not just a scientist. Accordingly, the public fascination by Einstein and the huge support to his theories became a factor in determining the fate of science because scientists are humans influenced by the public opinion[304] and hence they do not want to be deprived of public sympathy and support by opposing Einstein and his theories. This applies in particular to students and junior academics and researchers who through this "Einstein effect" are motivated or forced to support Einstein and his theories. The involvement of general media and propaganda machines have also contributed to this "Einstein effect" in determining the fate of science. Accordingly, even if Einstein theories (and general relativity in particular) are correct and ideal for the present state of science, this publicization and politicization of science will have negative impact on the progress of science in the long term because no one should believe that science has reached its ultimate destiny with Einstein and his theories and hence we should always look for the better and this cannot happen easily and naturally (as it should be) when we have an ultimate authority who very few have the will and courage to oppose and criticize.

The "Einstein effect" also extends to the funding bodies, academic and research institutes, refereed journals and in fact to all the institutions of modern science with no exception. So, if someone wants to get a research grant or a faculty post or publish a paper in a respected journal or avoid public humiliation by being mocked and described (legitimately and aptly!) as foolish crackpot then he must follow the line of Einstein and his theories. In fact, the easiest way to commit scholarly suicide or assassinate the character of a scientist and terminate his career is to make him an enemy of Einstein and portray him as an opponent to his great theories due (allegedly) to personal grudge or envy or lack of decency or deficiency in intellect and mental power. All these forms of collective attitude and mobbing conduct contributed badly to modern science and made it very different from the old rational science that is restricted to the intellectual elites who have the mental power and legitimacy to deal with science and manage its thorny issues by legitimate scientific means.[305]

In fact, these days even the illiterates and semi-literates enjoy chewing the "Einstein" word, with and without reason, to feel great and gifted by getting a link to "ingenuity" and gain public respect. Also, there is a distinctive feature in the modern scientific literature that is the personalization of science when it comes to the relativity theories where the tone changes abruptly to become a biography of Einstein and his achievements and personal traits more than a presentation of scientific theories and results. In fact, the scientific literature of the relativity theories (and even beyond) is full of this sort of bizarre manner of presenting science, most of which is based on personal emotions and tasteless exaggerations. So in brief, even if Einstein theories (and general relativity in particular) are correct and ideal from a theoretical perspective, with the involvement of these mobbing factors they become non-ideal for the progress of science in the long term because they make science like a religious order or cult or political party or ideological movement where emotions and prejudices play a major role in deciding what is right and what is wrong and determining the direction and destiny.

10.2.5 Controversies, Conflicts and Uncertainties

There are many aspects in the theory of general relativity that are not well determined and defined (see for example § 10.1.11) and hence they naturally become a source of controversies, conflicts and different interpretations and opinions which cast a dark shadow of uncertainty on the theory itself and its value

---

[304] This is particularly true with respect to the "public opinion" in educational, academic and research institutes although these are not "public" in the general sense.

[305] We should note that even in the old days of science there were some exceptions but they qualitatively and quantitatively differ from the miserable state of modern science.

and merit. In fact, a brief inspection of the literature (whether monographs or research and review papers or general articles) reveals that most aspects of general relativity are fertile fields for controversy and conflicting views and this should reflect an inherent haziness in the theory itself and its conceptual and logical structure. Although some of these controversial and obscure aspects belong to the interpretation they indirectly affect the formalism and how it is applied. This affects the value of the theory and casts doubts over its implications and alleged verifications because these verifications can be tailored according to certain interpretations to achieve the objective of endorsing the theory. Also, these gray areas of general relativity become a rich source of wild speculations, bizarre opinions and other sorts of fantasies and detachment from reality.

In brief, although the mess and confusion in the literature of the relativity theories (i.e. including special relativity) can be partly attributed to the exceptional complexity of these theories (mainly the general) and their unusual nature, if we dig deep we will find that the principal reason behind the mess and confusion (which are linked to the controversies, conflicts and uncertainties) is actually the lack of clarity and definiteness in the theoretical framework of these theories where many hazy concepts, interpretations and formulations[306] are employed in the construction of these theories. We can also add to this the lack of rationality in many of the arguments and reasoning which these theories are based on since irrationality naturally leads to controversies, confusion and mess.

10.2.6 Practical Value

We have a number of observations and reservations about the practical value of general relativity; some of these are outlined in the following points:

• As indicated earlier, general relativity has limited practical value due to its immense complexity and highly theoretical nature (noting that almost all the practical parts of the theory can be easily obtained from classical gravity). Analytical solutions (whether exact or not) are very few and are mostly related to simple or trivial cases whose results are already known from classical physics or observations and hence they are mainly for the purpose of checking and testing general relativity itself rather than producing novel results.[307] In fact, even numerical methods and other approximation techniques become highly complex and demanding when used with this theory.

• In many cases general relativity needs the guidance of classical theory and this highlights a limitation of general relativity since it does not work independently.

• The alleged practical results and benefits that are claimed to be obtained from general relativity such as the corrections to the global positioning system (GPS) can generally be obtained empirically or from classical gravity or from alternative theories and methods.

• General relativity is not tested (at least sufficiently) in extreme gravitational systems (and hence it may mislead science in the future in the investigation of these systems due to the reliance on the alleged success of general relativity so far).

• The incompatibility with quantum mechanics (plus the dogmatic persistence on keeping general relativity as the leading theory of gravity) hindered and will continue to hinder the progress of science especially in the field of finding a unified physical theory.

• There is hardly any tangible benefit from general relativity as a "General Theory".

• When we compare general relativity with other theories of modern physics (such as quantum mechanics and Lorentz mechanics) we find that the actual yield of general relativity is trivial. For example, quantum mechanics revolutionized not only modern science and technology theoretically and practically but it revolutionized even our daily life and our philosophy and conceptions about the physical world and the epistemological framework of science. On the other hand, general relativity (with all its lavish requirements and high cost and the generous support that it enjoyed since the first years of its appearance) made very little practical impact. For instance, almost all the serious astronomical investigations

---

[306] As seen, for example, in having several concepts for distance and having several interpretations for length contraction and having several formulae for Shapiro time delay.

[307] This is in sharp contrast to other scientific theories like quantum mechanics whose results cannot be replicated or substituted by any other theory.

and space adventures (such as sending spaceships and satellites to the outer space) can be made (and are usually made) using the classical gravity theory with purely classical analytical tools and methods. This also applies to the theoretical value of general relativity[308] because almost all the claims on which general relativity is based are redundant conceptually and theoretically and can be replaced by their classical counterparts or alternative theories. In brief, general relativity is based on a set of unverified and unverifiable ideas and framework (e.g. gravity is a distortion in spacetime and hence it is not a force like other forces) that may revolutionize the thinking of whoever believes in these claims and chooses to revolutionize his thinking in this way but this is not necessarily the case because we can keep our classical conceptions and framework (or even embrace alternative theories) such as thinking of gravity in terms of gravitational field and hence it is a force like any other force. This is unlike quantum physics, for instance, where we cannot keep our classical conceptions and framework (or use alternative theory) when we deal with quantum systems. The reality is that most of the significance of general relativity and its "huge" contribution to modern science are the result of the huge propaganda machine behind general relativity and the generous support that it enjoyed from the beginning. We strongly believe that if the classical theory or some of the alternative gravity theories are given the chance to evolve and are granted a tiny fraction of the generous and prejudiced support that is given to general relativity then they will make at least as much contribution as the alleged contribution of general relativity.[309] This is due in part to the huge complexity of this theory and its many problematic aspects (practical and theoretical) which diminish its actual yield despite the huge amount of resources that it consumes.

Exercises
1. Argue in favor of the redundancy of general relativity practically.

10.2.7 Theoretical Value

We also have a number of observations and reservations about the theoretical value of general relativity; some of these are outlined in the following points:

• General relativity is not intuitive or easy-to-understand at all. Interpreting gravity as distortion in spacetime may be mathematically elegant and appealing but it is far from being obvious or natural although it may seem so thanks to the systematic indoctrination which starts from the first years of science education. The notorious complexity and the lack of intuitivity of general relativity should be considered a negative factor rather than a positive factor as it might be imagined and claimed (see Exercises).

• Apparently, the most important contributions of general relativity are in the fields of cosmology and extreme astrophysical systems. However, cosmology is not a rigorous science since it is based in large part on philosophical contemplations and assumptions that degrade its status as a science. The reality is that cosmology (even in its modern version) is a philosophical discipline in essence rather than a physical science. Its scientific appearance is largely attributed to the involvement of general relativity and the consequent mathematization of this subject. Most of the claimed "facts" and results of cosmology are not verified by experiment or observation while many others are not verifiable at all because they are beyond the reach of science. Hence, cosmology is not a physical science in a strict sense and therefore the alleged achievements of general relativity in the field of cosmological theories and investigations (assuming that all these alleged achievements are as claimed) are not really a contribution of general relativity to science, although they contain valuable contributions to applied mathematics and philosophy as well as legitimate speculations and contemplations about the origin and destiny of the Universe. The situation of the general relativistic investigations in the field of extreme astrophysical systems (such as black holes) is

---

[308] In fact, the theoretical value of general relativity is discusses in § 10.2.7 and hence this part belongs more aptly to that subsection. However, we discuss it here to complete the comparison to quantum mechanics while avoiding discontinuity and interruption.

[309] It is noteworthy that the trivial yield of general relativity and its handicaps led to its decline in the 1940s and 1950s where it was dying but it was revitalized by orchestrated promotional campaigns from relativity-loving physicists like Wheeler. So, general relativity was "created" by the propaganda of Eddington and his alike and resurrected by the propaganda of Wheeler and his alike (noting that this propaganda is still going on) and hence general relativity did not survive and thrive naturally and by its own merit and strength.

10.2.7 Theoretical Value

very similar since these investigations are highly mathematical and speculative in nature. Moreover, the claimed results are mostly not verified at all (noting that even when they are "verified" the verification is generally insufficient). Hence, in most cases they cannot be regarded as authentic scientific achievements of the theory.

• We also refer the reader to the last point in § 10.2.6 where we briefly talked about the theoretical value of general relativity in the context of comparison to quantum physics.

Exercises

1. What is your overall assessment of general relativity?
2. What are the advantages of having intuitive and easy-to-understand scientific theory?
3. Is general relativity ideal?
4. Is general relativity correct?

Epilogue

- It may be difficult to reach a firm verdict about the validity of general relativity and if it is right or wrong, but we can say with certainty that general relativity is not an ideal theory. Hence, the search for a better alternative theory (both as a gravity theory and as a "General Theory") should be a priority for modern physics. The dogmatic embracement of this theory and accepting it as the perfect theory does not serve the progress of science.
- No one can claim that classical physics is a perfect and ultimate theory, but at least it is based on logical methodology and rational epistemology whereas modern physics is based in large part on illogical methodology and irrational epistemology and the main credit for this trend belongs to theories like general relativity. Thanks to the school of thought and methodology of the relativity theory and its alike we have things like wormholes, white holes, time machines, dark energy, and so on. In fact, this school of thought infested modern physics with distorted logic, metaphysical objects, creation theories, heavily mathematized physical models that are completely detached from physical reality, use of observations as supplementary tools to justify ready-made theories, ..., and the list extends to include many other crazes and heresies. Thanks to this school of thought we now read in our physics texts things like this (quoted with minor modifications):

"At $t = 10^{-43}$ s, the universe was extremely dense, small and hot. The universe spanned a region of 10^{-33} cm. Matter and energy were indistinguishable. The temperature of the universe was 10^{32} K. In a tiny fraction of a second, the universe expanded rapidly and doubled in size many times ... etc.".

This sort of nonsense is taught as solid science and the young minds are indoctrinated and brainwashed to believe that this is the true story of the Universe and its alleged creation. This sort of illusion forms the framework for huge amounts of academic work and research projects in the universities and scientific institutes around the world. In brief, this school of thought spoiled the beauty, rationality and realism of physics and reduced it to a fertile ground for hallucination, wild speculations and bizarre ideas.

- General relativity diverted modern physics far away from its natural and healthy route of development thanks to its unusual nature and methodology. What worsens the situation is that the undisciplined nature of this methodology encourages many incompetent or ambitious scientists to release their wild imagination in fancy scientific enterprises and ideas. Huge resources are spent for example on testing very trivial implications and consequences (or alleged implications and consequences) of this theory and this resulted in wasting huge amounts of precious resources in futile or trivial investigations instead of spending these resources in objective and useful investigations. Although we may be witnessing huge progress in modern science, modern science is highly inefficient since it consumes huge resources.[310] In other words, the scientific process can be highly optimized and the spent resources can be substantially reduced for the same amount of real progress.
- Although general relativity is commonly depicted in the modern literature as the perfect theory[311] and described as proven (or well established), it is far from being perfect or proven. It should be more appropriate to call it the imperfect theory. Only with bias, propaganda and brainwashing it can become the perfect theory. Regardless of its formalism and if it is right or wrong, the logical and epistemological foundations of general relativity can be questioned in many details. However, this is generally inhibited by the curtain of exaggeration that engulfs this theory as well as the complexity of its formal structure. In fact, there are many dark corners in general relativity (both in formalism and epistemology) that have

---

[310] This should be compared to classical science which is mostly developed by individuals with very little resources. The great achievements of classical science (despite its limitations) cannot be denied or disputed. In fact, these achievements (considering the limited spent resources) are much greater than the achievements of modern science.

[311] Interestingly, while the early physicists who witnessed the birth and early development of general relativity (and some have worked closely with Einstein) express their awareness of the defects and limitations of general relativity (as exemplified by the upcoming quote from Bergmann), the more recent physicists do not spare any opportunity to make absurd exaggerations in praising general relativity such as describing it the perfect theory or the most beautiful theory ever created.

Quote from Bergmann (see References): "The third part deals with several attempts to overcome defects in the general theory of relativity. None of these theories has been completely satisfactory".

not been investigated properly because modern physicists are more concerned about establishing the glory of Einstein and his theories than about establishing a reliable scientific theory, and hence amid the frenzy of this mystic devotion the truth is lost.

• Finally, the "eccentric" views that we expressed above (and throughout the book) may be challenged by the "huge" developments (particularly in astrophysical and cosmological studies) which are based on general relativity. However, we believe that most (if not all) of these developments are motivated and staged by general relativity in the sense that they are made with general relativity being in mind as the only viable theory. Any other alternative theory will make similar contributions within its framework and hence the general relativistic view that dominated and steered these investigations and developments is not a necessity. Yes, if these studies were based on objective observations and contemplations without a prior presumption and bias in favor of a certain theory, and general relativity was the only viable framework that can explain these observations (as it is the case for example in the development of quantum mechanics) then the credit for these developments can be given to general relativity. In brief, with the huge bias in favor of general relativity many of these developments are no more than re-production (or echo) of general relativity and hence they are not genuine scientific advances made by general relativity.[312] Moreover, we question modern cosmology as a credible, rigorous and precise science because it is largely based on theoretical speculations and mathematical models with very little observational and experimental substance or evidence. In fact, large part of modern cosmology is based on philosophical contemplations and assumptions as well as many unverified and even unverifiable premises. For example, large part of the theoretical investigations that are based on or related to general relativity, such as dark energy and wormholes, belong to metaphysics and hence they cannot be considered contributions of general relativity to science. In fact, general relativity sometimes plays a destructive role by re-directing huge resources to such futile investigations and taking science away from its natural route and objectives.

• Although we should do our best to understand the Universe and the laws that govern its behavior, there is no guarantee that we can reach this goal. The Universe may be too big and too complex to be fully understood or even understandable. Our physical limitations may impose boundary conditions on our ability to understand. This view seems to be very realistic from a scientific perspective although this should not impose any limit on the legitimacy of our philosophical and spiritual speculations, contemplations and experiences. It seems to me that large part of modern physics (especially in cosmology) belongs to this category of intellectual activities which are wrongly classified as science, and from this perspective general relativity may be a suitable theory for this kind of activities. Nevertheless, because of the many negative aspects of general relativity (e.g. theoretical and practical complexity, over-mathematization, lack of clarity and definiteness in many aspects, looseness and sloppiness in some formulations, epistemological defects, controversies, impracticality, etc.) and for the sake of the progress of real science, physicists should look for a better theory to replace this theory both as a gravity theory and as a "General Theory".

---

[312] In fact, most parts of astrophysics (and even cosmology) can be explained by classical Newtonian gravity even without any extension or modification. Moreover, most of the alleged necessities of employing general relativity in some details of these subjects are arguable and can be challenged.

References

P.G. Bergmann. *Introduction to the Theory of Relativity*. Dover Publications, second edition, 1976.

S. Carroll. *Spacetime and Geometry: An Introduction to General Relativity*. Addison Wesley, first edition, 2004.

O. Gron; S. Hervik. *Einstein's General Theory of Relativity*. Springer, first edition, 2007.

M.P. Hobson; G.P. Efstathiou; A.N. Lasenby. *General Relativity An Introduction for Physicists*. Cambridge University Press, first edition, 2006.

A.P. Lightman; W.H. Press; R.H. Price; S.A. Teukolsky. *Problem Book in Relativity and Gravitation*. Princeton University Press, second edition, 1979.

Open University Team (R.J.A. Lambourne *et al.*). *Relativity, Gravitation and Cosmology*. Cambridge University Press, first edition, 2010.

W. Rindler. *Relativity Special, General, and Cosmological*. Oxford University Press, second edition, 2006.

L. Ryder. *Introduction to General Relativity*. Cambridge University Press, first edition, 2009.

B.F. Schutz. *A First Course in General Relativity*. Cambridge University Press, second edition, 2009.

T. Sochi. *Introduction to Differential Geometry of Space Curves and Surfaces*. CreateSpace, first edition, 2017. **B2**.

T. Sochi. *Principles of Tensor Calculus*. CreateSpace, first edition, 2017. **B3**.

T. Sochi. *The Mechanics of Lorentz Transformations*. CreateSpace, first edition, 2018. **B4**.

T. Sochi. *Solutions of Exercises of Introduction to Differential Geometry of Space Curves and Surfaces*. CreateSpace, first edition, 2019. **B2X**.

T. Sochi. *Solutions of Exercises of Principles of Tensor Calculus*. CreateSpace, first edition, 2018. **B3X**.

T. Sochi. *Solutions of Exercises of The Mechanics of Lorentz Transformations*. CreateSpace, first edition, 2018. **B4X**.

I.S. Sokolnikoff. *Tensor Analysis Theory and Applications*. John Wiley & Sons, first edition, 1951.

B. Spain. *Tensor Calculus: A Concise Course*. Dover Publications, third edition, 2003.

J.L. Synge. *Relativity: The General Theory*. North-Holland Publishing Company, Amsterdam, first edition, 1960.

R.M. Wald. *General Relativity*. The University of Chicago Press, first edition, 1984.

S. Weinberg. *Gravitation and Cosmology: Principles and Applications of the General Theory of Relativity*. John Wiley & Sons, first edition, 1972.

Note: as well as the above references, we also consulted during our work on the preparation of this book many other books, research and review papers and general articles about this subject.

Index

3-
    operator, 17
    quantity, 17
    tensor, 17
    vector, 17

4-
    operator, 17
    quantity, 17
    tensor, 17
    vector, 17

Absolute
    derivative, 6, 49, 65, 66, 71, 75, 76, 78, 80, 156
    differentiation, 62, 64
    frame, 39, 131, 137, 141–146, 179, 213, 219, 227
    space, 145, 146
    spacetime, 145

Accelerated motion, 142, 144, 174

Accelerating
    frame, 13, 30, 33–36, 39, 131, 132, 142–145, 163, 167, 173, 213, 220
    observer, 213

Acceleration, 6, 9, 11, 13, 17, 19, 21, 24, 29, 34–37, 150

Accidental singularity, 42, 43, 110

Accretion, 11, 176, 178, 208

Action at a distance, 228

Adelard of Bath, 12

Advance of perihelion of Mercury, 157

Affine
    connection, 38, 58
    coordinate system, 45, 49, 58, 65, 78
    parameter, 8, 71, 72, 74, 76, 77, 112, 114, 117, 124, 155
    tensor, 49, 58, 60, 62, 65

Algebraic
    rules (of tensors), 49
    sum (of tensors), 50

Angle of
    deflection (of light), 8, 163
    precession, 8, 183

Angular
    momentum, 6, 7, 24, 25, 27, 28, 117, 119, 121, 129, 145, 150–152, 176, 178–180, 182, 183, 226
    speed, 8, 29, 128

Anti-
    matter, 9
    symmetric, 48, 50, 51, 79
    symmetric tensor, 48, 50, 51
    symmetry, 50, 51, 79, 83, 156
    symmetry properties (of tensor), 79, 83, 156

Aphelion, 7, 26, 28, 160, 162, 163
    distance, 7, 26, 28, 160

Areal
    speed, 6, 24, 25, 27, 28, 129
    velocity, 23, 27

Astronomy, 9, 13, 188

Astrophysics, 9, 13, 235

Attractive force, 9, 11, 17, 21, 225

Average
    density, 177, 178
    mass density, 178

Axial symmetry, 119, 180

Axially symmetric, 120

Azimuthal, 138, 184

Basis vector, 6, 21, 45–49, 55, 56, 59, 60, 62–65, 80

Bekenstein (Jacob), 182

Bergmann, 234

Bessel (Friedrich), 37

Bianchi identity, 79, 81, 83, 84

Big Bang, 13, 175, 176, 184, 206, 214, 215

Binary
    orbiting system, 127, 176, 201, 202
    pulsar system, 188, 202, 210
    system, 29, 187, 188, 202

Binomial series, 160

Birkhoff, 157, 187
    theorem, 104, 109, 111, 174, 175

Black
    body, 182
    body radiation, 180
    body radiator, 181
    hole, 10, 13, 39, 42, 175–182, 203, 204
    hole evaporation, 180

Block symmetry, 48, 79

Blue shift, 136, 137, 171, 173, 199

Boltzmann constant, 7, 181

Boundary condition, 105, 106, 108, 109, 235

Boyer-Lindquist coordinates, 6, 119, 120

Bucket experiment (of Newton), 145

Bullialdus, 12

Calculus of variations, 72

Calibration, 136, 148, 150, 218
    at infinity, 148

Cartesian, 8, 19, 20, 44–47, 49, 52–58, 62, 66, 77, 80, 83, 105, 134, 165
    -to-spherical transformation, 56
    quadratic form, 44

Cassini probe, 194

Causality, 41, 221, 228

Cavendish, 189, 191

Center of mass, 29, 174, 175

Central force field, 24, 114, 116, 126

Centripetal acceleration, 29

Chain rule (of differentiation), 25, 64, 73, 74, 77, 114, 115, 124, 159

Characteristic speed, 6, 16, 84, 85, 158, 163, 174, 176

Christoffel symbol, 6, 8, 46, 49, 52, 53, 55, 58–64, 71, 75–78, 80–83, 99, 106, 107, 111–113, 127

Circular
    motion, 29
    orbit, 26, 29, 183
    symmetry, 164

Circularity, 202, 219, 222

Circumferential, 138, 139

Classical
    black hole, 103, 181, 182
    equivalence principle, 36, 37, 39, 206, 212
    gravity, 10, 12, 19, 21, 29, 224

limit, 29, 31, 40, 94, 106, 109, 122–129, 158, 161, 163, 208, 224
mechanics, 19, 24, 28, 30–33, 39, 40, 114, 116, 130, 132, 134, 137, 140, 148, 150, 153, 157
physics, 11, 23, 32, 36–38, 40, 46, 49, 122, 127, 129, 147, 149, 157, 163, 168, 171–173, 189, 203, 213, 218, 229, 231, 234
Clock (i.e. time measurement device), 132, 168, 169, 171, 178, 192, 195, 196
Cluster, 9, 164
Co-moving frame, 89, 90, 122
Coefficient of the Field Equation, 8, 93
Commutative, 80, 81
Commutativity, 80, 81, 99, 100
Compact object, 11, 13, 127, 175–177
Composite rule (of differentiation), 25, 73, 74
Conformal tensor, 223
Conjugate semi-axis (of hyperbola), 6
Connection coefficient, 58, 61
Conservation of
    angular momentum, 24, 27, 28, 117, 145, 151, 182, 183
    energy, 24, 32, 34, 85, 93, 137, 148, 149, 151, 171–173, 191, 195, 196, 199, 201, 214–216, 221
    energy-momentum, 85
    momentum, 85, 91, 93, 145, 151
    momentum 4-vector, 91
Conservation principles, 24, 93, 151, 221
Conservative field, 23
Contraction (of tensor), 49, 51, 65, 86, 222
Contravariant
    basis vector, 6, 21, 45, 47, 55, 56, 59
    differential operator, 99, 124
    differentiation, 63
    metric tensor, 51, 55, 56, 58, 61, 98
    partial differential operator, 99, 124
    tensor, 48
Coordinate
    curve, 45, 46, 138
    singularity, 42, 43
    surface, 45, 46, 84, 138, 139
    system, 6, 17, 19–21, 24, 44–48, 130, 132, 133, 138, 139, 147, 148, 216
    time, 133–135, 147, 154, 155, 226
    time interval, 134, 135
    variables, 129, 132, 133, 140, 147, 162, 163, 169, 170, 177, 188, 194, 204, 205, 211, 216, 228
Copernican astronomical model, 11
Corpuscular theory of light, 190
Cosmic
    cataclysm, 201
    censorship (hypothesis or conjecture), 178
    dust, 120
Cosmological
    constant, 8, 92, 96, 97, 104, 214
    term, 93, 96, 97, 104, 105
Cosmology, 1, 9, 10, 13, 20, 42, 85, 90, 96, 97, 148, 175, 184, 214, 215, 228, 232, 235
Cotangent space, 45
Covariant
    basis vector, 6, 21, 45, 47, 55, 56, 59
    derivative, 6, 49, 55, 61, 63–65, 67, 79, 80
    differential operator, 80
    differentiation, 62, 64
    metric tensor, 6, 51, 55, 56, 58, 59, 61, 81, 93, 98
    tensor, 48

Creation, 151, 175, 214, 215, 234
    of the Universe, 175
    theory, 41, 214, 215, 234
Curl, 6
Curvature of
    space, 51–53, 66–68, 78
    spacetime, 10, 11, 29–31, 35, 52, 91, 95, 105, 133, 163, 176, 177, 182, 183, 188, 193, 194, 211, 217, 219, 220, 222–224, 226
Curvature vector, 8, 70
Curved
    geometry, 57
    space, 18, 19, 30, 45, 49, 52, 53, 56, 66–68, 70, 71, 75, 133, 170, 218
    spacetime, 10, 57, 90, 91, 95, 102, 133, 134, 144, 148, 150, 151, 162, 183, 207, 214, 216, 218–220, 223
Curvilinear coordinate system, 45, 47, 53, 59, 62, 63, 66, 134
Cylindrical coordinates, 8, 57, 58, 134

d'Alembertian operator, 6, 99, 124
Dark
    energy, 42, 96, 97, 177, 214, 218, 228, 234, 235
    energy density, 97
    matter, 42, 97, 214
de Sitter, 204
    effect, 182
Decelerating frame, 36, 39, 213, 219
Deceleration, 21
Deflection of radio waves, 189, 190
Density, 8, 10, 22, 42, 84, 86, 89, 90, 110, 176–178
Destiny of the Universe, 13, 232
Determinant, 6, 55, 81
Deviation vector, 6, 155
Diagonal
    elements (or components), 6, 17, 55, 107
    matrix, 6
    metric tensor, 17, 55, 57, 61, 106
Diagonality (of metric tensor), 61
Differential
    equation, 54, 76, 102, 103, 108, 111
    geometry, 1, 9–11, 13, 44, 46, 53, 71, 229
    operations, 49, 62
    operator, 6, 65, 80, 81, 99, 100, 124
    rules (of tensors), 49
Diffraction, 189
Dimension (or dimensionality) of tensor, 47
Dimensionless, 6, 95, 97, 108
Direct
    observation (of gravitational waves), 201–203
    transformation, 45, 46
Displacement, 6, 155, 164, 168
    vector, 6, 155
Distance, 6, 7, 10–12, 15, 17, 18, 20, 21, 23, 24, 26, 28, 29, 37, 39, 68, 69, 76, 105
Distortion of spacetime, 29, 142, 176
Divergence, 6, 22, 23, 42, 83–85, 91, 93, 94, 125, 126, 223
    -free, 93, 223
    theorem, 22, 23
Doppler frequency shift, 197, 198
Dot notation, 25
Dragging of inertial frames, 183
Dummy index, 47, 59, 63, 64, 73, 74, 100
Dust, 85, 87, 90, 120, 122, 176
    cloud, 87, 90, 122
Dynamic, 39, 89, 151, 193, 194, 220, 225

pressure, 89

Earth, 10–12, 18, 23, 37, 39, 128, 129, 143, 159, 164, 169, 170, 173, 174, 178, 182–184, 190, 192–196, 199, 204, 205, 218
    mass, 39, 182, 204
    radius, 173
Eccentricity (of orbit), 6, 24, 26, 28, 158, 159, 162
Eddington, 13, 15, 164, 209, 229, 230, 232
Eddington-Finkelstein coordinates, 104, 227
Effective
    mass, 16
    potential, 8, 153, 154
Einstein, 12–16, 33, 38, 96, 146, 164, 175, 185, 189, 223, 230, 234, 235
    cross, 164
    ring, 164, 189
    tensor, 6, 7, 20, 52, 78, 81, 83–86, 91, 93–95, 103, 222–224
Elastic body method (of gravitational waves detection), 202, 203
Electric
    charge, 7, 11, 35, 120, 121, 174, 176, 178–180, 182, 213
    force, 9, 11, 35
Electrically
    charged, 120, 121
    neutral, 120
    uncharged, 103, 104, 109
Electromagnetic
    emissions, 198
    energy, 171
    field, 7, 85, 91, 121
    field strength tensor, 7
    force, 10, 29, 150
    radiation, 20, 122, 149
    spectrum, 191
    waves, 136, 174, 175, 220
Electromagnetism, 11, 12, 15, 32, 33, 97, 175
Electron, 180
Ellipse, 6, 24, 26–29, 158
Elliptical orbit, 15, 23, 26, 161
Embedding space, 52, 68, 70
Energy, 6, 7, 9–11, 14, 16–18, 20, 23, 29, 30, 84–88, 96, 151, 180
    density, 85–87, 89, 90, 96, 97, 222
    flow per unit area, 85–87
Energy-momentum tensor, 7, 14, 57, 83–91, 93, 94, 96, 102–104, 120–122
Entropy (of black hole), 182
Eot-Wash experiments, 37
Eotvos (Lorand), 37, 39
Eotvos experiments, 39
Epistemological principle, 14, 32, 33, 40, 221
Equinox, 157, 188
Ergo-
    region, 177, 180, 181
    sphere, 177, 180
Erigena, 12
Escape speed, 8, 111, 176
Essential (or inherent) singularity, 43
Euclidean, 19, 29, 44–46, 53, 57, 58, 66, 68–70, 76, 78, 95, 133, 134, 138–140, 182, 183
Euler-Lagrange equation, 54, 72, 73
European Space Agency (ESA), 6
Event, 16
    horizon, 39, 110, 134, 154, 176–181, 203, 217, 226, 227
Evolution of the Universe, 13, 176
Existence of reality, 41
Expansion of the Universe, 13, 214, 234
Extra precession, 8, 157–159, 161–163, 188
Extrinsic property, 52, 145, 179, 180
Field
    Equation, 8, 13–16, 20, 30, 52, 57, 92–98, 100, 102, 103
    theory, 12, 21, 23, 94, 201
First
    Bianchi identity, 79, 81
    order approximation, 98, 123
    order deviation, 98
    order perturbation, 122
    order variation, 114
    postulate (of special relativity), 34
Flat
    geometry, 57, 97
    metric, 93
    space, 17, 19, 30, 44, 45, 52, 53, 66–68, 75, 77, 79, 80, 83, 84, 93, 120, 133, 134, 162, 170, 182, 188, 218
    spacetime, 19, 36, 57, 89, 131, 133, 134, 144, 147, 148, 156, 183, 188, 207, 222
Flatness (of space or spacetime), 31, 46, 52, 53, 92
Fluid, 85, 89, 90, 110
Flux, 22, 84, 88, 91
Focus (of orbit), 15, 23, 26
Force, 6, 9–11, 150, 224, 225
Form
    invariance, 31, 32
    invariant, 32, 33
Frame dragging, 180, 183–185, 205, 210
Free
    fall, 10, 11, 18–20, 36, 39, 94, 131, 132, 142, 144, 149, 150, 156, 163, 164, 178, 220, 224
    fall gravitation, 18, 19, 149, 150
    index, 49, 77
    object, 17, 19, 40, 69, 78, 102, 127, 151, 153, 154
    particle, 17, 19, 69, 197, 202
    particle method (of gravitational waves detection), 202
    space, 8, 16, 143, 144, 220, 227
    space constants, 227
Freely falling frame, 39, 131, 132, 142, 144, 220
Frequency, 8, 136, 149, 171–173, 197–199
    shift, 136, 137, 171–174, 178, 185, 191, 192, 195–201, 209, 210
Fundamental
    definition (of Christoffel symbols), 59, 60, 63, 64
    definitions (of geodesic), 71
    force, 10

Galaxy, 9, 12, 164, 175, 176, 217
Galilean transformations, 32, 33, 47
Galileo, 11, 12, 32, 34, 37
gamma ray, 198, 199
Gaussian curvature, 51–53, 67
General
    coordinates, 8, 20, 45, 54–58, 62, 63, 78, 83, 93
    relativity, 1, 9–16, 29–31, 100, 101, 234
    tensor, 49, 58, 62–64, 98, 216
    transformation, 17, 32, 38, 49, 93, 132, 147, 220
General relativistic time dilation, 192
General Theory, 9, 29, 30, 33, 35–38, 92, 100, 101, 142, 206–208, 210, 211, 215, 216, 226, 228, 231, 234, 235

Geodesic
    condition, 71, 72
    coordinates, 46, 47
    curvature, 8, 68, 70, 71
    curve, 17, 67–76
    deviation, 80, 150, 155, 156, 225
    equation, 20, 72, 75–78, 102, 111, 112, 117, 124–128, 151, 157, 163, 164, 190
    gyroscope precession, 182
    path, 10, 17, 66, 68–71, 76, 78, 93, 102, 149, 163, 204
    trajectory, 10, 40, 69, 78, 93, 164, 183, 204
    unit vector, 7, 70

Geodetic
    effect, 8, 182, 204, 205
    precession, 182, 185

Geometry of spacetime, 13, 36, 40, 93, 94, 98, 102, 103, 120, 121, 127, 141, 149, 174, 175, 184, 202, 223, 226

Gerber, 12, 157, 187

Global
    frame, 130, 137, 141, 217
    inertial frame, 141, 207
    positioning system (GPS), 7, 192, 195–197, 219, 231
    property, 46
    significance, 148–151, 214, 217

Gradient, 6, 17, 23, 45, 48, 125

Gravitated, 21, 141, 220, 222
    mass, 149
    object, 18, 21, 35, 36

Gravitating, 21, 141, 220, 222
    mass, 149
    object, 18, 21–24, 36, 39
    system, 18, 188

Gravitational
    acceleration, 6, 11, 29, 34, 37, 38, 220
    constant, 6, 20, 21, 24, 37, 94, 95, 157, 163, 181
    field, 10, 17, 18, 20, 22, 23, 29, 30
    field gradient, 35, 39, 155, 156, 178, 212, 213
    force, 6, 9–11, 22–24, 40, 70, 78, 150, 220, 224
    force field, 6, 22, 23
    frame, 13, 20, 30, 33, 35, 36, 130–132, 144, 163, 167, 173, 193, 220, 221
    frequency shift, 136, 137, 171–174, 178, 191, 192, 195, 197–200, 209, 210
    length contraction, 139–141, 174, 193, 194, 200, 201, 221
    lensing, 164, 168, 185, 189–191, 209
    mass, 7, 17–19, 34–38, 213, 220
    potential, 8, 23, 109, 153, 171, 191, 193–196, 199, 224
    system, 39, 142, 145, 167, 231
    time dilation, 103, 135–137, 140, 168, 171, 172, 174, 178, 185, 191–193, 195–199, 201, 209, 210
    waves, 97, 174, 175, 185, 201–203, 209, 210, 220, 221
    well, 135–137, 140, 168, 171, 172, 174, 193, 196, 199

Gravity, 9–13, 21, 29, 224
    of gravity, 175, 222

Gravity Probe
    A, 195, 199
    B, 205

Greek indices, 19
Grossmann, 14
Groups, 9
Gyroscope, 182–184, 204, 205
Gyroscopic precession, 184, 205

Hafele (Joseph), 192

Hafele-Keating experiment, 192, 193
Harvard University, 199
Hawking
    effect, 180, 181
    radiation, 180–182, 204
    temperature, 181
Hawking (Stephen), 180, 182
Hilbert, 14–16
Homogeneity, 151, 172
Homogeneous, 87, 122
Hooke, 12
Hubble law, 13
Hulse, 202
Huygens, 12
Hyperbola, 6

Ideal fluid, 85, 89, 90
Identity tensor, 55
Imaginary, 19, 39, 76, 176, 179, 218, 221
Improper
    mass, 150
    transformation, 47
Index
    lowering, 49, 51, 55, 56, 58, 60, 98
    raising, 49, 51, 55, 56, 58, 98, 100, 123
    replacement, 51, 56, 57, 59, 65, 73, 75
    shifting, 51, 56, 57, 65, 78, 79, 81
Indices of tensor, 47
Indicial notation, 20, 89, 94
Indirect observation (of gravitational waves), 201–203
Inertia, 17, 19
Inertial
    frame, 12, 13, 19, 30, 32, 33, 37, 46, 93, 131, 141, 142, 144, 148, 173, 180, 183, 184, 207, 220
    frame dragging, 183, 184
    mass, 7, 17–19, 34–38, 213
    observer, 177, 180
    rest frame, 131, 213, 220
Infinite red shift, 180
Inner
    event horizon, 181
    product, 18, 49–51, 59, 63–65, 67–69, 74, 218
Instantaneous
    inertial rest frame, 131, 213, 220
    rest frame, 131
Integral, 16, 17, 22, 23, 72, 161, 218
Integration, 106, 108, 126, 154, 162, 163, 170, 218
Interference pattern, 202
Interferometric techniques, 189, 201, 202
Interferometry, 8, 204
Interstellar medium, 198
Intrinsic
    derivative, 55, 63–66, 70, 71
    differentiation, 62, 64
    inhabitant, 68, 76
    property, 51, 58, 59, 76, 80, 145, 179, 180
Inverse transformation, 45, 46
Irreversibility, 110
Irrotational, 143
Isotropic, 90
    coordinates, 194
Isotropy, 151

Jacobian
    factors, 50

matrix, 34, 56
Jupiter, 159

Keating (Richard), 192
Kepler
    first law, 24, 26
    laws, 12, 15, 23, 29, 127
    second law, 27
    third law, 28, 29
Kerr
    black hole, 180, 181, 184
    geometry, 227
    line element, 120
    metric, 6, 119–121, 184
    parameter, 119
    problem, 119
    solution, 105, 119, 120, 180
Kerr-Newman
    metric, 121
    problem, 119, 120
    solution, 120, 121
Kinematic, 18, 36, 93, 94, 130, 137, 141, 148, 149, 169, 192, 196, 198, 225
Kinematical time dilation, 196
Kinetic energy, 171, 176, 180
Kronecker delta, 8, 44, 47, 51, 54–57, 59, 65, 73, 75
Kruskal-Szekeres coordinates, 217

Laplacian operator, 6, 23, 51, 125
Laser
    interferometric techniques, 201, 202
    ranging, 204, 205
LAser GEOdynamics Satellites (LAGEOS), 7, 205
Laser Interferometer Gravitational-wave Observatory (LIGO), 7, 185, 202, 203
Laser Interferometer Space Antenna (LISA), 7, 202, 203
Latin indices, 19, 44
Laue, 14
Leaning Tower of Pisa experiments, 37
Length, 7, 16, 17, 24, 37, 46, 54, 68, 69, 71–73, 84, 137–141, 148, 149, 174, 200, 201
    contraction, 139–141, 174, 193, 194, 200, 201, 217, 221, 226, 231
Lens, 164, 168
Lense-Thirring
    effect, 184, 205
    precession, 184
Lensed object, 168
Lensing object, 164, 168, 189, 190
Levi-Civita, 14
Light
    bending (by gravity), 103, 143, 145, 163–165, 167, 168, 185, 189–191, 193, 210
    cone, 69
    deflection (by gravity), 8, 163, 167, 168, 189–191, 209
Lightlike, 17–19, 69, 77, 118, 153
    geodesic, 69, 77, 118, 153
    interval, 17
    trajectory, 69
Line element, 6, 16, 17, 19, 44, 45, 55, 57, 76, 84, 104–106, 120, 138, 169, 170
Linear, 201, 203
    combination, 49–51, 59
    momentum, 150, 151
Linearity, 18, 23, 93, 103, 175, 219, 222

Linearization, 98, 100, 102
Linearized
    Field Equation, 97, 98
    general relativity, 201, 210
Local
    Cartesian coordinate system, 19, 44, 47
    flatness, 19, 30, 89, 92
    frame, 31, 130–132, 136, 137, 141, 217, 221
    inertial frame, 19, 37, 38, 46, 142
    property, 46
    significance, 53, 173, 204
    standard, 148, 149, 218
    stationary frame, 136, 168, 174
Locality, 19, 35, 36, 39, 49, 87, 93, 137, 141, 148, 163, 170, 188, 202, 204, 205, 207, 211–213
Locally
    Cartesian, 44
    Euclidean, 19, 44
    flat, 18, 19, 30, 36, 37, 45, 66, 89, 90, 130
    inertial, 37, 38, 142
Lorentz
    factor, 8, 84, 85, 87
    invariant, 216
    mechanics, 12, 13, 15, 20, 30–34, 40, 44, 92, 130, 141, 143, 150, 173, 209, 210, 215, 216, 219, 220, 231
    metric, 105, 106, 111, 119, 120, 139
    tensor, 34, 49
    transformations, 1, 15, 17–20, 30–33, 47, 49, 56, 92, 93, 207, 215
Lorentzian
    field, 97
    tensor, 49, 98, 216
Low-speed limit, 31, 124

Mach, 145
    principle, 143, 145, 146, 205
Machian
    paradigm, 146
    view, 145
Manifold, 16, 17, 30, 44, 53, 140
Marat (Jean-Paul), 190
Mars, 11, 159
Mass, 6, 7, 9–12, 16–21, 96, 149, 150, 180
    density, 22, 85, 87, 89, 94, 103, 122, 177, 178
    distribution, 23, 103
    energy, 18, 111
Mass-energy
    distribution, 40, 223
    equation, 10, 16, 84, 149, 150
    equivalence, 10, 20, 84, 104, 149, 180
    relation, 16, 149, 150
    source, 94, 95
Massive, 9–12, 16–21, 69, 77, 78, 93, 94, 102, 112, 117, 119, 127, 144, 149, 151–155, 163, 174–176, 183, 184, 193, 202
Massless, 10, 11, 16, 17, 20, 69, 77, 78, 93, 94, 102, 112, 118, 119, 149, 153, 154, 163, 164, 168, 211
Mathematical singularity, 42, 178
Matrix notation, 88
Matter, 9, 10, 17, 20, 29, 30
Maxwell, 32, 33
    equations, 12
Mean distance, 6, 15, 29, 158, 159
Mercury, 10, 12, 157–159, 162, 163, 185, 187–189, 193, 209–211

Metaphysical, 41, 97, 184, 206, 214, 234
Metaphysics, 214, 235
Metric
    connection, 58, 61
    gravity theories, 189, 198, 199, 203, 209, 210
    tensor, 6–8, 16–18, 38, 44, 45, 54–58
Metrical
    rules (of tensors), 49
    significance, 132–134, 216, 217
    variables, 133, 140, 177, 194, 228
Michelson-like laser interferometry, 202
Miniature black hole, 176
Minkowski, 14
    energy tensor, 14
    metric, 8, 18, 90, 98, 105, 122
    spacetime (or space), 13, 14, 17, 46, 78, 215, 223
Minkowskian spacetime (or space), 13, 16–18, 30, 69, 93, 97, 99, 105, 122
Mixed
    derivative, 80
    metric tensor, 51, 55
    tensor, 48, 51, 63
Modern physics, 1, 10, 41, 42, 97, 177, 181, 184, 210, 213, 218, 219, 229, 231, 234, 235
Momentum, 7, 84–88, 91, 150, 173
    flow, 85
    flow density, 85
    vector, 7, 84–88, 91, 173
Moon, 164, 204
Moon-Earth gyroscope, 204

nabla operator, 6, 45, 51
National Aeronautics and Space Administration (NASA), 7, 195
Natural parameter, 8, 71, 72, 74–76, 155
Nature, 32, 34
Nebula, 175
Neptune, 12, 159
Neutron, 176
    degeneracy pressure, 176
    star, 10, 13, 167, 174, 176, 197, 201, 202
Newton, 12, 15, 37, 190
    bucket experiment, 145
    first law, 197
    law of gravity, 12, 15, 17, 21, 23, 24, 29, 37, 151, 168, 213
    laws of motion, 23, 39, 151
    second law, 17, 24, 25, 34, 37, 125, 127, 213, 220, 225
    third law, 225
Newtonian
    absolute space, 146
    effective potential, 8, 154
    gravity, 10–12, 18, 20, 23, 31, 40, 46, 94, 95, 108, 111, 122, 127, 157, 163, 187, 208, 209, 235
    mechanics, 146, 154
    regime, 210, 226
No hair theorem, 121, 176, 178, 182
Non freely falling frame, 142, 144, 220, 221
Non-
    accelerating, 143
    classical, 122, 157, 185, 187–190, 192, 195, 196
    Euclidean, 78
    gravitational, 9, 17, 40, 78, 101, 137, 141, 148, 150, 189, 198, 219, 224
    homogeneity, 18

inertial, 13, 37, 93, 142, 144, 173, 207
isotropy, 18
linear, 76, 97, 98, 102, 103, 201, 202, 210
linearity, 18, 103, 175, 219, 222
metrical, 133, 140, 217
physical, 41, 134, 140, 214, 218
relativistic, 191, 192
rest mass, 87
Riemannian, 47, 223
rotating, 109, 131, 143, 178, 185
scalar, 45, 65, 218
scientific, 1, 184, 214
timelike, 72
uniform, 35, 36, 131, 145
uniformity, 17, 18
uniformly accelerating frame, 39
Nordstrom, 14
Normal
    curvature, 8, 70
    stress, 85
    unit vector, 7, 22, 70
Normalized spherical coordinates, 6, 21
Nucleus, 178
Null geodesic, 17, 19, 69, 70, 93, 94, 118, 153, 163
Number density, 7, 87
Number of
    components (of tensor), 48
    independent components (of tensor), 48

Optical, 189–191, 194
Orbit, 11, 12, 15, 23, 24, 26–29, 116, 157–159, 161–163, 174, 183, 188, 202, 205
Orbital
    motion, 9, 26, 127, 151, 153, 154, 157, 183, 187, 210
    period, 15, 23, 157–159, 202
    plane, 184, 205
    speed, 127, 129
Orbiting system, 127, 145, 174, 176, 188, 189, 194, 201, 202
Order of tensor, 47
Ordinary
    derivative, 65, 147
    differential operator, 65
    differentiation, 62–65
    partial differentiation, 62
    total differentiation, 63
Origin of the Universe, 232
Orthogonal coordinate system, 45, 55, 57, 61
Orthonormal, 8, 46, 54, 56–58, 84, 85
    Cartesian coordinate system, 8, 46, 54, 56–58
Outer product, 49, 51, 88
Over-mathematization, 43, 221, 228, 229, 235

Parallel
    propagation, 66
    translation, 66
    transport, 53, 66–69, 71, 75, 76, 78, 80, 150, 182, 183, 204, 218
Parallelism, 66, 75, 76
Partial
    derivative, 6, 49, 59, 63, 65, 76, 223
    differential equation, 102, 103
    differential operator, 6, 81, 99, 100, 124
    differentiation, 62, 64
Particle-antiparticle pair, 181
Pendulum, 37, 197

Perfect fluid, 89
Periastron, 157, 187, 188, 210
    precession, 157, 187, 188, 210
Perihelion, 7, 8, 10, 12, 24, 26–28, 157, 158, 160–163, 170, 185, 187, 188, 204, 205, 209, 210
    distance, 7, 26, 28, 160
    precession, 8, 10, 12, 157, 158, 162, 163, 170, 185, 187, 188, 204, 205, 209, 210
    precession of Mercury, 10, 12, 157, 158, 162, 163, 185, 187–189, 209, 210
Periodic time, 7, 136, 171, 195, 201
Permeability (of free space), 8
Permittivity (of free space), 8
Permutation (of tensors), 49, 51
Permutation tensor, 65
Perturbation, 12, 72, 73, 96, 122, 157
    techniques (or methods), 103
    tensor, 8, 18, 98, 99, 122
Philoponus (John), 37
Photon sphere, 179, 180
Physical
    dimensions, 24, 25, 57, 62, 81–85, 89–91, 95, 97
    significance, 84, 85, 133, 134, 147–150, 152, 204, 216, 218
    singularity, 42, 43, 104, 176–179
    time, 132, 133, 149
    variables, 129, 132, 133, 147, 163, 169, 170, 188, 194, 204, 205, 211
Planar, 24, 27, 28, 116
Planck
    constant, 7
    relation, 149, 171–173
Planet, 9, 11, 12, 15, 23–28, 103, 127, 128, 157–159, 161–163, 170, 187, 188, 193
Planetary
    motion, 6, 23, 24, 26, 127–129, 157, 158, 161
    orbit, 24, 26, 27, 158, 162, 188
Pluto, 159
Poincare, 32, 33
    mass-energy equation, 84, 149
Poisson equation, 23, 122, 123, 125, 126, 224
Polar
    angle, 24, 27
    axis, 119
    coordinates, 7, 24, 26, 165, 167
    equation, 168
    orbit, 205
    orbital plane, 205
    plane, 184
Pole (of geodesic coordinates), 46
Position, 6–8, 22, 45, 66, 78, 114, 180, 195, 197
    vector, 6–8, 22, 45
Positive definite, 16, 45, 46, 55
Potential energy, 18, 23, 111, 149, 151, 154, 171, 172, 224
Pound (Robert), 198, 199
Pound-Rebka experiment, 198, 199
Power, 182, 204
    rule (of differentiation), 25, 73
    series, 160, 173, 191, 197, 198
    series approximation, 160, 173, 191, 197, 198
Precession of equinox, 157, 188
Pressure, 89, 90, 176
Primordial black hole, 176, 203
Principle of
    causality, 40
    conservation of angular momentum, 27
    conservation of energy, 32, 149, 171, 201, 214, 221
    consistency, 40
    correspondence, 40, 41, 94, 108, 122, 129, 208
    covariance, 31
    equivalence, 9, 12, 17, 30, 31, 33–40, 93, 96, 144, 206, 212–214
    general covariance, 31–33, 35, 37, 38, 40, 131, 147, 215, 216, 220, 221, 224
    geodesic motion, 40, 96
    geometric gravity, 40
    invariance, 30–34, 38, 39, 47, 92, 93, 96, 215, 216
    metric gravity, 40, 93, 96, 100, 134, 142, 147, 150, 214
    relativity, 33–35, 40
    superposition, 103, 186, 188
Principles of reality and truth, 40, 41, 229
Product rule (of differentiation), 25, 59, 60, 63–65, 73, 74, 114, 115
Proper
    density, 8
    length, 138–140, 170
    mass, 150
    mass density, 89
    number density, 7, 88
    time, 8, 77, 112, 117, 124, 129, 134–136, 147, 154, 155, 168, 169, 226
    time at infinity, 135, 226
    time interval, 134–136, 168, 169
    time parameter, 77, 112, 117, 124
    transformation, 47
Proton, 180
Pseudo-
    Cartesian, 46, 47, 80
    Minkowskian, 97
    Riemannian, 16–19, 45–47, 76, 222
PSR B1913+16 binary pulsar system, 202
Pulsar, 188, 202, 210

Quadratic form, 16–19, 21, 44–46, 55, 57, 69, 105, 106, 109, 111, 117–120, 134, 138, 139, 153, 154, 158
Qualified tensor, 49
Quantum
    entanglement, 228
    field theory, 219
    gravity, 218, 219
    mechanics, 40, 177, 180, 181, 209, 210, 218, 219, 231, 232, 235
    physics, 181, 219, 232, 233
Quasar, 164, 208
Quasi-
    Cartesian, 44, 46, 52
    Euclidean, 78
    gravitational, 143
    Minkowskian, 97, 98
    Newtonian, 210, 226
    Riemannian, 45
Quotient rule of
    differentiation, 74
    tensors, 50, 80

Radar
    distance, 218
    ranging, 194
Radial
    coordinate, 7, 103, 110, 139, 140, 151, 154, 155, 165, 167, 169, 170, 183, 184

distance, 7, 21, 24, 133, 154, 168–170, 174
motion, 153
trajectory, 154, 155
Radio
astronomy, 13
band, 191
interferometry, 204
waves, 189, 190
Radius, 6, 7, 22, 29, 39
Rank of tensor, 47
Rebka (Glen), 198
Reciprocity relation, 59
Rectilinear coordinate system, 45–47, 52, 55, 59, 62, 80, 134
Red shift, 136, 137, 171, 173, 180, 185, 198, 199
Reduced Planck constant, 7, 181
Reference frame, 12, 13, 20, 29, 31–34, 36, 37, 46, 92, 130–132, 142–145, 187, 217
Refraction, 189, 190, 194
Reissner-Nordstrom
metric, 120
problem, 120
solution, 120
Relativistic
cosmology, 1, 90, 96, 148, 184, 215, 228
effective potential, 8, 154
precession, 187
Renaissance, 11
Repulsive force, 9, 11
Rest
energy, 6, 152
frame, 29, 89, 90, 131, 213, 220
mass, 16, 87, 149
Retarded gravitational potential theory, 12, 157, 187, 209, 210
Reversibility, 105
Reversible, 58, 104–106, 109, 110, 120
Revolution, 7–9, 11, 15, 24, 28, 69, 157–159, 161–163, 170, 183, 204
Ricci curvature
scalar, 7, 82–84, 91
tensor, 7, 52, 81–83, 86, 91, 106, 107, 111, 222, 223
Riemann-Christoffel curvature tensor, 7, 52, 53, 78–84, 86, 95, 156, 222–224
Riemannian
curvature, 53
geometry, 223
metric, 83, 93
space (or spacetime), 16–19, 29, 30, 36, 44–47, 55, 57, 66, 69, 70, 84, 89, 93, 95, 182, 222
Ring singularity, 181
Rolling balls experiments, 37
Rotating
black hole, 177, 180
frame, 36, 39, 213, 219
Rotation, 21, 103, 119, 143, 145, 170, 175, 177, 180, 181, 183, 184, 188
parameter, 119
Ruler distance, 218

Satellite, 7, 11, 13, 183–185, 194–196, 205, 232
ranging, 195
Saturn, 159
Schwarzschild
black hole, 110, 179, 180, 227
coordinates, 6, 104, 109–111, 130, 132, 138, 139, 147, 169, 170, 194, 217, 226, 227

geometry, 109, 154, 170, 174, 183
line element, 105, 106, 169
metric, 6, 7, 57, 103–112, 118, 121, 125, 127, 130, 133, 139, 140, 147, 151, 152, 157, 158, 163, 164, 169, 170, 174, 175, 177–179, 210, 226
object, 104, 106, 119–121, 127, 170
problem, 102, 104, 105, 109, 110
quadratic form, 105, 111, 117, 118, 134, 138, 154, 158
radial coordinate, 139
radius, 6, 7, 39, 108, 110, 111, 119, 152, 154, 167, 176–179
solution, 102, 103, 109, 110, 119–121, 125, 179, 217, 224, 226
spacetime, 111, 127, 132–135, 137–139, 149, 151, 155, 168–171, 174, 175, 194
Second
Bianchi identity, 79–81, 83, 84
order absolute derivative, 80, 156
order derivative, 223
order differential equation, 76
order partial differential equation, 103
order perturbation, 99
Sectorial
speed, 23
velocity, 23
Semi-
major axis (of ellipse), 6, 28, 29, 158, 162
minor axis (of ellipse), 6, 28
Semicolon, 63
Separation vector, 155
Shapiro (Irwin), 193, 194
Shapiro time delay, 185, 193–196, 231
Shear stress, 85
Simultaneity, 148
Singularity, 39, 42, 43, 104, 110, 111, 120, 134, 147, 170, 176–179, 181, 203, 214, 217, 221, 226, 227
Sirius, 198
Smithsonian Astrophysical Observatory (SAO), 7, 195
Snider, 199
Solar
corona, 194
eclipse expedition (of year 1919), 13, 15, 164, 191, 209, 229
mass, 173, 178, 202
system, 9, 127, 145, 157–159, 163, 174, 187, 188
Soldner (Johann), 189, 191
Space, 6, 8, 10, 13, 16–21, 44–47, 51–53, 140–142, 215
interval, 7, 16–18, 77
Spacelike, 18, 19, 69, 77, 110
geodesic, 69, 77
trajectory, 69
Spacetime, 6, 8–11, 13, 14, 16–21, 29, 57, 130, 131, 134, 137, 147, 154, 227, 228
interval, 8, 17, 19, 77, 93, 194
Spatial coordinate, 16, 19, 34, 57, 84, 105, 119, 133, 134, 138, 139
Special
relativity, 1, 11–20, 30, 31, 137, 140, 141, 207, 215
tensor, 49
Special relativistic
length contraction, 141
time dilation, 137
Spectrum, 189, 191
Speed, 6–8, 150
limit, 211

of light, 6, 16, 33, 84–87, 122, 145, 150, 158, 163, 173–176, 180, 181, 193, 194, 228
Spherical
    coordinates, 6, 7, 21, 22, 24, 53, 57, 58, 104, 106, 109, 120, 133, 134, 138, 139, 184
    symmetry, 23, 103–106, 109, 110, 117, 127, 170
Spherically symmetric, 23, 103, 104, 109, 110, 119, 120, 157, 163, 170, 175
Star, 9, 10, 12, 13, 103, 145, 157, 164–168, 173–176, 187, 191, 197, 201, 202
Static, 39, 89, 96, 103–105, 109, 110, 119–121, 157, 163, 179, 180, 220, 225
    limit, 180
    pressure, 89
Stationary, 72, 89, 104, 110, 120, 134, 135, 137, 139, 141, 144, 177, 180, 192, 226
    frame, 132, 134–139, 144, 168, 174, 193, 220, 221
    gravitational frame, 131, 193, 220
    limit, 180, 181
    observer, 178, 213
Stefan-Boltzmann constant, 8
Stellar
    black hole, 176, 180, 203
    core, 175
Stevin (Simon), 37
Stick (i.e. length measurement device), 132, 174
Stress, 84, 85
Stress-energy tensor, 84
Stress-energy-momentum tensor, 84
String theory, 218
Strong
    equivalence principle, 31, 33–39, 100, 131, 132, 147, 206, 208, 210–214, 220, 226
    force, 9, 10
    gravitational field, 10, 18, 30, 39, 122, 163, 176
    gravity, 10, 170, 202
Sum rule (of differentiation), 65
Summation, 16, 19, 86, 90
    convention, 16, 19
Sun, 9, 11, 15, 18, 23–27, 128, 157, 158, 161–166, 168, 170, 173, 178, 185, 191, 193, 204, 209, 218
Super-massive black hole, 176, 203
Supernatural, 41, 214
Supernova, 175
Superposition, 23, 103, 186, 188, 192, 193, 196, 205
Symbolic notation, 88, 89
Symmetric tensor, 48, 50, 51, 83, 84, 94, 102
Symmetry, 23, 48, 50, 51, 58, 60, 61, 73, 74, 79, 81, 82, 87, 98, 100, 103–106, 109, 110, 117, 119, 121, 127, 147, 161, 164, 170, 174, 180, 225
    properties (of tensor), 48, 51, 58, 60, 79
Synge, 35, 187, 188, 212, 223

Tangent
    space, 45, 59, 218
    vector, 7, 19, 63–65, 67, 69–72, 74–76, 151
Taylor, 202
Temperature, 7, 181, 182, 234
    of black hole, 181, 182
Temporal coordinate, 7, 16, 91, 135, 138, 154
Tensor, 47, 54, 57, 62, 78, 81–84, 222
    calculus, 1, 9–11, 13, 33, 44, 103, 229
    derivative, 147
    differential operator, 65
    differentiation, 55, 62, 64–66, 71, 80

Test
    object, 18, 151
    particle, 18, 19, 21, 22, 36, 111, 219, 220, 222, 225
Thermodynamic effects, 89
Thermodynamics, 182
Thermostatic effects, 89
Third law of thermodynamics, 182
Tidal force, 6, 11, 17–19, 35, 39, 155, 156, 212, 213, 224
Time, 7, 8, 13, 15, 16, 21, 134–136, 140–142, 148, 149, 168–170, 191–195, 215
    dependence, 110, 122, 126, 174
    dependent, 174
    dilation, 31, 103, 135–137, 140, 141, 168, 171, 172, 174, 178, 185, 191–193, 195–201, 209, 210, 217, 226
    independence, 104, 106, 110, 123–126
    independent, 29, 103, 104, 109, 110, 120, 122, 126, 134, 174, 175
    interval, 27, 134–136, 138, 148–150, 168, 169, 171, 174, 195
    machine, 42, 177, 218, 234
    of observation, 137
    of occurrence, 137
    period, 28, 29, 162, 170, 188, 197
Timelike, 17–19, 69, 72, 75, 77, 110, 117, 152
    geodesic, 69, 77, 117
    interval, 17
    trajectory, 69, 75, 152
Torque, 24, 182, 183
Torsion
    -free space, 61, 71
    balance techniques, 37
    tensor, 61
Total
    derivative, 49, 63, 65, 66, 107
    differentiation, 62–64
    energy, 89, 152, 154, 201
Trace, 6–8, 20, 84, 91, 94, 95, 99
Transformation (of coordinates), 13, 17, 29–31, 44–51
Transformation rules, 47–51, 56
Translation, 21, 66, 103
Transposition, 7, 88
Transversal, 138
Transverse semi-axis (of hyperbola), 6
Twin paradox, 193, 196
Tycho Brahe, 12

Ultimate speed, 211, 228
Uniform
    acceleration, 35, 36, 143
    density, 110
    motion, 137, 141
Uniqueness of
    absolute frame, 143
    reality, 40, 41, 229
    truth, 40, 41, 229
Unit
    tangent vector, 7, 67, 70, 71, 75, 76
    vector, 7, 22, 24, 70
Unity tensor, 59
Universe, 9, 11–13, 85, 96, 145, 146, 175, 176, 180, 184, 214, 232, 234, 235
Uranus, 159

Vacuum
    energy, 96, 97

　　　　problem, 91, 105, 108, 224
　　　　region, 86, 109, 175, 223, 226
　　　　solution, 86, 91, 95, 104, 119–121, 170, 224
Value
　　　　invariance, 32, 50, 80
　　　　invariant, 51
Variance type, 48–51, 55, 56, 63–66
Variational principle, 54, 72, 75, 206
Velocity, 7, 8, 12, 13, 16–18, 21, 23, 24, 27, 77, 84, 86–89,
　　　　114, 122, 124, 150, 151, 155, 173, 176, 180, 195
　　　　vector, 7, 8, 18, 77, 85–89, 122, 124
Venus, 159, 193
Very Long Baseline Interferometry (VLBI), 8
Virtual particle-antiparticle, 181

Wavelength, 8, 173, 175
Weak
　　　　equivalence principle, 33–39, 100, 131, 190, 206, 212
　　　　force, 10
　　　　gravitational field, 18, 29, 97, 111, 122, 125, 126, 167,
　　　　　　226
　　　　gravity, 10, 89, 167, 170
Weyl tensor, 223
Wheeler (John), 13, 232
White
　　　　dwarf, 173, 175, 198
　　　　hole, 177, 234
Whitehead, 157, 187
Work, 151
World line, 10, 11, 19, 69, 93, 151, 174, 183
Wormhole, 42, 177, 184, 206, 218, 229, 234, 235

X-rays, 176

Zero tensor, 50, 51, 80

Author Notes

- All copyrights of this book are held by the author.
- This book, like any other academic document, is protected by the terms and conditions of the universally recognized intellectual property rights. Hence, any quotation or use of any part of the book should be acknowledged and cited according to the scholarly approved traditions.
- This book is totally made and prepared by the author including all the graphic illustrations, indexing, typesetting, book cover, and overall design.

www.ingramcontent.com/pod-product-compliance
Lightning Source LLC
Chambersburg PA
CBHW060411220526
45465CB00008B/2838